THRIVING IN AN ACADEMIC CAREER

This book provides an invaluable guide on how to achieve a successful and fulfilling academic career. Academics must balance multiple roles and responsibilities, between teaching, research, and offering services to the department, university, and broader community. This book provides practical, research-based guidance on how to adopt a healthy and balanced perspective that accounts for these interconnections.

Research shows that faculty who achieve early balance in their academic responsibilities and home life are more likely to succeed in all aspects of their career, while strengthening the quality and climate of their programs and campuses. This book's chapters accordingly feature case studies and examples that dig deeper into strategies and principles of holistic and balanced career practice and planning. This book assists readers in understanding the relationships between their individual talents as teachers and scholars; the obligations of their department as a community nested with others contributing to the university mission; and the role and responsibility of their university and discipline in the wider society. The themes of balance and harmony underpin this book's approach to faculty development.

Thriving in an Academic Career is for anyone beginning their academic career in geography and related social and environmental sciences, at all types of higher education institutions. This book will be of particular interest to graduate students and early career faculty in geography and nearby social, environmental, and natural sciences.

Michael Solem is a Professor in the Department of Geography and Environmental Studies at Texas State University. He has been principal investigator on several large-scale, federally funded projects supporting geography at all levels of education. Since 2002, Michael has contributed to the AAG's annual Geography Faculty Development Alliance workshops for early career faculty. He has twice received the Journal of Geography in Higher Education's award for promoting excellence in teaching and learning for his research on faculty development and graduate education in geography. Michael is the 2015 recipient of the AAG Gilbert Grosvenor Honors in Geographic Education.

Kenneth Foote is a Professor of Geography, Sustainability, Community and Urban Studies at the University of Connecticut. His research focuses on historic preservation, heritage tourism, and the commemorative landscapes of the US and Europe, especially the way events of violence and tragedy are memorialized. His recent research has also focused on the spatial dimension of racialized and gendered violence in the US. In addition, Foote is also known for his work in geospatial technologies, especially geovisualization, as well as his efforts to improve professional development opportunities for early-career academics and department leaders. Foote is a past president and fellow of the American Association of Geographers (2010-11) and a past president of the U.S. National Council for Geographic Education (2006). He has received major national and international awards for his research, teaching, mentoring, and service from the American Association of Geographers, National Council for Geographic Education, University Consortium for Geographic Information Science, and the Royal Geographical Society.

Shannon O'Lear is a Professor of Geography at the University of Kansas, is a political geographer with interests in environmental geopolitics, critical geopolitics, the South Caucasus, and Science and Technology Studies (STS). Her recent work includes Environmental Geopolitics (2018), A Research Agenda for Environmental Geopolitics (2019), and A Research Agenda for Geographies of Slow Violence (2021). She co-authored a report for the National Science Foundation, Environmental Change and Human Security: Research Directions. She has served as the councilor for the Great Plains Rocky Mountain Region of the American Association of Geographers (AAG). She holds a BA in Geography and Russian and a master's degree in Geography from the University of Colorado, Boulder, and a PhD in Geography from Syracuse University.

LaToya E. Eaves is an Associate Professor of Geography and Sustainability at the University of Tennessee, Knoxville. She has been instrumental in increasing the visibility of Black geographies in addition to her research, which centers on questions of race, Blackness, gender, sexuality, and place – especially in terms of the US South and Southeast. Recipient of numerous awards, including the 2019 Ronald F. Abler Distinguished Service Honors for her transformative impact on the American Association of Geographers through her commitment to Black geographies, she serves as PI on a half million dollar NSF grant for a collaborative project on museums, public pedagogy, and Black geographies in the US. Her profoundly interdisciplinary contributions reflect her experiences, training, and teaching across Africana studies, women's and gender studies, and geography departments.

Jongwon Lee is a Professor Social Studies Education at Ewha Womans University in Seoul, South Korea. Specializing in teacher education, his work focuses on the development of innovative design and instructional materials for technology-enhanced learning environments, particularly those involving geospatial technology. His approach emphasizes active learning and inquiry-based learning. Since 2016, Jongwon has also held the position of Secretary for the Commission on Geographical Education under the International Geographical Union, contributing to the field of geographical education on an international scale.

THRIVING IN AN ACADEMIC CAREER

An International and Interdisciplinary Guide for Early Career Faculty

Edited by
Michael Solem, Kenneth Foote, Shannon O'Lear, LaToya E. Eaves and Jongwon Lee

Routledge
Taylor & Francis Group
LONDON AND NEW YORK

Designed cover image: fizkes/Shutterstock

First published 2025
by Routledge
4 Park Square, Milton Park, Abingdon, Oxon OX14 4RN

and by Routledge
605 Third Avenue, New York, NY 10158

Routledge is an imprint of the Taylor & Francis Group, an informa business

British Library Cataloguing-in-Publication Data
A catalogue record for this book is available from the British Library

Library of Congress Cataloging-in-Publication Data
Names: Solem, Michael, editor. | Foote, Kenneth E., 1955– editor. | O'Lear, Shannon, editor. | Eaves, LaToya E., editor. | Lee, Jongwon, editor.
Title: Thriving in an academic career : an international and interdisciplinary guide for early career faculty / Edited by Michael Solem, Kenneth Foote, Shannon O'Lear, LaToya E. Eaves and Jongwon Lee.
Description: Abingdon, Oxon ; New York, NY : Routledge, 2025. | Includes bibliographical references and index.
Identifiers: LCCN 2024042174 (print) | LCCN 2024042175 (ebook) | ISBN 9781032379944 (hardback) | ISBN 9781032379951 (paperback) | ISBN 9781003342953 (ebook)
Subjects: LCSH: College teaching—Vocational guidance. | College teachers—Professional relationships. | Geography—Study and teaching (Higher) | Universities and colleges—Faculty. | Career development.
Classification: LCC LB1778 .T47 2025 (print) | LCC LB1778 (ebook) | DDC 378.1/24—dc23/eng/20241002
LC record available at https://lccn.loc.gov/2024042174
LC ebook record available at https://lccn.loc.gov/2024042175

ISBN: 9781032379944 (hbk)
ISBN: 9781032379951 (pbk)
ISBN: 9781003342953 (ebk)

DOI: 10.4324/9781003342953

Typeset in Galliard
by codeMantra

Dedication

As geographers, we approach faculty development mindful not only of the many ways that individual teaching, research, and service activities intersect but how these core academic practices are situated in diverse institutional settings. While the work of individual faculty can often seem solitary, the truth is all faculty must conduct their scholarship in a web of social relations linking them to colleagues near and far, campus administrators, journal editors, program officers, professional societies, students, and members of the local community. Professional life in the ivory tower is often stereotyped as occurring outside the "real world," but, in reality, faculty practice cannot be separated from the dynamics of race, gender, class, sexuality, and other social characteristics that, while unrelated to academic ability, are too often associated with unequal career advancement and opportunity.

Perhaps no one in our discipline understood this better than Jan Monk, Professor Emerita at the School of Geography, Development and Environment at the University of Arizona. A cherished colleague, mentor, and friend to so many, Jan dedicated much of her career to the needs of graduate students and early career researchers of all backgrounds. She gave special attention to the gendered nature of faculty work and life. Jan was one of the original leaders of the Geography Faculty Development Alliance (GFDA) workshops and contributed numerous research articles for the Enhancing Departments and Graduate Education (EDGE) project, both initiatives of the American Association of Geographers (AAG) supported with funding from the U.S. National Science Foundation.

For helping countless new faculty thrive, we dedicate this book to Jan. Yet we also wish to acknowledge the very spirit of faculty development that Jan cultivated, one that recognized the "all hands on deck" approach that over time has made rich professional development part of the social fabric and programming of academic departments and scientific and educational societies to which we and so many others belong. We accordingly salute the hundreds of junior and senior faculty who have contributed to the vision of the GFDA through their participation in workshops, conference sessions, online webinars, and informal conversations on and off their respective campuses.

To the readers of this book: each time you pick it up and select a chapter, a baton passes from the authors to you. We hope this facilitates a productive and enjoyable career and inspires a commitment to lifelong professional development and mentoring.

AAG
AMERICAN ASSOCIATION of GEOGRAPHERS

CONTENTS

CONTRIBUTORS

Derek H. Alderman, University of Tennessee, Knoxville, TN, USA

Dagmar Budikova, University of Michigan, Dearborn, MI, USA

Dydia DeLyser, California State University Fullerton, CA, USA

LaToya E. Eaves is an Associate Professor of Geography and Sustainability at the University of Tennessee, Knoxville.

Dafna Fisher-Gewirtzman, Technion-Israel Institute of Technology, Haifa, Israel

Kenneth Foote is a Professor of Geography, Sustainability, Community and Urban Studies at the University of Connecticut.

Eric Fournier, Washington University in St. Louis, MO, USA

Douglas W. Gamble, University of North Carolina, Wilmington, NC, USA

Brandi Gaertner, Pennsylvania State University, University Park, PA, USA

Adrienne Goldsberry, Pennsylvania State University, University Park, PA, USA

Meg Gregory, Washington University in St. Louis, MO, USA

Herbert Q. Hambati, University of Dar es Salaam, Tanzania

Peter J. Haproff, University of North Carolina, Wilmington, NC, USA

Robin Matross Helms, Association of Community College Trustees, Washington, DC, USA

Miguel Á. Hernández, San Francisco State University, San Francisco, CA, USA

Jennifer Ho, University of Colorado, Boulder, CO, USA

Keiko Ikeda, Kansai University, Osaka, Japan

Joshua Inwood, Pennsylvania State University, University Park, PA, USA

Zuhri James, University of Cambridge, Cambridge, UK

Andrew Jolly-Ballantine, University of Connecticut, Storrs, CT, USA

Joseph J. Kerski, Esri, Louisville, CO, USA

Beth Fletcher King, Pennsylvania State University, University Park, PA, USA

Francis T. Koti, Middle Tennessee State University, Murfreesboro, TN, USA

Hilda E. Kurtz, University of Georgia, Athens, GA, USA

Gil Latz, Waseda University, Tokyo, Japan, and Indiana University, Indianapolis, IN, USA

Jongwon Lee is a Professor of Social Studies Education at Ewha Womans University in Seoul, South Korea.

Juliana Mansvelt is a Professor of Geography at Massey University/Te Kunenga ki Pūrehuroa, New Zealand/Aotearoa

Deborah G. Martin, Clark University, Worcester, MA, USA

Chris McMorran, National University of Singapore, Singapore

Michael E. Meadows, Nanjing University, China and University of Cape Town, South Africa

Emily Murai, University of California, Santa Cruz, CA, USA

Shannon O'Lear is a Professor of Geography at the University of Kansas.

Patricia Price, Queens College, City University of New York (CUNY), NY, USA

Sarah A. Radcliffe, University of Cambridge, Cambridge, UK

Lynn M. Resler, Virginia Polytechnic and State University (Virginia Tech), Blacksburg, VA, USA

Michael Solem is a Professor in the Department of Geography and Environmental Studies at Texas State University.

Patricia Solís, Arizona State University, Tempe, AZ, USA

Daniel Sui, Virginia Polytechnic and State University (Virginia Tech), Blacksburg, VA, USA.

Dan Trudeau, Macalester College, St. Paul, MN, USA

Antoinette M.G.A. WinklerPrins, Johns Hopkins University, Washington, DC

Chen Wu, Institute of Tibetan Plateau Research, Chinese Academy of Sciences, Beijing PRC

May Yuan, National Science Foundation, Alexandria, VA USA, on leave from the University of Texas at Dallas, TX, USA

PREFACE

This book was created to support aspiring academics in geography and related social and environmental sciences. Building on the editors' experience over 20 years leading faculty development workshops and research initiatives, this book's chapters collectively aim to help early career faculty and graduate students get a quick start on their academic careers. Even the best graduate training can leave new faculty with questions about how to succeed in new jobs: What are the expectations for research, teaching, and service, and what are effective strategies for meeting and exceeding these expectations? This book helps answer these questions with concise, to-the-point chapters focusing especially on the issues that are often the greatest source of stress for new faculty, including the ways academic responsibilities sometimes intersect with our personal lives.

We prepared this book in the spirit of building supportive academic cultures in which early career faculty thrive and become successful in their roles as teachers, scholars, and colleagues. The movement to make faculty development an ongoing focus of academic careers can be traced to Earnest Boyer's seminal book *Scholarship Reconsidered* (1990). Boyer argued for sweeping change in academic culture to support research into and critical discussion of faculty teaching and educational practices. The argument came at a time of growing concern in academia that members of the professoriate, especially early career faculty, were struggling to maintain healthy and productive professional lives in a "sink or swim" work environment.

Scholarship Reconsidered led to a wave of research in higher education focusing on faculty experiences in graduate school and the first few years of their academic appointments. A key finding was that very few graduate programs offered training in teaching and learning processes. As a consequence, undergraduate teaching became a primary source of stress, in many cases undermining the ability of faculty to conduct quality research, find time to secure external funding, maintain a consistent publication record, and so forth. And because few colleges and universities at the time had centers for teaching and learning, most faculty had nowhere to turn for help and professional development support.

Things started to improve with the growth of the "scholarship of teaching and learning" movement, which slowly changed academic culture to reward faculty for researching their teaching, openly presenting their courses and efforts to improve student learning outcomes, and publishing their pedagogical experiments in the educational literature. Still, much of the advice available to faculty was general in nature. Little attention was given to teaching and learning issues that were often unique in the context of different academic fields and disciplines.

Although graduate programs in the social and environmental sciences do a fine job in developing research specialists, few programs prepare individuals for the full range of responsibilities that characterize life and work as an academic. *Thriving in an Academic Career* therefore complements, but considerably expands, the topical foci of existing titles in faculty development, approaches to graduate preparation, and campus support systems by introducing readers to discipline-specific strategies for succeeding in an academic position.

Organization of the book

The themes of balance and harmony underpin this book's approach to faculty development. Many existing titles focus on teaching and learning issues, tips for grant writing and publishing, preparation for tenure and promotion, and other forms of specific advice for academic professionals. These approaches are valuable, but their narrow topical focus sometimes misses the holistic and intersectional nature of academic practice. While many experienced faculty members understand this, research shows that faculty who achieve early balance in their academic responsibilities and home life are more likely to succeed in all aspects of their career, while strengthening the quality and climate of their programs and campuses.

The chapters in this book are organized into four thematic sections. The first sets the tone with big picture ideas for getting off to a fast and successful start in an academic position. Emphasis is placed on the importance of time management, building collegial relationships and professional networks, and finding ways of connecting our professional work as faculty to our personal interests and aspirations.

Section II addresses many core topics of teaching and learning that often are the most challenging issues facing faculty at the early stage of their career. The chapters offer practical tips for deploying instructional technologies, designing face-to-face and online courses, teaching in the field, and supporting critical thinking. This section also addresses the importance of teaching for diversity and with an equitable mindset, with special attention given to decolonizing knowledge in the curriculum and advancing social justice through our teaching practices.

The focus of Section III is on research and scholarship. Chapters address strategies for building a research agenda and trajectory with goals set in the short and long terms, and the crucial role of publishing and promoting academic work in that process. Many faculty are expected to seek and secure external funding to support their research and teaching, and this section offers guidance for writing high-quality proposals.

The final section of this book examines topics that are fundamental to a successful academic career over the long run. This includes taking measures to prioritize our physical and mental health and how to cope with adversity. Other chapters provide tips for preparing a competitive dossier for tenure and promotion and the value of service for one's career, department, and academic community.

ACKNOWLEDGMENTS

This book was produced with the support of Gary Langham, Executive Director of the American Association of Geographers (AAG). The editors are grateful for the assistance provided by Nirvana Heidarian, Department of Geography and Environmental Studies, Texas State University.

SECTION I

Getting off to a good start

Kenneth Foote

The chapters of this section address a paradox of contemporary graduate education in many parts of the world -- that some of the knowledge, skills, and strategies needed to thrive in an academic career aren't actually taught in graduate school. Graduate programs provide excellent training in research, writing, and publishing skills and, typically, also offer some teaching experience. However, as new faculty open the door to their first position, they often find themselves facing issues and challenges they've never faced before. Suddenly they are handling several teaching preparations each semester, trying to meet expectations for research and publication, serving on departmental committees, advising students, and working toward tenure or a long-term appointment (or, if in a temporary appointment, keeping active in the job market). Many early career faculty are fortunate to have colleagues that provide help and mentoring on these issues, but other new hires are left to fend for themselves.

In the chapters in this section, we push back against this sort of do-it-yourself professional development. The topics addressed include service, collegiality, collaboration, care, career planning, and other interpersonal skills important to success. In her book *Making the Implicit Explicit* (2007), Barbara Lovitts has termed this the implicit curriculum of graduate education, to distinguish it from the seminars, classes, and labs that constitute the explicit curriculum of graduate education. One of the most important points Lovitts makes is that access to this implicit knowledge is often privileged and accessible to some students more than others.

We have chosen to open this book with a set of chapters aimed at Lovitts's goal of making the implicit knowledge needed to thrive in an academic career, an explicit part of professional development for early career faculty. Even if

DOI: 10.4324/9781003342953-1

they weren't part of your graduate training, we believe these chapters will provide you with some important strategies for career success. We also hope they will reduce some of the stresses of starting an academic life and encourage you to be proactive in planning your work and career.

"Getting Started in Academic Life" by Koti and Hambati focuses on some of the effective strategies for launching a successful academic career. Many new hires are surprised by the greater range of responsibilities they face in their new positions, often far beyond their experiences in graduate school. For some, the pressure of these new responsibilities in teaching, research, and service is compounded by a lack of clarity about expectations and limited mentoring. Drawing on their own experiences and published literature, Koti and Hambati argue that newly appointed faculty should themselves take the lead in defining and refining their new roles. Boiling down their advice to some key points, they stress, among other points, the importance of honing our teaching craft, knowing we always have room to improve; setting realistic goals and being patient with ourselves and our progress; tracking and documenting the professional growth; getting research started as soon as possible; and being strategic in taking on service responsibilities. They understand that starting an academic career can be stressful but argue that there are many positive steps a person can take to reduce stress and work toward success.

"Building momentum: Developing research and teaching trajectories," by Sui and Fisher-Gewirtzman, takes the next steps beyond those recommended by Koti and Hambati. They note that even though beginning an academic career is an exciting time, setting some plans and goals early in a new position can help us make good use of new resources and opportunities for collaboration in research and teaching. Sui and Fisher-Gewirtzman are particularly effective in suggesting ways of building career momentum through career planning. They suggest developing plans to secure early wins; seeking synergy between research and teaching; and building a record of research publication and exposure, while also embracing serendipity and keeping an open mind about unexpected opportunities.

Whereas the first two chapters focus largely on our professional work, Mansvelt's "Aligning Personal Aspirations and Professional Goals: Finding the 'Sweet Spot'" argues that long-term success and career satisfaction hinges on considering how our academic responsibilities and personal lives intersect. She takes a broad, integrated approach to career development, accepting that conflicts do sometimes arise between our personal and professional lives. Her chapter focuses on identifying goals and aspirations and argues that a value-based approach can help you navigate both your personal and professional worlds. She addresses the myth of separating work and personal goals by outlining some practical strategies for managing the areas where they intersect. A particular focus is on considerations around part-time employment with the need to assess benefits and pitfalls carefully. She also advocates intentional reflexivity and self-assessment across one's career path to assist us in considering our personal aspirations and professional goals together. In doing so we can make better choices around short-term and long-term goals that can help us avoid potential conflicts before they arise.

Murai's chapter focuses on some of the opportunities and challenges of pursuing a career as a non-tenure track faculty (NTTF) member. These non-tenure track positions have been growing over the past few decades in higher education owing to budget cuts and rising expenses. Though NTTF are often associated with exploitation, de-professionalization, and invisible labor, non-tenure track positions encompass a wide range of roles and experiences, and NTTF experiences are highly variable. Murai argues that under fair labor conditions and one's own expectations, it is possible to thrive on the non-tenure track. She makes the case that, given the current realities of the academic job market, these positions offer worthwhile career pathways, even if work needs to continue on making them more respected and inclusive. Drawing on her own career experiences as an NTTF, Murai puts her finger on issues to consider if you're thinking about an NTTF opportunity and offers some strategies to help you thrive if you're in one already.

Time management is often reported as one of the greatest stresses faced by early career academics. But, as Price points out, effective time management is best conceived as a process of setting goals and allocating time to those goals, rather than letting clocks and calendars set our priorities. Goals shape intended actions, and time management is simply a set of repeated practices that, if not aligned with goals, lead nowhere. She stresses the value of setting realistic goals and aligning daily actions with those goals, making sure to always keep an eye on activities that are important, even if they aren't urgent. Her chapter covers a set of tried-and-true time management techniques, and tips for using technology. Time management concerns your own time, as well as the time of others. As a faculty member, you will interact with colleagues, administrators, students, staff, and members of the public. It is important to be able to manage these interactions in a way that respects your time and the time of others. Additionally, structural considerations such as gender, race, age, and other aspects of our social location can shape an individual's time management calculus. The good news is that time management is a skill that can be learned and improved, and developing good habits early in your career will set you up for success.

In "Creating communities of care and caring for community: Building community through work-life integration," Trudeau and DeLyser further develop some of the ideas raised in the preceding chapters, particularly those by Mansvelt and Price. Like Mansvelt, they argue that the major facets of academic life – scholarship, teaching, and service – are often treated as separate domains of our work and that any one, or all of these, can even be at odds with aspects of our private lives – family, community, and leisure. This can lead to situations where we feel we must scramble to cover disparate, even apparently conflicting, parts of our lives, leading to feelings of loneliness, depletion, and stress. Trudeau and DeLyser make the case that we can live otherwise, and it's up to us to make that happen. In this chapter, they argue that it is possible to move toward a more conscientious and collaborative formation of the academy by prioritizing care for ourselves and for others and centering reciprocity, collaboration, and support in our relationships. A propensity for circulating these types of care can help us connect with others, find support for ourselves, and create circumstances in which we and others thrive. To explore this idea, Trudeau and DeLyser consider how some academics have integrated different facets of their work and lives through community-engaged scholarship and community geography.

The final chapter in this section, "Slowing down and saving our humanity," ties together many themes discussed by the other authors, particularly the many stressors that new faculty face as they embark on an academic career. Like the other authors, O'Lear and Foote recognize the excitement of getting started in an academic career, but they also offer perspectives and actionable suggestions for navigating challenges that arise as we move forward in our careers. Drawing on other chapters in this volume that look at self-care, time management, and mentoring, as well as research on academic work, this chapter focuses on understanding a broad and realistic view of stressful aspects of an academic career and key strategies aimed at supporting physical well-being and mental and emotional health, and slowing down to identify ways of saving time, energy, and effort. From cultivating supportive networks to winning at email, this chapter encourages new academics to maintain a sense of perspective and a sense of humor.

Reference

Lovitts, B. E. 2007. *Making the implicit explicit: Creating performance expectations for the dissertation.* Sterling, VA: Stylus.

1

GETTING STARTED IN ACADEMIC LIFE

Advice for new hires

Francis T. Koti and Herbert Q. Hambati

Introduction

Diving into an academic career is both exciting and stressful for a variety of reasons (Dennihy 2020; McDonnell 2020). New hires are suddenly expected to contribute a greater range of tasks than they've had to face as graduate students, including teaching, research, and service. The balance of these responsibilities tends to vary by the program a person joins and the terms of their appointment. While new hires are facing these tasks, time always seems to be running out as they now face regular reviews for salary increases, reappointment, promotion, tenure, funding, as well as immigration and citizenship for foreign-born faculty. Additionally, the early career period is often a time when key life decisions are being made including but not limited to marriage, partnering, starting or growing families, and the care of parents and members of extended families.

Unfortunately, graduate programs don't always provide formal training in all the professional responsibilities that will be important in their new positions (McDonnell 2020; McLeod et al. 2014). Except for some relatively limited training in course preparation, curriculum, and pedagogical issues, the level of experience needed to pursue competitive funding and get published and how to be an effective colleague and academic "citizen" are novel territories for many new hires.

This chapter fills that void. We offer advice to new hires on effective ways to successfully launch an academic career. We draw from personal and professional experiences and the literature to highlight effective strategies in the transition to academic life. Throughout the chapter, we try to shed light on the following questions: (1) What are some baseline things that – if we get

DOI: 10.4324/9781003342953-2

them right at the beginning – can lead to a successful liftoff of our academic careers? (2) What opportunities, resources, and tools are available for early career faculty to help reduce the stress that comes with getting started? (3) What are some common challenges to be expected at the launch of academic careers, and what are effective strategies to overcome them? We have organized our recommendations into seven broad strategies that can lead to the successful launch of an academic career.

Intentionality as key to a successful takeoff: take the lead in charting your own course – know your job, the people, and places

We identify seven ways you can take the lead in charting your own course.

Clarifying your roles and responsibilities: keep asking questions

When Francis started his first appointment, he found out that one of the most important and challenging things was getting a clear sense of his roles and responsibilities in his new department. In order to get a clear picture of your roles and responsibilities, and how they fit within the broader university mission, we recommend taking the time to ask and find answers to important questions related to the job. The following are examples of questions that can be asked: What were they looking for when they hired me? What exactly was I hired to do? What am I expected to do in the first year? How will I know I have met these expectations? What resources will I need in order to fulfill my duties in the first year? An effective way to answer these questions is reviewing and documenting important aspects of your position as listed in key documents such as your appointment letter, job advertisement for your position, or the Faculty Handbook (mainly for universities in the United States). Create a folder on your computer that contains all these documents, and regularly refer to them to evaluate your progress. Because they are your immediate supervisor, schedule periodic meetings with your department chair to get clarification on your specific expectations at the department level (Dennihy 2020).

Organizing your workspace from Day 1: treat it as your identity and inspiration

Your office or lab space is part of your identity, and how well you organize it reflects on you as a professional. Be intentional about organizing it as part of the "getting-started" process. A well-organized office workspace increases your efficiency and can also be a source of inspiration for you. In their study, "*How Your Office Space Impacts Employee Well-Being*," Kohll (2019) estimated that we spend a third of our lives in the office. Based on this assertion, it makes perfect sense to make organizing your workspace a priority during the "getting-started" process. Furthermore, there are documented benefits of neatly organized office spaces including but not limited to reduced stress levels, better time management, ability to remember important tasks, increased productivity, and better impression on your constituents (Colenberg et al. 2021; Forooraghi et al. 2021).

Study and feel the campus climate and situate your activities within the "big picture"

From the moment you set foot in an institution, treat it as your job to understand how the university works – including the chain of command, who controls the budget, who makes major decisions, as well as who authorizes things (Marquette University, 2022). Furthermore, every university has its own culture – what is acceptable in one place may be unacceptable elsewhere. In the first year, pay attention to the unwritten rules in your university. For example, what are some things most people in your university would frown upon. At the same time, make sure what you do advances a broader purpose of the university. When Herbert started his appointment as a lecturer, he thought he would focus on his teaching, research, and consulting responsibilities but quickly learned that what he was doing as a lecturer had to align with a broader mission, vision, and values of his university. As part of getting into the academic life mindset, always ask yourself, "How does what I do support the broader mission of the university and my unit?" Fortunately, many universities, colleges, and departments have well-documented strategic plans that clearly spell out their missions, visions, and values (Ishumi and Anangisye 2021). These may be published in print or simply posted on the university website. Incorporate the university's mission, vision, values, and goals in your work (McLeod et al. 2014).

Develop a working relationship with your colleagues

One of the most important investments for new hires is building trust. Find ways to develop a working relationship with your colleagues. Small acts of kindness, such as always being optimistic about things, treating colleagues with respect, and saying hello in the hallways, will lead to trust over time. Dennihy (2020), for example, recommends talking to your colleagues about their own teaching experiences, specifically "the types of assignments they use, lesson plans they have found particularly successful, challenges they have faced in the classroom and the like." When Francis joined his department, he always treated each of his colleagues like there was something he wanted to learn from them. What you are striving for as a new hire is for your colleagues to describe you with positive adjectives such as curious, interested, respectful, positive, uplifting, enthusiastic, encouraging, collegial, energetic, and supportive.

Know your way around the campus, physically and virtually

Knowing your way around the campus, asking questions, and knowing where to get help from are your biggest investments in getting started as a faculty member. In addition to locating venues where your classes meet, locate other offices and contacts on your campus that may help make your life easier. The University Library is a key resource for your teaching and research needs, and knowing where it is, how to get there, and important contacts there is as important as knowing the location of specific resources for your needs. A faculty member with heavy research expectations should locate and introduce themselves to a contact in the grant's office as early as possible. Other strategic offices and places may include gyms and wellness centers, the human resource office, information technology division, the dean's office, and on-campus daycare facilities for people with families. Try to set aside time every few weeks to visit and meet people in these units. These contacts can be very helpful when you have questions or need help.

Maintain a list of strategic contacts on campus and beyond

There is a group of people on your campus who are going to be important as you launch your academic career, and we break those down into three categories: *allies, mentors, and sponsors.* Your *allies* are people that make your life manageable every day. The administrative assistant in your department, the staff in the grant's office, information technology staff, the librarians, the custodial staff in your building – all these folks make your daily life possible. Your *mentors* are trusted people you turn to with questions about, and advice on, specific areas of your work and your own progress. From your department chair, senior colleagues, peer mentors, to mentors outside the university – maintain a list with their phone numbers and emails. Vick and Furlong (2010) recommend talking to other new faculty members in departments across the university, arguing that "while expectations may vary from department to department, chatting with others who are 'learning the ropes' will help you learn as well as think of questions to ask in your department." Finally, it is important to forge positive relationships with people who make decisions that will impact your career. For starters, develop a working relationship with your senior colleagues and your immediate supervisor. They will be first in the line of decisions being made for your promotion and long-term appointment. In addition to writing your annual evaluation for instance, your department chair or your immediate supervisor is your biggest resource for information, and perhaps one of the most effective person to introduce you to senior administrators and also across campus. In her keynote address to the 2021 Kenya Scholars and Studies Association conference, Dr. Sherine Obare noted, "everyone needs to have someone who speaks on their behalf when they are not in the room." She called that group "your *sponsors.*"

Don't wait to be asked; get involved on campus from the beginning. It is the most effective way to know people and places

As part of taking the lead in charting your own course, make it part of your work culture to get involved on campus. Among the opportunities are faculty recreational activities like tennis or basketball, intramural sports, online games, faculty learning communities, student-led activities, brown bag lunches and colloquia, community gardening, and so on. To give you a personal example, during the second year of Francis's appointment, he joined a weekly faculty and staff exercise group without knowing any of the members. Fast-forward several years later, he found he had made friends with the provost's personal assistant, the university registrar, a librarian, accounting staff, new faculty members, a nurse at the infirmary, and the principal at his son's school. That one small social network helped Francis meet people who could support him professionally and personally in subsequent years as his roles changed from new hire to tenured faculty member, and then to study abroad director, department chair, and interim associate dean.

Teaching your first class: hone your craft but leave room to improve; mastery of teaching takes time

Teaching is often not the reason most people choose to become college professors (Fink 2013), yet it is central to the mission of most universities and colleges (Kniel 2008). Because teaching is going to be so important, endeavor to hone your craft from the beginning. In addition to preparing your lectures and labs, invest time to research effective ways to connect with your students. To make your classes exciting, learn and use teaching aids and tools appropriate to your area. Regardless of the class size and level, invest the time needed to prepare and present material in challenging and engaging ways.

Be aware, however, that no one expects you to start off as an expert. But there are ways to overcome self-doubt and build confidence. For example, inviting a senior faculty colleague to sit in on your classes to provide feedback may sound intimidating, but it is a great way to boost your confidence and build trust with colleagues. As mentioned in chapters in the second section of this book, there are plenty of resources for course design and for improving your teaching skills (Fink 2013; Gooblar 2019; Nilson 2016). If there is one skill we would highlight, it would be to master the Learning Management System (LMS) being used in your institution. It will save you time and headaches. Additionally, attend as many faculty development workshops you can fit into your schedule to learn the latest and greatest teaching techniques. Take advantage of any course improvement grant programs your campus might offer. Also, teaching assistants can help to prepare and improve materials, as can students in your classes who can provide feedback and good ideas for improvement.

Scholarship and creative activities as your bread and butter: get started as soon as possible

One may wonder if research is the only area in which most new PhDs have the most training and background, how come it is also their big source of anxiety as new hires? Well, research is the only thing they were doing as graduate students, but as new faculty members, there is suddenly an expectation to contribute a greater range of tasks beyond research — including teaching and service. In their work, Medina and Baert (2013) highlight the inherent precariousness in the early stages of scientific careers by shining a light on the contextual and temporal dimensions of academic life – further drawing our attention to the multiple roles early career academics must juggle. Developing a research trajectory is covered in other chapters of this book, but here we emphasize getting a research agenda started as soon as possible.

But where to begin? We recommend doubling down on your dissertation research during the first year or two, mining as many publications as possible from this work or from closely related topics. In the meantime, learn about finding intramural and extramural research funding opportunities, gain a sense of research and publishing expectations at your institution, and develop a plan for building your research trajectory. This may involve attending grant-writing workshops at your university or professional meetings, joining research and writing accountability and support groups with peers, and seeking collaboration with other more experienced researchers in your area. Keep in mind that internal grants from your university may provide just the right amount of seed funding to pursue larger projects and grants (Graul 2022).

Service responsibilities are important, but be strategic and selective

While teaching and research requirements for faculty appointments are usually spelled out in appointment letters and the Faculty Handbook, service responsibilities are rarely as clearly defined. If this is the case for you, we recommend meeting with your department chair to clarify expectations and to learn what types of service opportunities are available and valuable to you, the department, and university. We urge you to be strategic and selective in your service activities. While McDonnell (2020) recommends practicing to say "no" to service requests, McLeod et al. (2014), on the other hand, advise to "do service sparingly." However, both Monk (2004) and Hanson (2007) emphasize that service is also central to a successful career. It is how we build networks of contacts, collaborators, and friends who can help us with our research and teaching.

Set realistic and practical goals for Year 1 and be patient with yourself

Prevailing circumstances can pressure new hires to set unrealistic goals for themselves in their first year on the job, and here is a personal example. In his first semester, Francis started off teaching four classes, which included three different course preparations, one of which he had never taught before. Francis's appointment contract also required maintaining an active research agenda. On the other hand, the area of greatest need in the department for which he had been hired was to build undergraduate research. Due to increasing enrollment, Francis's colleagues also expected him to help with academic advising, something he said he "loved to do" during his interview.

Fortunately, Francis took the advice of his experienced department chair who recommended that he avoid burnout by setting some modest and realistic goals, rather than do everything at once. For example, one of his Year-1 research goals was to *publish a chapter from his doctoral dissertation in a peer-reviewed journal,* work he already had in motion. Likewise, one of his Year-1 teaching goals was to revise the syllabus for his urban geography course to incorporate a geographic information systems (GIS) component – a class he had already taught in graduate school. The key point here is that setting modest goals as a new hire is a positive step forward, rather than one that reflects badly on you. After all, McLeod et al. (2014) caution new faculty to "recognize that your first year is about learning the ropes and building capacity."

Mapping your progress: create a professional development plan to keep yourself focused and within schedule

Professional development planning is discussed elsewhere in this book, but we want to emphasize its value here, too. Think of a plan as a table or chart of your career development goals and strategies to achieve them (Stalnaker 2018; Whitaker 2018). Create one that, at the very least, includes professional goals (categorized into teaching, research, and service), strategies and activities to be undertaken to meet those goals, the knowledge and skills as well as other resources needed to accomplish the activities, criteria to be used to measure success of the goal, and year-end status of the goal (Table 1.1).

A professional development plan is useful for several reasons. Most importantly, it serves to align professional activities with career goals and university mission. Second, year-to-year plans can help you to focus on the goals and track your progress. Finally, a well-structured plan can keep you from taking on too many or unnecessary additional responsibilities when faced with the temptations of, for example, joining interesting projects with senior colleagues or investing time in an article or project related to your expertise, but tangential to your plans.

TABLE 1.1 A sample of Professional Development Plan with Year 1 goals

Assessment Area/Aspect	*Teaching*	*Research*	*Service*
Goals	Goal 1: Revise the syllabus for the "Urban Geography" course to incorporate a GIS component	Goal 1: Publish paper from dissertation chapter, "Assessing the downstream impacts of COVID-19 pandemic on urban livelihoods in Sub-Saharan African cities"	Goal 1: Organize game night for graduate students
Tasks/ Activities/ Strategies	(1) Obtain old syllabus from archives; (2) review and revise learning outcomes; (3) incorporate GIS activities and assessment strategies; (4) submit to the department curriculum review committee	(1) Compile and clean data; (2) review literature; (3) analyze data; (4) write; (5) seek informal peer review; (6) revise; (7) submit.	(1) Meet with Geography Club officers; (2) propose game night and set date; (3) obtain funds from the department to purchase board games; (4) book venue
Knowledge/ Resources needed to achieve this goal	Six hours; GIS software; GIS textbook	36 hours; GIS; remote sensing; funding	Six hours; funding to buy board games; venue
Success criteria (evidence of completion)	Revised syllabus; submission of syllabus for review	Paper submission	Hosting of game night during spring semester
Year-end status	Syllabus review completed and approved by the department Curriculum Review Committee	Paper submitted to *"Africa Today"* journal on March 23; pending reviewer comments	Game night hosted on March 22

Document your professional growth from Day 1 in a portfolio

From Day 1 of your first appointment, collect evidence of your accomplishments. You can assemble them into a portfolio that documents evidence of your accomplishments across all of your responsibilities. This portfolio can be very useful if you need to prepare for annual or periodic merit reviews, and is equally helpful for reappointment, promotion, and tenure reviews. Because academic portfolio development is a gradual process, the collection of materials should be an ongoing process. From Day 1 of your job, scan and save evidence of accomplishments both great and small, such as thank-you notes from administrators, colleagues, and students; invitations to speak in classes; requests for interviews; and programs from conference presentations (Kniel 2008). The structure of the academic portfolio may vary from one university to another or by discipline. Some portfolio may focus primarily on teaching (Seldin et al. 2010) while others may cover all three areas of evaluation (Pinto Zipp and Simpkins 2010).

A conclusion, or just a beginning...

The following chapters highlight other equally important issues to getting off to a good start in an academic career. In our experiences, we would certainly highlight the same issues. We feel it is particularly important to focus on Juliana Mansvelt's chapter on how to realize our professional goals while being mindful of our lives and families outside our office, lab, or classroom. Many of our suggestions dovetail nicely with those of the authors of chapters in the next section focusing on "Becoming a curriculum leader," or look ahead to topics related to research, grants, publishing, and building toward long-term success in the third and fourth sections of this book.

In this chapter, we have tried to provide some guidance on your first steps toward thriving in your academic career. We have drawn from our own professional and personal experiences as well as from published literature to highlight key strategies, resources, opportunities, and tools for getting started. Throughout the chapter, we have noted that starting an academic career can be stressful. At the same time, we have tried to provide steps to help reduce this stress by suggesting the value of building working, productive relationships, setting attainable goals, honing teaching abilities, getting started with research early, enjoying life outside of work, practicing self-care, and getting involved around campus. We also emphasize the value of taking the lead in working toward these goals, while also being strategic in our choices and being careful to document our work and accomplishments. We see all these as steps toward a successful academic career.

References

Colenberg, S., T. Jylhä and M. Arkesteijn. 2021. The relationship between interior office space and employee health and well-being – A literature review. *Building Research and Information* 49 (3): 352–366. doi: 10.1080/09613218.2019.1710098

Dennihy, M. 2020. Starting a tenure track career. *Inside Higher Ed.* September, 21. Accessed December 28, 2022. https://www.insidehighered.com/advice/2020/09/22/tips-how-succeed-during-your-first-year-tenure-track-opinion

Fink, D.L. 2013. *Creating significant learning experiences: An integrated approach to designing college courses*, rev. ed. San Francisco, CA: Jossey-Bass.

Forooraghi, M., E. Miedema, N. Ryd, and H. Wallbaum. 2021. How does office design support employees' health? A case study on the relationships among employees' perceptions of the office environment, their sense of coherence and office design. *International Journal of Environmental Research and Public Health* 18 (23): 12779. doi: 10.3390/ijerph182312779

Gooblar, D. 2019. *The missing course: Everything they never taught you about college teaching.* Cambridge, MA: Harvard University Press.

Graul, A. 2022. Improving performance through strategic teaching documentation: The continued importance of Seldin, Miller & Seldin's "The Teaching Portfolio" in the face of a new reality. *Journal on Empowering Teaching Excellence* 5 (3): 4–8. doi: 10.26077/94ca-3df5

Hanson, S. 2007. Service as a subversive activity: On the centrality of service to an academic career. *Gender, Place & Culture* 14 (1): 29–34.

Ishumi, A., and W. Anangisye. 2021. *Fifty years of education in Tanzania (1961–2011): A Historical account and review.* Dar es Salaam: Dar es Salaam University Press.

Kniel, K. 2008. Getting started in your academic career. In *Careers in food science: From undergraduate to professional*, ed. R. Hartel and C. Klawitter, 305–315. New York: Springer. doi: 10.1007/978-0-387-77391-9_29

Kohll, A. 2019. How your office space impacts employee well-being. *Forbes*, January 24. Accessed July 26, 2023. https://www.forbes.com/sites/alankohll/2019/01/24/how-your-office-space-impacts-employee-wellbeing/?sh=37b31d7264f3

Marquette University, Department of English. 2022. *Starting your academic career.* Accessed December 29, 2022. https://www.marquette.edu/english/documents/academic-career-advice.pdf

McDonnell, J. 2020. *Navigating an academic career: A brief guide for PhD students, postdocs, and new faculty.* Hoboken, NJ: John Wiley & Sons and the American Geophysical Union.

McLeod, S., G. López, K. Murtadha-Watts, D. Quinn, M. Tschannen-Moran, and P. Tucker. 2014. *Now what? Making the transition from student to faculty member.* Accessed December 28, 2022. https://www.researchgate.net/publication/237790618

Medina, L., and P. Baert. 2013. Local chairs vs international networks: The beginning of the scholarly career in a peripheral academic field. *International Journal of Politics Culture and Society* 27 (1): 93–114. doi: 10.1007/s10767-013-9150-y

Monk, J. 2004. Valuing service. In *Presidential musings from the meridian*, ed. M.D. Nellis, J. Monk and S. Cutter, 69–71. Morgantown: West Virginia University Press.

Nilson, L. 2016. *Teaching at its best: A research-based resource for college instructors*, 4th ed. New York: Wiley.

Seldin, P., J. Miller, and C. Seldin. 2010. *The teaching portfolio: A practical guide to improved performance and promotion/tenure decisions*, 4th ed. San Francisco, CA: Jossey-Bass.

Stalnaker, R. 2018. The professional development plan process in an educational setting: The need for revision. *Arts and Social Science Journal* 9 (6): 1000424. doi: 10.4172/2151-6200.1000424

Vick, J., and J. Furlong. 2010. From graduate student to faculty member. *The Chronicle of Higher Education,* August 26. Accessed July 28, 2023. https://www.chronicle.com/article/from-graduate-student-to-faculty-member/

Whitaker, S. 2018. *Professional Development Planning.* Paper presented at SPYCH Café, Minneapolis, MN, March. doi: 10.13140/RG.2.2.20330.06084/1

Pinto Zipp, G., and S. Simpkins. 2010. The role of the academic portfolio in documenting faculty development. *Journal of College Teaching & Learning* 7 (10): 1–6.

2

BUILDING MOMENTUM

Developing research and teaching trajectories

Daniel Sui and Dafna Fisher-Gewirtzman

Launching an academic career offers exciting opportunities for early career faculty. The benefits include access to resources, opportunities to collaborate with other scholars in teaching and research, and the ability to begin new, cutting-edge research in your field. In this chapter, we will provide some advice for building momentum in all these areas.

Colleges and universities differ greatly in the stress they place on research, teaching, and service. Many are looking for people who are good at all three, but some put the highest expectations on research, others seek more balanced contributions, and some focus on teaching excellence.

These differences often make a difference in the resources that will be available to faculty. Also, many universities support a range of faculty positions. Some are traditional tenure-track positions offering the potential of long-term employment, but colleges and universities are employing increasing numbers of full-time and part-time teaching and research faculty who may have short- or long-term contracts without granting tenure.

So, it is important to think carefully about your priorities and ask questions about the blend of responsibilities that are most attractive to you. Then, look for institutions that suit these career plans and priorities. Looking for a good match involves considering the emphases and types of support that will be available at a particular institution, and hence you will be able to start the position with your eyes wide open.

As the first step, it is very important to find a university that is a good fit for your research interests and teaching goals. Research universities are known for their diverse and highly specialized faculty, which makes it possible to find a community of scholars who share your research interests and can provide valuable feedback and support. Additionally, research universities often offer a wide range

DOI: 10.4324/9781003342953-3

of resources and facilities to support research, including libraries, laboratories, and institutional support for faculty to pursue external funding opportunities.

Career planning – setting goals (short-, medium-, and long-term)

Career planning is an important aspect of professional development. Setting goals, both short- and long-term, can help guide and gauge your career progress. Immediate tasks often claim much of our attention, but looking ahead can help to keep us on track. Thinking ahead to where you would like to be in your career and life a few years in the future can be a good way to develop momentum in your research and teaching. Drawing from our own experience, we'd like to share the following advice to help you plan and advance your career:

Define your interests and goals: Clarify your research interests and goals and develop a plan for achieving them. This may include identifying funding opportunities, publishing in top-tier journals, collaborating with colleagues, and developing a research agenda. And revisit your goals periodically to gauge progress, but also to revise your goals as you move forward.

Seek mentorship and feedback: Seek mentorship and feedback from more experienced faculty members in your department or field. They can provide advice and guidance on navigating the academic job market, developing research, and teaching strategies, and preparing for tenure.

Develop a strong teaching portfolio: Teaching is an important component of your career as a faculty member. Develop a strong teaching portfolio by seeking out professional development opportunities, creating effective course materials, and gathering feedback from students and colleagues.

Engage in service and outreach: Engage in service and outreach activities that align with your research and teaching goals. This may include participating in professional organizations, serving on committees, and sharing your research with the broader community.

Network and collaborate: Networking and collaborating with colleagues and researchers in your field can help you stay up to date with new developments and opportunities. Attending conferences, workshops, and other events will help you build connections and find potential collaborators.

Balance work and life: Balancing your work and personal life is important for maintaining your well-being and long-term success. It is very important during your early days of your career to prioritize self-care, develop time management strategies, and seek out a supportive network of colleagues, friends, and family.

Remember that career planning is an ongoing process that requires continuous reflection, adaptation, and effort. This involves revisiting goals regularly, perhaps every few weeks of short-term goals, and at long intervals for longer-term projects. By setting clear goals, seeking out mentorship and feedback, developing a strong teaching portfolio, engaging in service and outreach, networking and collaborating, and balancing your work and personal life, you can successfully plan and advance your career.

Just like we all need a road map for the trips we take, it is also critically important to develop a career development "map" when you embark on your academic journey. Typically, academic career planning entails short-, medium-, and long-term goals. It is recommended that you should share these goals with your family and partner, as all sides are influenced by these plans and decisions.

Short-term (first three years after your Ph.D.): post-doc or other temporary research/teaching positions or opportunities; securing your first permanent position (tenure track position or equivalent); publications (dissertation work & beyond), securing your first grants (internal vs. external), and getting your classes in order.

Medium-term (5–6 years after your Ph.D.): obtaining tenure, promotion, or a continuing contract; sustaining your research productivity; graduate students' supervision and advising.

Long-term (post-tenure): becoming a leader/master in your field; exploring new terrae incognita and new careers on the horizon.

Executing your plan and secure early wins – stay focused, secure early wins, seek mentorship, and build social capitals

Executing your career plan and securing early wins can set you on a path toward success. We can't emphasize more how critically important it is to stay focused, secure early wins, seek mentorship, and build social capital when you are executing your plan:

Stay focused: Developing a clear plan for your research and teaching goals can help you stay focused on your priorities. Set ambitious but realistic goals and establish a timeline to achieve them. Adopting a "divide and conquer" strategy whenever you can: break down your goals into smaller, achievable steps and track your progress.

Secure early wins: Identify achievable goals that can provide early wins in your research and teaching activities. This can help you build momentum and confidence in your abilities. Celebrate your achievements and use them to motivate yourself to continue making progress in areas and activities that you truly excel.

Seek mentorship: Seek out senior faculty members who can provide mentorship, guidance, and moral support. A good mentor can help you navigate challenges, provide feedback on your work, and help you develop the skills and knowledge necessary to succeed in your field.

Build social capital: Attend conferences and workshops, join professional organizations, and network with colleagues in your field. Building social capital can help you establish collaborations, find new opportunities for research and teaching, and stay current with trends in your field. Building a trusting relationship and being trustworthy are essential in building and expanding your social capital. Some additional tips are provided in the chapters on cultivating a collegial workplace and cultivating supportive networks and collaborations.

Remember that you have many competing demands on your time and energy. It is important to prioritize your goals and stay focused on what is most important to you, which means from time to time you must have the discipline to say no diplomatically to lots of activities and tasks. By securing early wins, seeking mentorship, and building social capital, you can establish a solid foundation for your research and teaching career.

Focus your time and energy on research and teaching, but also be mindful of the importance of your service roles in the department and in the profession. Balance "what's in it for me" with "how can I help or contribute to something that is larger than my own research and teaching?" Selfless service (in committee roles, professional organizations, etc.) can be strategically important in building connections, networks, and support that enhance one's research and teaching career and generate broader impacts.

Research to teach and teach to research: seeking synergy between research and teaching

Balancing research and teaching responsibilities can be challenging. Here are some tips to help you find balance:

Set clear priorities: Identify your priorities and set realistic goals for your research and teaching activities. This can help you manage your time effectively and avoid feeling overwhelmed.

Create a schedule: Develop a schedule that allows you to allocate time for both research and teaching activities. Be sure to include time for administrative tasks, such as grading and responding to emails.

Delegate tasks: Consider delegating tasks, such as grading or leading discussion sections, to graduate students or teaching assistants. This can help free up time for your research activities. Collaborate with colleagues: Collaborating with colleagues can help you balance your workload and enhance your research and teaching activities. Consider co-teaching a course or working on a research project with a colleague. We also recommend you to collaborate with a colleague from another field for interdisciplinary research, making sure that the contribution from your field of research is clearly articulated. In addition, you should consider collaborating with colleagues at different stages of their career. It is helpful to work with young colleagues as well as more advanced/established colleagues so that your contribution will be acknowledged.

Seek feedback: Seek feedback from your colleagues and students to identify areas for improvement in your teaching activities. This can help you enhance your teaching effectiveness and reduce your workload over time.

Protect your research time: Set aside dedicated time for research activities, such as writing, data analysis, or lab work. Consider blocking off this time on your calendar and communicating your availability to colleagues and students.

Stay organized: Stay organized by keeping track of deadlines, meeting dates, and other important tasks using a planner or online calendar.

By setting clear priorities, creating a schedule, delegating tasks, collaborating with colleagues, seeking feedback, protecting your research time, and staying organized, you can balance your research and teaching responsibilities effectively. Remember that finding balance is an ongoing process, and it may take time to develop a routine that works for you.

Integrating research and teaching responsibilities can enhance the quality of both your research and teaching activities. Here are some tips to help you integrate research and teaching:

Incorporate your research into your teaching: Identify opportunities to incorporate your research findings or methodology into your teaching activities. This can help you engage students in your research and provide them with practical examples of how research is conducted.

Use student projects as research opportunities: Consider using student projects as an opportunity to collect data or test hypotheses related to your research interests. This can help you generate new research ideas while providing valuable research experience to your students.

Foster research-oriented discussions in class: Encourage discussions in your class that relate to your research interests. This can help you stay current with research trends in your field and provide opportunities for your students to engage in cutting-edge research.

Collaborate with students: Consider working with students on research projects or papers. This can help you generate new research ideas, provide students with valuable research experience, and enhance your teaching effectiveness.

Attend conferences and workshops: Attend conferences and workshops in your field to stay current with research trends and network with other researchers. This can provide you with new research ideas and opportunities for collaboration with other researchers.

Use your research to inform your teaching: Use your research findings to inform your teaching activities. This can help you provide students with practical examples of how research is conducted and enhance their learning experience.

By incorporating your research into your teaching, using student projects as research opportunities, recruiting research students to your lab/team, fostering research-oriented discussions in class, collaborating with students, attending conferences and workshops, and using your research to inform your teaching, you can effectively integrate your research and teaching responsibilities. Remember to be flexible and open to new ideas as you navigate the integration of your research and teaching activities.

Research and teaching should not be two competing priorities – seek synergy between teaching and research whenever you can.

Research to teach – Tips: cover your own research discovery or involve your students in your research whenever you can:

a Design a research seminar on a topic you want to explore.
b Design a research workshop where students actually participate in your experiments in a pilot study or the actual study and ask them to express their opinion (helps also in recruiting future research students or research assistants).
c Develop a design studio (architecture) or senior capstone class where the students design proposals or work topics related to your on-going research.

Teach to research – Tips: encourage your students to answer critical questions and convert some of the questions into research projects whenever you can:

a Design a course so that the students are working on case studies for your research. Could work as a pilot study or more.
b Suggest assignments that could become preliminary tools for your research.
c In your class, always encourage your students to think critically and challenge what you teach them, admit candidly what you don't know and conduct research (broadly defined) to follow up.

Classroom and teaching innovations can be evaluated and written up for publication in the *Journal of Geography*, *Journal of Geography in Higher Education*, *Professional Geographer*, etc.

Research publication and exposure

Maximizing the quantity and quality of your research publications helps build a strong research profile and advance your career. Here are some tips to help you achieve this:

Set clear research goals and priorities: Define your research interests and goals and prioritize the projects and publications that align with these goals. This will help you focus your efforts and make the most of your time and resources.

Seek feedback and collaboration: Seek feedback from colleagues and mentors on your research projects and publications. Collaboration with other researchers – Preferably from other universities – in your field can also help you enhance the quality and impact of your work as well as build a strong support network of people who can help answer questions and address challenges on other professional topics.

Identify and target high-impact journals: Aim and publish your research in high impact and flagship (Q1) journals in your field or in the broader scientific/scholarly community; Identify high-impact journals in your field and tailor your research projects and publications to meet their criteria. Be strategic in selecting journals, but also consider submitting to open access journals or other alternative publication models. There should be a fine balance between articles submitted to high impact journals and those that are lower and be aware to your university's policy and your departmental culture/expectations. Also, pay attention to those articles that you mention and quote in the literature review section of your articles. Chapters in books are evaluated as less significant at some universities and in most disciplines – so make sure you are investing your time wisely.

Develop a strong writing and editing process: Develop a systematic approach to writing and editing your research publications. This may include outlining, drafting, revising, and proofreading your work. Seek out resources and support for improving your writing and editing skills. The chapters on writing strategies and the publishing process discuss these issues in detail.

Stay up-to-date with trends and developments in your field: Stay informed about the latest trends, developments, and controversies in your field, and incorporate these insights into your research projects and publications. Attend conferences, workshops, and other events to stay connected with your colleagues and stay up-to-date with the latest research. Make your best effort to attend the best/highly ranked conferences and seminars; get invited to speak at invitation-only events speaks volumes.

Use social media and other online tools to promote your work: Promote your research publications through social media, academic networking sites, and other online tools. This can help increase the visibility and impact of your work, as well as facilitate connections with other researchers in your field. Derek Alderman discusses these strategies in more detail in his chapter on promoting your work.

Remember, the quality and the impact of your research publications are not solely determined by the quantity of your output. Rather, it is important to focus on producing high-quality, innovative, and impactful research that contributes to the advancement of your field. By setting clear research goals, seeking feedback and collaboration, targeting high-impact journals, developing a strong writing and editing process, staying up-to-date with trends and developments in your field, and promoting your work online, you can maximize the quantity and quality of your research publications and make a significant impact in your field and beyond.

Harnessing the power of uncertainty: embracing serendipity and open-mindedness

Embrace uncertainty and be open-minded about unexpected opportunities. Some situations and circumstances are beyond our control. Life doesn't always go as planned, but it is also often full of unexpected surprises. To prepare for the unexpected, you need to harness the power of uncertainty and embrace serendipity:

Be open to unexpected results and opportunities: In research, unexpected results can sometimes lead to new and exciting discoveries. Similarly, unexpected teaching moments can often lead to valuable learning experiences for both you and your students. Embrace these opportunities and be open to exploring new avenues of research and teaching.

Foster a culture of curiosity: Encourage your students and colleagues to be curious and ask questions. This can help spark new ideas and collaborations and encourage a culture of open-mindedness in your research and teaching community.

Build a diverse network: Cultivate relationships and collaborations with people from diverse backgrounds, disciplines, and perspectives. This can help expose you to new ideas and approaches and foster a culture of openness and innovation. Such efforts can be particularly important for international faculty, faculty of color, women, and LBGTQ individuals as a way to build strong networks in support of personal and professional growth.

Take calculated risks: Sometimes, taking risks and pursuing unconventional ideas can lead to breakthroughs in your research and teaching. Don't be afraid to take calculated risks and try new things.

Embrace failure: Failure is an inevitable part of any research or teaching career. Rather than viewing failure as a setback, embrace it as an opportunity to learn and grow. Use failure as a chance to reflect on your approaches and strategies and consider how you can improve in the future.

Stay up-to-date with emerging trends: Keep up-to-date with emerging trends and new technologies in your field. This can help you identify new research opportunities and teaching strategies and stay ahead of the curve in your field. Keep asking questions even when you think you know the answer.

By embracing uncertainty and open-mindedness, you can foster a culture of innovation and creativity in your research and teaching career. By being open to unexpected results and opportunities, fostering curiosity, building a diverse network, taking risks, embracing failure, and staying up-to-date with emerging trends, you can maximize your potential and make a significant impact in your field.

Preparing for the unexpected: What happens in case you do not get tenured? Continue the academic path or explore non-academic careers? In addition to academic careers at other universities, you should be open-minded about a wide variety of career paths in government, industry, and non-profit organizations. Companies from various sectors have significant research departments that lead the development of knowledge and tools in the various practices. They are also looking for leading researchers in the relevant field.

Open to new possibilities: Security comes with tenure, but it may also create complacency; to avoid becoming a dead wood after tenure, you need to set new career goal(s) and make it a moving target.

Aiming for the long-haul: You need to aspire to keep your professional life in balance with your personal life. Our professional and personal lives are interconnected, and these interconnections need to be acknowledged rather than pushed aside. Seeking higher purpose and deeper meaning can make both your professional and personal lives fun; gratification, satisfaction, and fulfillment come only when we are in both good physical and mental health.

Some (gender and) family perspective: The years of building an academic career are often also the years of building relationships and a family. The apparent issues may seem competitive, but it is important to build them simultaneously and with the spouses/partners in mutual considerations. When one of the spouses/partners is doing academia and the other non-academic work, and when both spouses are in academia, these two different situations lead to different complexities that require joint planning (such as applying for post-doctoral positions or universities for sabbatical years as well as housing location) and figuring out a fair distribution of the burden.

Summary and conclusions

In this chapter, we've tried to provide a range of useful strategies for building momentum in an academic career. We emphasize the importance of your career planning – setting your goals on a short, medium, and long term and how to execute your plan in ways that help you gain social capital while also building your academic profile. Seeking synergy between your research and teaching will be beneficial to all: your research will be challenged, new ideas may grow, your students will benefit from cutting-edge knowledge, and they will spread the word. Research publication and exposure are key mechanisms for academic development. We suggest how to choose and balance. Last but not least, we suggest ways to harness the power of uncertainty, which is very typical of an academic career.

We can't stress it enough that it is crucial to prioritize your research, teaching, and service activities in order to make the most of your time and resources so that they support each other as much as possible. Universities can have high expectations for both research productivity and teaching effectiveness, and it can be challenging to balance these demands. It is important to develop a plan for your research and teaching activities that consider your interests, skills, and available resources. We recommend that you focus on one or two areas of research and not spread too thin over many areas in order to develop a national (eventually global) reputation in your specialty area. It is also highly recommended that you align your teaching duties with your research areas as closely as you can.

It is also critically important to build relationships with colleagues and mentors who can provide guidance and support as you launch your research and teaching career. Universities offer many opportunities to connect with other scholars, including departmental seminars, workshops, and conferences. It is also important to seek out mentorship from more experienced faculty members who can provide advice on research and teaching strategies, as well as guidance on navigating the academic job market and tenure process. Other strategies are discussed in the chapter of the book entitled *Cultivating a Collegial Department Workplace*.

Launching a research and teaching career is a challenging and rewarding opportunity. By finding a university that is a good fit, prioritizing your research and teaching activities, building relationships with colleagues and mentors, and staying committed to your goals, you can successfully launch your research and teaching career and make meaningful contributions to your field of study. Launching a career benefits from dedication, persistence, and a willingness to take risks and try new approaches. By staying focused on your goals and seeking opportunities to learn and grow as a scholar and educator, you will find academic life to be one of the most gratifying and fulfilling careers in this world.

Acknowledgment

The authors are grateful for the comments made by Ken Foote and Lili Wang on an earlier draft.

Further reading

Boice, R. 2000. Advice for new faculty members. Needham Heights, MA: Allyn and Bacon.

Buller, J. 2009. The essential college professor: A practical guide to an academic career. San Francisco, CA: Jossey-Bass.

Caplan, P. 1993. Lifting a ton of feathers: A woman's guide to surviving in the academic world. Toronto: University of Toronto Press.

Childress, H. 2016. The PhDictionary: A glossary of things you don't know (but should) about doctoral and faculty life. Chicago, IL: University of Chicago Press.

Furstenberg, F. 2013. Behind the academic curtain: How to find success and happiness with a PhD. Chicago, IL: University of Chicago Press.

García, M., ed. 2000. Succeeding in an academic career: A guide for faculty of color. Westport, CT: Greenwood Press.

Gasman, M. 2021. Candid advice for new faculty members: A guide to getting tenure and advancing your academic career. Gorham, ME: Myers Education Press.

Gray, P. and D. Drew. 2012. What they didn't teach you in graduate school 2.0: 299 helpful hints for success in your academic career. Sterling, VA: Stylus Publishing.

Hanson, S. 2007. Service as a subversive activity: On the centrality of service to an academic career. *Gender, Place and Culture* 14 (1): 29–34, DOI: 10.1080/09663690601122176

Lang, J. 2005. Life on the tenure track: Lessons from the first year. Baltimore, MD: Johns Hopkins University Press.

Lucas, C., and J. Murry, Jr. 2011. New faculty: A practical guide for academic beginners, 3rd ed. New York: Palgrave.

Monk, J. 2004. Valuing service. In *Presidential musings from the meridian,* eds. M. D. Nellis, J. Monk and S. Cutter, 69–71. Morgantown: West Virginia University Press.

Rockquemore, K. and T. Laszloffy. 2008. The Black academic's guide to winning tenure—*Without* losing your soul. Boulder, CO: Lynne Rienner Publishers.

Solem, M., K. Foote, and J. Monk, eds. 2009. Aspiring academics: A Resource book for graduate students and early career faculty. Upper Saddle River, NJ: Prentice-Hall.

Toth, E. 2008. *Ms. Mentor's new and ever more impeccable advice for women and men in academia.* Philadelphia: University of Pennsylvania Press.

Wankat, P. 2002. The effective, efficient professor: Teaching, scholarship and service. Boston, MA: Allyn and Bacon.

3

ALIGNING PERSONAL ASPIRATIONS AND PROFESSIONAL GOALS

Finding the 'sweet spot'

Juliana Mansvelt

Introduction

If you'd asked me what my personal aspirations and professional goals were when I first started as a lecturer, I'd probably have answered, 'to survive my first year'. I felt so grateful to have obtained an academic job, so concerned whether I could teach, do research, and live up to the expectations of those who had employed me, that I couldn't think beyond completing the immediate tasks at hand. I was grateful that my university didn't have tenure – a trial period before one's job is made permanent – but at the same time, the expectations placed on me as part of my appointment were very clear. I was very much aware that my personal aspirations (to do my best, to succeed, and to not be sacked!) were absolutely bound up in my career performance.

Though my personal aspirations and professional goals were aligned at the time, it was an alignment that meant my sense of self was entirely dependent on my work performance. As someone who has long suffered from imposter syndrome, I believed I had to work as hard as I could to prove I was good enough, smart enough or productive enough to succeed in a university career. It drove me to work crazy hours, and, ironically, I was nowhere near as productive in my teaching or research as I hoped to be, which made me feel even worse about my performance! I wish I had confided in colleagues about how I was feeling or sought out an academic mentor (having since learnt the valuable perspectives more experienced academics can bring), but I was reluctant to reveal myself as incompetent. In short, the focus on all work and no play (or very little) meant I began to resent the time I was investing in an academic career, so much so that in my third year of employment, I applied for a nine-to-five-position in a local voluntary organisation

DOI: 10.4324/9781003342953-4

(which I failed to obtain!). Feeling exhausted and becoming pregnant with our first child caused me to re-evaluate my personal and professional goals; I couldn't see how I was going to manage both work and family!

I decided to take a year's parental leave, not convinced that I would go back to university. It was during this time that I really did some serious thinking and created a list of positives and negatives around my choice to return, stay at home with our baby, or find another position. I realised that it wasn't the job I disliked but the pressure I put myself under to succeed. I reflected on how much I had learnt from my teaching successes and failures, and how I enjoyed the relationships I had with my students and colleagues. Though I was not the most innovative or productive researcher, I wanted to continue to do research on topics I felt were rewarding and fascinating. This propelled me to return to my academic career when my leave period ended.

Taking the time to reassess my priorities and re-evaluate my personal aspirations was a worthwhile exercise, but personal and career choices are not made in a vacuum. Having another full-time income in our household, being in a permanent position with parental leave provisions, having a supportive head of school and partner, and benefitting from a buoyant job market meant I had a range of choices available to me. However, the complex intersections between structural, political, social, economic, cultural and personal circumstance can circumscribe options and subvert one's ability to plan personal and career trajectories. Many universities are now also subject to neoliberal rationalities (Ball 2012). These can result in increased managerial surveillance, a strong emphasis on metric-based notions of productivity and ranking, increased competition between universities, units and individual staff, reduced academic autonomy and the treatment of staff as human capital (Berg et al. 2016; Davies et al. 2006; Maisuria and Helmes 2019). As Berg et al. (2016) argue, these factors can increase staff anxiety as they feel the pressure to meet key performance indicators for teaching, service and research. Neoliberal rationalities may also have profound effects on early career academics through increasing the precarity of and competition for employment and placing additional expectations on junior staff to manage workloads and maximise performance in order to achieve tenure.

Universities are populated by diverse people and positions, including academics with varying research, teaching and service capabilities, roles, interests and personalities. As institutions, they can be spaces that foster creativity, innovation, knowledge production, conscience and critical thinking, but like any other workplace, they can be subject to social, political, cultural, economic inequalities and negative practices. While the ability and opportunity to have choices and control the effects of career decisions may be limited, self-critical introspection of one's goals and aspirations remains important to shape one's unique trajectory, identity and place in the academy. Reflexivity is a key to managing the complexity of our personal and professional lives across all stages of one's career. Reflecting on the alignment of one's values, short- and long-term goals and aspirations can identify the paths, plans and options through which these can be achieved, and head off potential conflicts before they arise.

Observations

- Reflecting on personal aspirations and professional goals during one's career can help you navigate the complexity of professional and private life, assist you in making decisions about short- and long-term priorities and help you decide what matters and why with respect to the choices you are able to make. Reflexivity with respect to aspirations and goals can also assist you in developing a career trajectory and in finding your unique 'place' as an academic.

Values as a foundation for personal and professional activities and aspirations

Some years ago, I attended a teaching and learning workshop on workload management. One of the first tasks we did was to undertake a values exercise (see Box 3.1), which I found incredibly revealing. This was the first time that I had consciously reflected on those principles which are important to how I lived my life. This exercise enabled me to understand what I appreciated about my academic and personal life and what frustrated me.

Values can be thought of as internalised principles and social representations which guide and provide the rationales for action (Oyserman 2015). They are abstract ideas which have affective/emotional dimensions, but their effects depend on how they are interpreted in everyday contexts (Maio 2016). Consequently, values are socially and culturally constructed. They derive from family, group, community and societal socialisation and the ways we interact with and understand the non-human world too.

BOX 3.1 IDENTIFYING YOUR VALUES

Undertaking an exercise to identify my core values was something of revelation to me, enabling me to reflect on how these values supported the activities I was currently engaged in and to evaluate future personal and professional plans and priorities. I include here several websites which have various exercises to enable you to identify your core values. Though these websites all draw on values which are often seen as universal, it is important to recognise that values are still relative and relational. Values are formed and articulated differently in different spatial, social and cultural contexts, and they may change over time. For example, common lists of values don't include the Māori concept of manaakitanga, a value that is important to my personal and academic practice. An English translation of manaakitanga might centre on showing hospitality, but an understanding of manaakitanga as involving mana*-enhancing care more comprehensively captures the sense of giving, generosity, kindness and support which comes from hosting and hospitality. [*Note: mana is a supernatural force in a person, place or object which enables authority, leadership and power (Moorfield, 2003–2023)].

https://www.berkeleywellbeing.com/core-values.html

It contains lists of core values and a quiz, reflecting on what values are and whether personal values may align with those of your workplace.

https://www.fearlessculture.design/blog-posts/what-do-you-stand-for-an-exercise-to-discover-your-values

A value-identification exercise that encourages you to develop your own values rather than choosing from a specified list.

https://www.think2perform.com/values/#start

An online values card exercise that enables you to sort and reduce your list of values to a list of five core values.

A values exercise involves identifying a small number of core values. Through engaging in this exercise, I recognised my core values underpinned much of my personal and professional being. The realisation that different values might be stressed and enacted in different contexts was also helpful. Creativity for example is a core value of mine that is important in my teaching and research. However, this value was perhaps most cogently expressed through my involvement in sculpting clay – attending pottery night classes for many years has been vital for me to switch off from work demands and to achieve a state of flow (Csikszentmihalyi et al. 2014). I also recognised valuing 'relationships' was critical to the pleasure I found in co-leading interdisciplinary teaching and learning mentoring circles in my institution. 'Integrity' was another core value, and recognition of this prompted me to discontinue a frustrating research relationship. 'Service' has always been important to me and has propelled me to contribute to my local community, my discipline and university, defining the kind of 'public academic' I am. Understanding the congruence of my values with those whom I interact enabled me to identify what I found satisfying and/or challenging about the different situations and contexts I found myself in. Though I really enjoy service activities, I endeavour to ensure I balance service commitments with research and teaching to meet the priorities of my tertiary institution and facilitate a pathway for promotion!

Academic workplaces do not just provide the opportunity to express our values. They can also pose considerable challenges to how we enact/represent them (Maio 2016). It may be difficult, particularly early on in an academic career, to exercise agency around the activities you engage in. If our practices and the contexts in which we operate don't reinforce or align with our values, we can feel frustrated and unfulfilled. Even now, as I write this, as an academic in the later stage of my career, I find myself wrestling with a misalignment of personal values with some of the values of my institution as I see them practised (rather than professed!). It is the recognition that many of my core values continue to be supported through collegial relationships, through research collaborations and in my teaching which have sustained me through rounds of institutional restructuring and change.

It is important to consider the relationship between one's values and one's goals. My drive to serve meant that in the early years of my career I felt obliged to say yes to everything (not something I would recommend!), with the consequence that I soon felt overcommitted. I didn't necessarily prioritise those things that would enable me to achieve my goals, for example, so keen was I to help a colleague (and with the added incentive of the availability of external funding) that I agreed to lead a large research study on e-learning and professional development. While I learnt a lot and it gave me a sense of achievement (another of my values), taking the time to research and write in an unfamiliar disciplinary field was something of a distraction from my existing research on ageing, consumption and place, and it did not hold the same sense of excitement (another significant value for me). I eventually learnt that I could prioritise the kind of activities my institution values, but in ways and forms which aligned with my personal values. Research productivity is an expectation of my employment contract, but I have choices in terms of involvement in research teams and topics. Consideration of whether or not my intended activities align with my personal values and both short- and longer-term goals and aspirations has also enabled me to say no and to feel ok about doing so.

Thinking about one's values can also help you to make positive choices – to contribute to personal/work projects and activities that you enjoy, feel comfortable about and that will give you intrinsic satisfaction. Research also indicates that when people think more deeply about their values, these can become more powerful motivators of actions, affirming them and rendering them more tangible and concrete (Maio 2016). Consequently, reflecting on the values we view as foundational can provide a basis for managing expectations of ourselves and others. While values may assist us to achieve a sense of balance and alignment in our personal and professional lives, it is important to consider how these values might operate in relation to our goals and aspirations, and to our passions, expertise and strengths.

Observations

- It's a good idea to consider what you want to achieve personally and professionally prior to applying for academic, industry or private sector, or NGO positions, and whether obtaining the advertised position aligns with your values and will enable you to reach your goals. Reflecting on whether your core values align with those of your colleagues, programme or unit, institution or workplace can provide opportunities for reflection and change.

- If there is dissonance between your values and those of others and/or your institution or intended workplace, what options are available to you? Do you want or need to accommodate this dissonance? How might you do this without compromising your personal values? Are there any trade-offs you can make (e.g. complying with some requests while exercising more autonomy in other areas)? Are there any changes you can make or advocate for in your workplace and/or personal life to see your values more clearly expressed in your own practice?

Identifying personal aspirational and professional goals

While our aspirations – the dreams, ideals or long-term desires – can influence career and personal pathways, goals are generally much more tangible and can be short or long term. Monk and Jacoy (2009, 18) provide a list of questions that can help determine personal and professional goals. They argue that it is useful to reflect on these before applying for your first academic job:

- How are my personal and professional lives interdependent?
- Is my career linked with that of a partner/family?
- How do policies and practices of institutions or organisations constrain choices?
- How might I/we negotiate within the institution or organisation? How might I work to change it?
- What do I do well? How can I identify my strengths? What can I do to improve my qualifications?
- How flexible/broad/specialised do I want/need to be?
- With whom do I wish to communicate?
- What are my expectations and desires with regard to contributing to institutions, organisations and society?

Setting goals can help identify the purpose of your personal and or professional endeavours, and the steps needed to achieve these goals including prioritising relationships, time and activities. Management consultant George Doran (1981) established the idea that goals should be SMART: *specific* (i.e. clear and sufficiently detailed to establish what is necessary to accomplish them), *measurable* (so you will know when you have accomplished them), *attainable* (they are actually goals which can be achieved), *realistic* (they are achievable with the resources, timeframes, actions you have available) and *timely* (they can be accomplished within a specific time frame). While the establishing smart goals seems reasonable, it is worth noting that they focus primarily on outcomes (e.g. I want to publish a paper in in a specific tier one journal this semester) rather than on process (e.g. learning how to write a publishable article), and they tend to ignore the wider context in which goals are established (balancing personal and professional commitments, the journal submission/rejection rates). Regardless of the degree of specificity of your goals, it is useful to consider what you want to achieve, why and how, and to reflect on the extent to which these goals are based on realistic expectations. Considering how your goals might intersect with other structures, practices, relationships and resources can help shape reasonable assessments of one's own performance or actions and avoid a crushing sense of failure if things don't pan out as you planned! Regular evaluation and revision of your goals can also enable you to assess your progress, manage your time more effectively, reassess your priorities and aspirations and help you to move forward if you encounter challenges.

In my academic position, my annual performance appraisals with my manager are focused on annual and longer-term goal setting, the identification of resources, skills and time needed to meet these goals and discussion of how or why these goals have been or not been met. When I haven't achieved my goals, I find it easy to see this as a personal failure. Notions of meritocracy can underpin staff promotion structures and metrics, where success or failure is seen as a product of individual effort, education, talent and expertise. But, while individual thought, effort and achievement is important, I have learnt that achieving goals and aspirations is also dependent on numerous relationships, events and practices, for example, institutional and personal support, the work/input of other colleagues into your research and teaching, finance, workloads, access to resources, grant availability and funding priorities. Uncritical acceptance of meritocracy obscures the wide social and spatial context which may give rise to inequalities, and which influence the ways in which individual work is enabled and constrained (Liu 2011).

There may be material consequences of not reaching one's goals, particularly in the early stages of a career, but in hindsight, I can see how my own setbacks (though unwelcome) have been learning opportunities that provided moments of reassessment and reprioritisation. They provided me with the motivation to advocate for resources and discuss realistic work expectations with employers. It was a sense of frustration at my teaching load and not getting to research writing that led me to request increasing my part-time percentage from 0.5 to 0.6 FTE to reflect the time I was actually investing in paid work.

Achieving or failing to achieve one's goals can be an intensely emotional and affective experience. Establishing good relationships with work colleagues (particularly trusted colleagues more senior than myself) and cultivating meaningful relationships, interests and activities with people outside of academic work provided support and enabled me to bring some perspective to the inevitable peaks and troughs of academic life.

Observations

- Developing specific goals and plans based on reasonable workload expectations and manageable time frames can help in establishing realistic career aspirations. Engaging in periodic 'reality checks' regarding your progress towards your goals can help you to determine whether or not to persist when things are not going as planned, if it is possible to make other choices with respect to your personal life or career or if your initial goals were realistic and need to be re-evaluated.
- While promotion and appraisal structures emphasise individual achievement, this is often dependent on personal and professional networks and is influenced by political, economic and social structures. Developing meaningful and supportive relationships with significant others in our personal and professional lives can assist in managing the complexities of everyday life and help bring perspective to successes and failures.

Managing the interconnected nature of personal and professional life – the myth of separate compartments

I have always found the concept of 'work–life balance' deeply unsatisfactory for a range of reasons. The term seems premised on the notion that paid and unpaid work, as well as professional and personal life, can be easily separated, quantified and that some magic balance can be achieved. Uneven intensities and seasons in work and personal demands have characterised my career, and although overall I might feel like I have a kind of balance, often these intensities skew my attention and practice one way or the other. An understanding employer has allowed me to work flexibly to manage differing work and home priorities. Even with this advantage, there have been times (such as when I was caring for my mother-in-law after a fall and heart attack) that I had to accept that I did not have the emotional, cognitive or physical ability to maintain my usual capacity to work in both my private and professional roles (Box 3.2).

BOX 3.2 THE PERSONAL AND THE PROFESSIONAL: THE MYTH OF SEPARATION

For me it was having children that was the death knell of the assumed separation of my home and work life. One of the most memorable examples of this was when I had our third child. I had taught right up until the week before I gave birth (there were still three weeks of semester remaining), and Grace was born at home. About an hour after giving birth the phone rang, and my husband answered. It was one of my students calling with a query about his assignment. Neil proceeded to tell him, 'Juliana's just had the baby', 'That's alright', said the student, 'I will ring back in half an hour!' – at which point I decided I may as well answer his query and be done with it! While I have always tried to be available to my students – this provoked a further realisation that one can be too available, and some renewed personal and professional boundaries were in order!!

Another incident at work reminded me of the fallacy of the assumed separation between work and personal life. I needed to run a three-day field trip and take the baby with me as I was breastfeeding. A geography technician who would normally assist on fieldtrips refused to attend lest they ended up holding the baby (this was never my expectation!). This incident provoked me to realise that managing the intersection between personal and working life was as much about developing strategies to manage others' expectations, as it was about managing my own. As our children grew older, the intensity and practical manifestation of my parenting identity in the workplace lessened. However, the emotional investments, spaces and relationships that went with caring for significant others and the pressures of my other roles and identities in my community continued to blur home and work boundaries. These interactions have required regular reassessment and renegotiation of my own and others' expectations.

Caring for significant others, sickness, bereavement, financial concerns, relationship upheavals, living distant from one's workplace, community commitments and working through a pandemic are just a few of the life circumstances and events that can provide challenges and opportunities in managing and meeting personal and professional goals. Similarly, changes in institutional structures, work expectations and workload can create new challenges for managing academic tasks – for example, moving to online and blended learning has been shown to increase workload for academics (Tynan et al. 2015). Interruptions in the continuity of employment, fractional contracts, inequitable work allocations and inaccurate workload calculations can also disadvantage research activity and promotion, particularly for women (Barrett and Barrett 2011; Rajagopal and Farr 1992).

Working part-time for all but three years of my academic career has intensified the recognition of the leakiness of personal and professional identities, roles, goals and boundaries. While I still work more hours than my part-time allocation, over the years I have developed strategies that help manage the activities I am engaged in, and the expectations placed on me by myself and others at work and in my private life. While each of you will need to find your own ways of managing the intersection between personal and professional, I hope some of the insights shared here might be of use.

After giving birth to the first of our three girls, I negotiated to reduce my contract to part-time. Though notionally a choice, my partner and I both lived distant to the university (and worked in different cities), and there initially was no full-time formal childcare provision in the small town I resided. An increasingly casualised tertiary workforce may mean many people have no choice but to work part-time. Research has indicated that women more commonly than men are also likely to report chronic stress (Schulte 2014, cited in Mountz et al. 2015, 1245), and these stresses may be magnified with part-time work. Though I haven't regretted working part-time, there have been trade-offs. My teaching workload has consistently been higher than my FTE component, and with service and teaching commitments I have found it challenging to carve out time at work for research and publication and to manage student contact on the days I am not employed. I have had few periods of sabbatical (for both family and financial reasons), and it has also taken me longer to progress through an academic career track, even in an institution that recognises fractional appointments as part of its promotional processes.

After recognising I was way too available to my students, I began to put more boundaries around student contact – publicising my office hours, putting messages on distance teaching sites about my availability and when students might expect a response to their queries (specifying I have weekends off). In recent times, I have also used automatic replies and email signatures to note when I am employed by the university and in the office and when I am likely or able to respond to messages. I have begun to say no to requests which don't align with my teaching or research interests. I think much more carefully about taking on work, reflecting on the extent to which it aligns with my values and my personal and professional goals and whether I can fulfil my promised obligations. I use my calendar to track and block out time for research, teaching and private activities. I keep a list of current tasks and prioritise these, where possible, to focus on the important, not simply the urgent. I prioritise writing, reading and thinking time, and I endeavour to nurture the things, people and activities I value. For things I love less (exercise classes at the gym, administration, marking assignments), I try and devise rewards for achievement – these vary in size, from a holiday at the end of the semester to a meal out with friends or a walk around our university campus after a sustained period of intensive work! The merits and challenges of slow scholarship have been much debated in the academy (Hartman and Darab 2012; Mendick 2014; Mountz et al. 2015; Vostal 2021). Slow scholarship has emerged as a response to the acceleration and intensification of academic work and proposes changing our work time. Taking the time to think, to read, to listen to others and to expand our horizons by engaging outside of our offices and institution can assist in expanding our collective capacities and finding joy in our learning researching and teaching (Mountz et al. 2015).

While I still endeavour to be as collegial as possible, recognising the importance of journal refereeing, contributing to textbooks and reference material, thesis examination and societal participation play in my discipline, I do consider each invitation or request on its merit (including whether I have the time to devote to doing it properly) before I decide to take anything on. Again, such choices may become easier as one's career advances, but choosing strategically and making choices to the extent you can, according to your passions, expertise and aspirations means that the tasks you do commit to are likely to be more rewarding. For example, though I enjoy it, university service is an expectation at my institution. I volunteered for the library committee (having enjoyed working in a library prior to undertaking a PhD) and got involved in co-leading teaching and learning mentoring circles (an initiative that helped my own professional development and learning), rather than find myself 'volunteered' for committees which I found much less enjoyable or productive!

As I reflect back on my career, there have been benefits of working part-time – less commuting, more time to engage in activities with my family and greater involvement in education in my local community. When I reflect on my values – service and a belief in equity – being involved in school governance, facilitating educational communities of learning across learning institutions in my community and supporting the equitable provision of digital devices for young people have been all hugely rewarding. When I applied for promotion to Professor – which required a comprehensive overview of my entire career's contribution – I was able to reflect on the ways in which my personal and professional goals and my work and voluntary activity helped shape my identity and contribution as an academic in my community.

As academics positioned differently with respect to goals, aspirations, values, work demands, career journeys and institutions, the options, choices and resources we may have available to us to manage work and professional life can differ greatly. Career decisions may be influenced by such things as the labour market, financial considerations, obligations (rent/mortgages, debt levels, household income), health and well-being, social and cultural responsibilities, as well as the goals set by yourself and your employer. Personal, social and economic considerations can influence decisions, trade-offs and negotiations you might engage in with your employer to manage workload and personal obligations, responsibilities and priorities.

Observations

- The complete separation of personal and professional life is a fallacy, although finding ways to manage the intersection is critical. This is even more important for those on part-time contracts.
- Research on part-time employment and gender reveals the disadvantages that can accrue across one's working career. However, part-time work can

also have many rewards. It is important to assess the pitfalls and possibilities of fractional appointments prior to undertaking these contracts.
- The kinds of roles and identities each of us assume as an academic will differ. It is worth considering your unique blend of personal and professional expertise, skills, interests and desires and how these relate to your intended career trajectory and goals.

Concluding thoughts: finding the sweet spot

After many years of feeling I lurched from one deadline to the next, I have found ways of working which help me manage my work and personal life. A key to enhancing scholarly and personal development is reflexivity – intentionally reflecting on and across one's career journey, interrogating your sense of self and experiences in relation to values, goals and aspirations and the political, economic and cultural contexts in which you are embedded. Doing this should help you assess the limitations and possibilities of working in your chosen institution and enable you to achieve a blend of personal and professional activity that is comfortable for you.

I have learnt over the years to recognise and cope better with the veritable rollercoaster of workload and the intensity of professional and private activity. Most of my learning has come from my mistakes rather than from my successes. I have developed strategies which help me to feel more on top of my workload, enabling me to enjoy both my work and non-work time to a greater extent. While I still struggle with managing the shifting boundary between personal and professional life, I do feel like I have found the sweet spot – a broad arrangement of time and activities which works for me. I continue to reflect on what matters, what I need to do and my aspirations – this self-critical introspection has necessitated a change in my own perspective as much as a change in practice.

I have learnt (again made easier as I progressed through my academic career) that achieving my goals is dependent on a whole raft of other people and things, though this does not discount the value and importance of individual effort. I often used to compare myself with others who seemed far more intelligent, productive and articulate, and I felt incredibly intimidated. Not surprisingly, this was a fruitless and unrewarding exercise! As academics, we each bring our own unique blend of experience, expertise, enthusiasm, strengths and weakness to the job. I eventually realised that in order to fulfil my personal and professional goals, I needed to be my authentic self in and outside of work, and accept that I could be successful as 'me', a quiet person (Cain 2013). Baking cakes and biscuits (sadly, this still tends to be a highly gendered activity in New Zealand) has not simply been a source of enjoyment in my spare time, but has come to represent my expression of manaakitanga, a mana-enhancing demonstration of care for my students, friends, colleagues and community contacts. For me, baking is a symbol of the 'sweet-spot' which represents a fulfilling intersection of my personal and professional life. I hope that you, the reader, will have the opportunity to craft your own unique identity and practice as an academic – and that you too will find some sweet spots – the points where academic and personal goals align in ways which are immensely rewarding.

References

Ball, S. J. 2012. Performativity, commodification and commitment: An I-Spy guide to the neoliberal university. *British Journal of Educational Studies* 60 (1): 17–28. doi: 10.1080/00071005.2011.650940

Barrett, L., and P. Barrett. 2011. Women and academic workloads: Career slow lane or Cul-de-Sac? *Higher Education* 61 (2): 141–155. doi: 10.1007/s10734-010-9329-3

Berg, L. D., E. H. Huijbens, and H. G. Larsen. 2016. Producing anxiety in the neoliberal university. *Canadian Geographies / Les géographies canadiennes* 60 (2): 168–180. doi:10.1111/cag.12261

Cain, S. 2013. *Quiet: The power of introverts in a world that can't stop talking*. New York: Crown.

Csikszentmihalyi, M., S. Abuhamdeh, and J. Nakamura. 2014. Flow. In *Flow and the foundations of positive psychology: The collected works of Mihaly Csikszentmihalyi*, Vol. 2, ed. M. Csikszentmihalyi, 227–238. Dordrecht: Springer.

Davies, B., M. Gottsche, and P. Bansel. 2006. The rise and fall of the neo-liberal university. *European Journal of Education* 41 (2): 305–319.

Doran, G. T. 1981. There's a S.M.A.R.T. way to write management's goals and objectives. *Management Review* 70 (11): 35–36.

Hartman, Y., and S. Darab. 2012. A call for slow scholarship: A case study on the intensification of academic life and its implications for pedagogy. *Review of Education, Pedagogy, and Cultural Studies* 34 (1–2): 49–60.

Liu, A. 2011. Unraveling the myth of meritocracy within the context of US higher education. *Higher Education* 62 (4): 383–397. doi: 10.1007/s10734-010-9394-7

Maio, G. R. 2016. *The psychology of human values*. London: Routledge.

Maisuria, A., and S. Helmes, S. 2019. *Life for the academic in the neoliberal university*. London: Routledge.

Mendick, H. 2014. Social class, gender and the pace of academic life: What kind of solution is Slow? Forum Qualitative Sozialforschung Forum: Qualitative Social Research 15 (3): Article 7. doi: 10.17169/fqs-15.3.2224

Monk, J., and C. L. Jacoy. 2009. Career planning: Personal goals and professional contexts. In *Aspiring academics: A resource book for graduate students and early career faculty*, ed. M. Solem, K. Foote, and J. Monk, 16–31. Upper Saddle River, NJ: AAG and Pearson.

Moorfield, J. C. 2003–2023. *Te Aka Māori dictionary*. https://maoridictionary.co.nz/

Mountz, A., A. Bonds, B. Mansfield, J. Loyd, J. Hyndman, M. Walton-Roberts, R. Basu, R. Whitson, R. Hawkins, and T. Hamilton. 2015. For slow scholarship: A feminist politics of resistance through collective action in the neoliberal university. *ACME: An International Journal for Critical Geographies* 14 (4): 1235–59.

Oyserman, D. 2015. Values, psychology of. In *International encyclopedia of the social and behavioral sciences*, 2nd ed., ed. J. D. Wright, 36–40. Oxford: Elsevier.

Rajagopal, I., and W. D. Farr. 1992. Hidden academics: The part-time faculty in Canada. *Higher Education* 24 (3): 317–331.
Schulte, B. 2014. *Overwhelmed: Work, love and play when no one has the time.* New York: Sarah Crichton Books.
Tynan, B., Y. Ryan, and A. Lamont-Mills. 2015. Examining workload models in online and blended teaching. *British Journal of Educational Technology* 46 (1): 5–15.
Vostal, F. 2021. Four reasons slow scholarship will not change academia. *Impact of Social Sciences Blog,* 11 May 2021. Accessed November 3, 2024. https://blogs.lse.ac.uk/impactofsocialsciences/2021/05/11/four-reasons-slow-scholarship-will-not-change-academia.

4

NAVIGATING NON-TENURE CAREER TRACKS

Emily Murai

As a non-tenure track lecturer within a large, research-intensive public university, I'm often asked by graduate students about my experiences and how being in a non-tenurable position differs from the work of their advisors and dissertation committee members. They have often found great satisfaction working as teaching assistants (sometimes for me!) and want to know more about teaching careers in higher education. They ask me how I got to be where I am, if I'm happy in my position, and what my work–life balance is like. Others share uncertainties about being in the graduate program. They feel pressured to emulate their advisors – to vy for external funding, publish in highly ranked journals, and aim for a tenure-track job at R1 universities or other prestigious institutions – and are not sure it is the right path for them. Many of them did not know what academic careers entailed before enrolling in the PhD program.

As a graduate student similarly coming out of a research-intensive program, I too felt these pressures and internalized the standards and expectations of the faculty around me. Yet, it was in classroom spaces where I felt I most thrived and made a tangible difference, especially for students of color and first-generation students with whom I identified. Toward the end of my graduate studies, it became clear to me that if I was going to pursue a career in academia, it would be on a teaching-centered pathway in a place I wanted to live and where I could have work flexibility so I could contribute to raising a family. I no longer wished to pursue a high-pressure academic career, with the "publish or perish" mentality, rigid status hierarchies, and overidentification with work. As my peers moved across the country (or in some cases halfway across the world) to pursue post-docs or temporary or tenure-track positions, I moved back to my hometown where I planned to stay for the rest of my adult life and hoped for the best.

DOI: 10.4324/9781003342953-5

Back then, I did not have much insight into navigating non-tenure track positions, and frankly, even if I identified them as career options, my advisors most likely would not have known how to guide me. I'm not sure that, like the whispered conversations I have with grad students in my department behind closed doors, I would have spoken about it openly with them. Now, after 14 years being in non-tenure track teaching roles and observing the academic job market radically shift towards casualization, I feel a greater sense of urgency that we, in the academic community, have these conversations. Given these labor market realities, we must discuss a wide-range of career options available to graduate students and how to make themselves marketable for them. Not every graduate student has the resources or desire to pursue coveted tenure-track positions, and we do a disservice if we solely prepare them for these kinds of roles. Realistically, they will at some point consider or find themselves in a non-tenurable position, whether as a post-doc, a contracted teaching job, a temporary visiting professor position, or another "soft-money" funded position. We need to prepare them for what to expect and how to navigate these roles in ways that enable them to thrive. Beyond this, we also need to discuss how we can collectively improve the stature, inclusivity, and levels of respect for these positions. In this chapter, I discuss teaching-heavy pathways in particular, although there's a wide range of non-tenure track positions with their own uniquenesses. I discuss key considerations when thinking about a non-tenure track role and what to do if you're in one or want to leave one. Much of what I write about here is derived from my own experiences and from observing others in non-tenure track roles.

First, why the non-tenure track? Where does it come from?

The rise of non-tenure track faculty (NTTF), along with rising tuition costs and levels of student debt, correlates with the decline of public funding for higher education. Over the past several decades, state funding for public higher educational institutions has dropped nationwide, with steep drops after the 2008 recession (Cook 2017; Mitchell et al. 2018; Pew Trusts 2019). In the humanities especially, budgets – and in some cases, entire departments – have been slashed due to declining enrollments (Hartocollis 2023; Heller 2023). Meanwhile, the production of PhD degree holders has not slowed down, creating a surplus of qualified candidates for a shrinking number of secure, well-paying jobs (Carey 2020a, Kolata 2016). Cutting permanently funded faculty lines and paying for instructional staff on a part-time or "as needed" basis is one way funding-strapped universities are able to cut labor costs and save money (Washburn 2006), particularly as higher educational institutions face even greater budget cuts post-pandemic (Hubler 2020; Carey 2020b). Whereas NTTFs were once a minority within the academy, the American Association of University Professors (AAUP) now notes that over two-thirds, or 68%, of higher education faculty hold a contingent, non-tenure track appointment, almost a half of them (47%) part-time (Colby 2023; Schuster and Finkelstein 2006). And the growth of NTTF has impacted all sectors of higher education – even though associate-degree granting institutions employ the highest overall percentages of NTTF, the largest growth of NTTFs over the past 20 years has been at public and doctoral granting institutions (McNaughtan et al. 2017).

A recent article in the *Chronicle of Higher Education* notes that the public generally favors high-quality teaching and learning, and that high-quality classroom instruction can be a major driver of student academic success and achievement in higher education (McMurtrie 2023). Yet, a dependence on contingent and part-time adjuncts often leads to poorer teaching and learning outcomes for students and faculty alike (Kezar et al. 2019), while permanent, tenure-track faculty are expected to pick up the slack in departmental and institutional administration and governance, leading to their own levels of burnout and overwork (Kinman and Jones 2008). The overinvestment in high-level executives and administrators, research output, and luxury facilities at the expense of high-quality instruction and student learning (Tuchman 2009) partially explains a cynicism and deep public mistrust of higher educational institutions (McClure 2021) and why academic workspaces have become highly toxic and workloads untenable (Smyth 2017). These factors have also contributed to the sharp increase in the number of academic workers leaving the profession in recent years (Dolezal 2022; Ellis 2021; Gerwin 2022).

One of the biggest challenges in discussing NTTF is treating it as a singular, unified category. There is a great deal of variability in titles, roles, experiences, and levels of responsibility encapsulated in the category of "non-tenure track," even among those solely focused on teaching. The AAUP, for instance, differentiates between full-time NTTFs, which they call "contract renewable" or "term" faculty, while part-timers are referred to as "adjuncts" (Colby 2023). In the UK, they are referred to as "teaching fellows" or "teaching-only" staff (Peters and Turner 2014). Other common NTTF titles include visiting assistant professor, teaching professor, professor of practice, lecturer, or instructor. In addition, graduate student instructors and teaching assistants are also considered NTTF. Positions also vary across institution types, geographic regions, and disciplines/fields. Some non-tenure track faculty voluntarily teach a class or two a year as a side gig, while others have a full-time load and stable professional pathway. Others still cobble together multiple low-wage, part-time gigs at different institutions to create a full-time load (e.g., the freeway flyer). Some NTTF have master's or professional degrees in their fields, while others have PhDs (American Federation of Teachers [AFT] 2022). Some disciplines, for example, composition/writing, music, or foreign languages, depend upon NTTF to teach their service-heavy, skills-based courses and are at the forefront of recognizing and redressing the inequities of these two-tier career tracks (Dorfeld 2022; Fink 2016). In other disciplines and institutions, the marginality and invisibility of NTTF suggests that they may as well not exist. Demographically, the AAUP found that as of fall 2021, women and underrepresented minority (URM) faculty members held both contingent and part-time appointments in greater proportions (Colby 2023).

Similar to the difficulties in universalizing the types of NTTF positions, it is equally difficult to universalize NTTF's experiences. Much has been written about the contingency, precarity, and labor exploitation faced by NTTF after years of specialized academic study and personal sacrifice; indeed, there is a large body of academic "quit-lit," with contingent faculty among some of the most prominent voices (Childress 2019; Purcell 2007). While it is undeniable that there is a great deal of room to improve equity and inclusion within the academy on multiple levels (Zahneis 2023), it is also true that others, like me, have found great satisfaction in these roles, finding positions that offer enough pay, stability, and flexibility to meet other demands in our professional and personal lives (Kelsky 2019; Trower 2021). Improvements in labor conditions, driven by institutions, academic labor unions, and other social justice movements, are making NTTF positions increasingly viable, stable, and respectable, although there is plenty of research that suggests that making more improvements is necessary for the longevity of the profession (AFT 2022).

As a result, while I don't want to minimize the disparities and inequities of a two-tier labor system, I also don't want to define all NTTF positions as ones of doom and gloom. Given the variability of NTTF positions and working conditions, each position needs to be examined in relation to one's own professional needs and wants. And given the evolving nature of both NTTF and TT roles, it is possible an NTTF position could better align with one's lifestyle than a TT position (Kelsky 2019). With an adequate set of expectations and a supportive working environment, it is possible to thrive in such a role. In the following section, I give some practical suggestions and insights into how to do so.

Thinking about non-tenure track jobs?

One of the most important things one can do before entering the job market – for any type of job, tenure-track, or non-tenure track, inside or outside the academy – is to be clear about your own value system and personal priorities when looking for a job. You need to be clear about what you want and value most in your life, and then to think about how a job will complement or supplement that, and not the other way around. Too often, feelings of desperation or anxiety, or the unreflective "hoop jumping" we've been trained to do in graduate school, propels us to search for and accept any job we are offered. Without careful consideration about *your* wants and needs, you may find yourself frustrated, resentful, and regretful for the choices you made down the road.

In this way, I strongly endorse the values clarification and goal-setting exercises discussed in Juliana Mansveld's chapter, "Aligning Personal Aspirations and Professional Goals: finding the 'sweet spot.'" Undertaking a deep analysis of your own values and priorities will clarify which jobs to apply for, which to accept (and avoid), and how to best organize work around other important aspects of your life. This, in turn, will give you more control and enable you to take more responsibility over the job search process. By centering on your own needs and wants, you'll increase the likelihood of being satisfied in a particular job, since you're applying for jobs based upon its fit with your value system and other priorities in life.

If you decide that NTTF positions might be a good fit, do as much research as you can beforehand to know what a specific NTTF job entails (Boxes 4.1 and 4.2). Read the job listing carefully to understand what the job is for, and what it does and does not offer. Is it a one-time contract to cover someone's sabbatical or personal leave, or is it a recurring and possibly long-term commitment? Is there the possibility to increase or decrease your workload in the future without penalty? Look closely at the pay and read the institution's labor contract for your particular role/classification (if they have one). What are your responsibilities, and how much will you be compensated for this work? Does the pay seem fair? How are additional responsibilities outside the labor contract compensated and handled? What is the process if you have a work-related grievance? Does the position offer opportunities or funding for professional development? Talk to other NTTF in the department (or institution) about their experiences and treatment by the department AND the institution. If a program's cohort of NTTF feel marginalized or ignored, department morale is low, and institutional support is threadbare or non-existent, think seriously if you will be able to thrive in such working conditions. Always weigh these considerations against other factors in your life (cost of living, schedule flexibility, family considerations) that are important in deciding if a specific position is right for you.

BOX 4.1 JOB FACTORS TO CONSIDER IN AN NTTF POSITION

- Pay
- Schedule/Flexibility
- Benefits and retirement
- Number and types of courses you'll teach
- Short- and long-term stability
- Employment in subsequent terms
- Type of institution and institutional fit

- Types of students you'll be working with
- Teaching format – remote, hybrid, or in-person
- Collegiality and reputation of the department
- Opportunities for professional development and institutional advancement
- Available resources at the institution you can access (tuition reduction, library, career services)
- Union representation and union activity

BOX 4.2 PERSONAL QUESTIONS TO ASK YOURSELF WHEN CONSIDERING AN NTTF POSITION

- What are my goals/expectations in taking this job?
- What are the short- and long-term costs/benefits analysis of taking this position?
- Could this position lead to other opportunities? Are these opportunities ones that I want?
- Could this position help me develop skills/contacts that can transfer to a different position?
- Are there other benefits to this position besides pay (e.g., location, networking)?
- What other job options do I have at the moment?
- Does this job make sense financially?
- Will I incur other costs (e.g., moving) in taking this job?
- Are there any hidden "opportunity costs" in this position?
- Will I be appropriately supported and/or set up for success in this position?
- Will I be able to support myself (and any dependents) financially in this position?

I have found that a NTTF position can be a good fit if you value some combination of three different factors that can be hard to meet in a tenure-track job: (a) you enjoy teaching (or at least as much as or more than research); (b) you prioritize living and working in a particular geographic location or don't have the freedom to relocate; (c) you need more schedule flexibility than a traditional nine to five job or want the option of having summers off. Most NTTF I've spoken to have been willing to trade some level of stability and higher pay for flexibility in these other areas.

All three of these reasons also explain why I pursued an NTTF path. I love teaching and currently teach at my undergraduate alma mater, where I feel a connection with the undergraduates, many of whom have backgrounds similar to mine. I started teaching multiple sections of a freshman seminar when I was in the throes of writing my dissertation and decided to continue once I finished my degree, partly because I enjoyed helping students transition into higher education. My spouse and I live close to our immediate families, high priorities for us, and I had enough flexibility in my schedule to be our daughter's primary caregiver when she was born. The position has offered many wonderful options: As an instructor, I am mostly responsible for teaching my own classes and am not required to do any additional uncompensated service work or teach in the summer; I have avoided getting enmeshed in draining departmental politics and institutional bureaucracy; I can receive merit raises for voluntarily engaging in scholarly work or service without the high-stakes pressures of "publish or perish"; I've observed NTTF colleagues start businesses, take on other paid work, pursue hobbies, and have a life outside of work – all things I hope to do as my child gets older.

There are, of course, trade-offs: my salary is considerably lower than those of tenure-track (TT) colleagues; I'm ineligible for perks like on-campus housing, spousal hire, or relocation funds; I often feel isolated with no opportunities to participate in institutional governance or departmental life; and for a few years when I was first getting started, I was financially very precarious: I had to put together a few different teaching gigs at different institutions until something more permanent came along, a common experience among NTTFs. Yet for me, the trade-offs were worth it. It is important to note, however, that some of this speaks to the relative privilege of my positionality: I am not the primary breadwinner in my family, and I have the option to choose a flexible work schedule in exchange for taking on more unpaid caretaking responsibilities. It also speaks to the privilege of my working conditions relative to other NTTFs: at my institution, NTTFs have the option of full-time contracts with a professional pathway, including full benefits, a pension, and strong union representation. In contrast, as reflected in NTTF employment statistics, the local community college caps teaching loads at part-time to avoid paying higher salaries and benefits. If my institution did not offer professional grade working conditions, I would not have continued teaching here.

This leads to my next point: when thinking about the range of NTTF positions available to you, think about your long-term career trajectory and how a particular job fit into that. Avoid taking jobs that move you away from those goals. This will help you avoid getting pigeonholed into roles that aren't helping you in the long-term. If you'd like to pursue research, don't take a teaching gig at a community college. If you want a stable, full-time position, don't scrape a bunch of part-time, one-off gigs together for too long, as it makes you look like you're willing to settle for part-time work. I often tell undergraduates that jobs beget jobs – being employed makes them more attractive to potential employers. But future employers are most attracted to (and want to poach) workers in related contexts, so their first job should be in the realm of what they want their second and third job to be. The same is true here. Unless you're planning to make a major career change in advance, align the jobs you accept with your broader career path.

Following are the takeaway points if you're thinking about non-tenure track positions:

1. Identify your own value systems and priorities and organize your career and work–life balance around them.
2. Find out as much as you can about a particular NTTF position to see if it aligns with your short- and long-term personal and professional goals.
3. Consider what an NTTF job has to offer in terms of flexibility in time, location, and scope of job responsibilities as well as possible job security and institutional advancement, and ask yourself if these are worthwhile trade-offs.

I'm in a non-tenure track job. How do I thrive?

Survival in any job involves recognizing, accepting, and, where possible, changing workplace contradictions and discrepancies. The non-tenure track, of course, has its own unique built-in contradictions given the devaluation of our labor. We should neither ignore nor blithely accept these conditions. But surviving – and even thriving – on the non-tenure track means managing workplace conditions as best you can to avoid feeling endlessly frustrated and consumed by them (e.g., Purcell 2007), while also working to improve the things that can be changed. Some suggestions about building collegiality and community can be found in Dagmar Budikova and Shannon O'Lear's chapter, "Cultivating a Collegial Department Workplace."

One of the most important things I think NTTF can do for themselves, which is arguably difficult to maintain in the face of our collective precarity, is to own your own value. Don't expect others to do this for you or to even notice what you are doing. In many cases, administrators care more that a class *is* being taught than who, what, or how it is being taught. This often gives instructors a lot of latitude to teach a class the way they want, but the trade-off is a general disinterest in our work. It is incumbent upon you to document your skills, training, education, and any other unique contributions and professional achievements you bring to your role. Having everything documented makes all the work you do, paid and unpaid, visible to the institution and justifies your requests when it comes time for a promotion or raise, or even to ask to get your teaching contract renewed for another term. Perhaps more importantly, documenting your work makes all of your labor and skills visible to *you*. Sometimes, given how easy it is to internalize institutional messages around our own devaluation, NTTFs can be our own worst enemy in acknowledging the value and scope of our labor. Keeping track of your skills, breadth of knowledge, and accomplishments will ultimately enable you to know your worth.

Secondly, set boundaries around your work and try not to overinvest in your job. This can be challenging given the emotional elements of our jobs, especially if an instructor identifies with their students or wants to help mitigate the other aspects of their lives (housing, employment, family, etc.). In addition, teaching and academic labor often get conflated with aspirational notions of a vocation or a calling, which perpetuates an overidentification with and culture of overwork. But NTTF positions are usually governed by job contracts that spell out the specific terms of employment, and we should use this to our advantage – aim to fulfill outlined tasks and responsibilities competently and do no more (and no less), unless it is for some other reason, like personal satisfaction or a potential salary increase. At the same time, however, I recognize the inherent contradiction of this point: in many instances, in order to keep their jobs, NTTF often have to demonstrate "professional excellence," which ends up involving invisible, unpaid labor. In my case, lecturers are expected to demonstrate "teaching excellence" and currency in our subject matter, which requires ongoing professional development and awareness of the research trends in the discipline, all of which go uncompensated. The broader point here is do your job well enough to keep it, but put boundaries around it and don't expect rewards, promotions, or accolades for excelling or going beyond.

Third, following the suggestions given by Dan Trudeau and Dydia DeLyser in their chapter, "Cultivating supportive networks and collaborations," I strongly suggest building communities in your workplace, and communities of two specific kinds. First, build a community of fellow NTTF in your department or institution. There are many reasons for this. Many NTTF report feeling lonely and isolated in their roles (Chatterjee 2023), and building a community can help you find a sense of belonging among faculty with experiences similar to yours. You can commiserate, find support, build solidarity, and share information. Second, aim to build a community of allies who can advocate on your behalf. These can be NTTF, but they are ideally more permanent, higher-ranked faculty and administrators who will "sponsor" you – that is, champion you in closed door spaces and possibly protect you from criticism (Hewett 2019). Creating this kind of advocacy network may feel infantilizing, but the reality is you may not have access to closed door spaces where decisions are made and need representatives to speak on your behalf. Also keep in mind that in most of academia, as in most professional fields, one's reputation and loose network of connections carry a lot of weight. Taking the time to establish yourself as a reliable colleague and to build these connections may pay off in unforeseeable ways in the future, both inside and outside the workplace. Similarly, as you grow in stature, becoming an ally, mentor, or eventually a sponsor to others can also add meaning to your work and help you pay it forward.

If and when you can, push for institutional changes to make the system more just. There's a wide variety of activities one can engage in to promote greater equity in the workplace: form or become active in a union, serve on institutional or professional association committees that focus on inclusivity and fostering equitable workspaces, or engage in other scholarly activities around NTTF/academic labor. Of course, getting involved in these ways may involve more unpaid work, and you might find it difficult to organize or collaborate with colleagues who are similarly not incentivized to take on additional work. And making large-scale professional change is slow, so you might not see the rewards of your labor very quickly. At the same time, and this is yet another potential frustrating and contradictory element of institutional change, the system of unpaid, invisible labor will probably remain the status quo until more people put additional unpaid work into changing it. I have found a sense of belonging and larger purpose in doing this work, which has made it personally rewarding. Also, getting involved in labor politics and workplace matters may not be feasible when you're just starting out, but it may be something to consider when you feel more stable and secure.

As you grow into your position and become more aware of your specific working conditions, keep your career options open. Maintain your professional networks and take advantage of opportunities to up- or transfer skills. I knew when I pursued a full-time lecturer position that it had an institutionally designed glass ceiling – even though it has a professional pathway, NTTF are not typically given the same opportunities for institutional advancement as tenured faculty, and I didn't want to keep teaching the same class(-es) over and over again for my entire career. Consequently, I've purposefully maintained very broad professional networks and have kept my eyes and ears open for different opportunities. I often ask people about their jobs and careers and have thought about the ways my skills might transfer to different roles. Similarly, create and keep an updated and active LinkedIn profile alongside your Google Scholar profile. Academia tends to train us to be very narrow content specialists, which may make it difficult to see ourselves outside of this role. Keep your options open, network broadly, and have your list of transferable skills and knowledge updated in case you ever want to exit.

Ultimately, take care of yourself. If you're teaching a wide variety of courses and/or working at different institutions, you may feel frazzled and mentally and emotionally drained. Build in enough downtime to refresh yourself and pursue other interests and activities outside of work. This may be hard to do, but it reinforces the importance of putting boundaries around your work and not overinvesting in a job that won't love you back (the students might, but the job never will). Don't overcommit, as easy as that is to do. If the workload starts to feel overwhelming and untenable, it might be time to look for things to cut or to possibly switch roles.

Takeaway points to help you thrive in a non-tenure track position:

- Acknowledge and work to improve the contradictions inherent in this role.
- Self-advocate: Document the work you are doing and share this information with others
- Avoid overinvesting in your job
- Build allies and networks
- Maintain professional networks outside of academia and actively consider how your skills can transfer to other roles
- Work to make the system more just

My non-tenure track job has become a dead-end. How do I leave?

If the position you are in is no longer offering you the kind of satisfaction or opportunities you're seeking, it might be time to think about transitioning jobs or careers. I don't mean for this to sound simple or blasé. Transitioning jobs and professions can be a monumental task, with life-changing implications . In most cases, it requires a great deal of planning and may not always be feasible (Adams 2013). But one of the major blockages academics face is actually believing that (a) changing careers/professions *is* possible with the skills and work experience we have; (b) we might actually be happier in a different role or profession; (c) we deserve to have happiness, work–life balance, and job satisfaction in our lives. Being an NTTF may have undermined our self-worth and led us to believe that we somehow deserve to be in jobs that rely on our constant exploitation or that we're not able to do anything other than teach. Overcoming these internalized messages may be the most difficult – and first step – in working toward change.

The second major block many NTTF often need to work through is the idea of the sunk cost fallacy – the idea that if you switch careers, you'll have wasted the time, personal resources, and finances you spent pursuing this profession. Do not let this keep you tied to a bad situation. First off, the fallacy is just that: an illogical fallacy. There is no experience or training that is not worthwhile. If your education or job were serving your needs at a particular time in life, you haven't lost anything. You will be able to apply these skills or experiences in a different way in the future. You may need to repackage and showcase different skills and achievements, but everything you have learned up until this point is valuable in any other job you have or pathway you pursue. I often counsel students who feel they pursued the wrong major or chose the wrong internship and are convinced that the rest of their life is overdetermined by that choice. I offer them the same advice I offer here: there is no loss of experience. You bring those experiences with you, and they may offer some of the most valuable insights down the road. You may not be able to see it yet, but it is there.

One major lever to help overcome these mental blocks is to take stock of your accomplishments and be proud of your work. Leave your job on a high note. Remember that nobody can take these achievements away from you. Again, this may mean that you have to actively resist institutional or professional messages that devalue your labor. Neoliberal economic policies and the institutions that propagate them devalue the very entities that are important in teaching and make life meaningful: developing caring relationships; cultivating passion, curiosity, and empathy in others; and engaging in transformative teaching, learning, and mentoring (Sennett 1998). If you awakened these qualities within students in any way, you achieved a greater good that goes way beyond an institution's metrics that measure learning outcomes. This is something to be celebrated and honored. It is a cruel reflection of the increasingly corporatized nature of higher education that individualistic achievements in research and publication are prized and rewarded over more collective educational endeavors that reduce social inequalities and vastly improve human life. But understanding the system makes it easier to take pride in your work and to walk away with a feeling of accomplishment when you feel that work is complete.

All of this points to attending to the mental, emotional, and psychic elements of leaving an NTTF position. These elements are often less discussed in conversations around the practical steps to finding a new job. Changing jobs, professions, or careers may feel like an enormous loss as your entire sense of self undergoes some changes. Make space to grieve the loss of identity and shift in goals (Simula and Elue 2023). As Ebru Ustundag and Asli Duru note, prioritize your physical and mental well-being. Give yourself the time and space to work through this process. Get professional help in the form of a therapist or career coach if you can afford it and build a support network. Give yourself opportunities for self-care. Working through your grief will make it easier to let go of your expectations, better accept the changes to come, and make room for and bring positive energy to allow those changes to take place.

Once you've made the mental and emotional shift to move on, the practical steps you need to take to get there will become clear. There is a plethora of resources to guide you through this (see Box 4.3 for a list of resources that may be helpful); some are free, while some cost money or time. Just remember that just as when you were first starting out and looking at NTTF jobs, you should do your values clarification exercises to see the ways your priorities or values have shifted, which will enable you to focus your job search on positions that meet your criteria and lifestyle. In addition, taking basic job search steps, such as scanning job ads, doing informational interviews, creating a LinkedIn profile, or converting your CV into a resume, will allow you to gain more confidence and imagine alternate pathways for yourself.

BOX 4.3 RESOURCES FOR TRANSITIONING JOBS/CAREERS

- Consulting and job-seeking resources/networks:
  - The Professor is In blog and consulting services
  - The Professor is Out Facebook group
  - Resources on LinkedIn
- Books (check your institution or local library for free access):
  - Karen Kelsky, *The Professor is In: the Essential Guide to Turning Your PhD into a Job*
  - Christopher Caterine, *Leaving Academia: A Practical Guide*
  - Sarah Jaffe, *Work Won't Love You Back*
  - Henry Cloud, *Necessary Endings*
  - Annie Duke, *Quit: The Power of Knowing When to Walk Away*
- Resources for upskilling, career changes, and professional development:
  - Your institution's career center or professional development resources (free tuition or free course auditing)
  - Reputable online certification courses (Google, Amazon, LinkedIn)
  - Reputable online classes (e.g., Coursera, Harvard/MIT free courses)
- Common "alt-ac" and "ac-adjacent" industries and careers worth exploring
  - Ed tech
  - Project/Program management
  - K-12 education
  - Public policy
  - Freelance/Self-employed writing, editing, consulting services
    Start a business
- First steps
  - Network, network, network. Tell everyone (neighbors, friends, family, and, where possible, professional networks) that you're looking for a change. Use LinkedIn to build professional connections in possible careers, industries, or employers you'd like to pursue.
  - Convert your CV into a resume if you're looking for jobs outside of academia
  - Start scanning job sites/boards to see what appeals to you
  - Do a wide variety of informational interviews to help you see what is possible and out there.

Takeaway points on leaving an NTTF position:

- Work to overcome your mental blocks around your value and around the sunk fallacy
- Take pride in your achievements
- Understand the institutional factors that shape our working conditions, and do not take them personally
- Take time to properly grieve the loss of identity and shifting goals
- Use available resources to help you move forward.

* * *

As a graduate student, it can be hard to get good insight into the experience of NTTF, especially since many of the NTTF at your institution are busy, on the margins of your department, and not necessarily compensated to engage with graduate students in this way (and could possibly be asked not to). That said, most NTTF are delighted when others take interest in their work and would probably be willing to open up about their experiences if asked. Consequently, I would find out about NTTF working conditions at your institution and converse with NTTF when the opportunities arise for your own professional benefit and to glean insight to see if this pathway is for you. It could also be valuable for forging cross-institutional labor networks that may become valuable in the future.

Overall, I believe that the elements that give NTTF positions the worst reputation of being an "adjunct trap" can be mitigated if you anticipate and plan around these things in advance and set appropriate expectations. My own experiences and those of my own NTTF community demonstrate this. Put your own values, needs, and priorities first, and organize work around that. Do not give more than you get, and do not pursue jobs that are less than what you are worth. Such advice seems easy to dispense and harder to follow, given the way we are trained to prioritize our studies over all else in grad school. But remember, having successfully completed a lengthy, rigorous training program, you deserve a livable wage, professional respect, and reasonable work–life balance. If certain jobs you're looking at are not offering these things, they are not worth it. Push back against institutional messages that say otherwise, and trust your inner voice when it tells you to stay or move on. It is possible to thrive on the NTTF, but you need to make it work for, and not against, you.

References

Adams, J. 2013. Switching Sectors: Transitioning into and among Business, Government, and Nonprofit Careers. In *Practicing Geography: Careers for Enhancing Society and the Environment*, eds. M. Solem, K. Foote, and J. Monk, 27–40. Upper Saddle River, NJ: Pearson.

American Federation of Teachers. 2022. An Army of Temps: AFT Contingent Faculty Quality of Work/Life Report. Accessed November 30, 2023. https://www.aft.org/sites/default/files/media/documents/2023/Contingent_Faculty_Survey_2022_interactive.pdf

Carey, K. 2020a. The Bleak Landscape of 'Adjunctopia' for PhDs. *New York Times,* March 5. Accessed November 30, 2023. https://www.nytimes.com/2020/03/05/upshot/academic-job-crisis-phd.html

Carey, K. 2020b. The 'Public' in Public College Could be Endangered. *New York Times,* May 5. Accessed November 30, 2023. https://www.nytimes.com/2020/05/05/upshot/public-colleges-endangered-pandemic.html

Chatterjee, A. 2023. Adjunct Professors face a 'Constant Struggle to not Give Up.' *Chronicle of Higher Education,* October 26. Accessed November 14, 2023. https://www.chronicle.com/article/adjunct-professors-face-a-constant-struggle-to-not-give-up-report-says

Childress, H. 2019. *The Adjunct Underclass: How America's Colleges Betrayed Their Faculty, Their Students and Their Mission.* Chicago, IL: University of Chicago Press.

Colby, G. 2023. Data Snapshot: Contingent Faculty in US Higher Ed. American Association of University Professors. Accessed November 30, 2023. https://www.aaup.org/article/data-snapshot-tenure-and-contingency-us-higher-education

Cook, K. 2017. Higher Education Funding in California [fact sheet]. Public Policy Institute of California. Accessed November 30, 2023. https://www.ppic.org/publication/higher-education-funding-in-california/

Dolezal, J. 2022. The Big Quit: Even Tenure-line Professors Are Leaving Academe. *The Chronicle of Higher Education,* May 27. Accessed November 30, 2023. https://www.chronicle.com/article/the-big-quit

Dorfeld, N., ed. 2022. *The Invisible Professor: The Precarious Lives of the New Faculty Majority.* Denver: The WAC Clearinghouse; University Press of Colorado. https://doi.org/10.37514/PRA-B.2022.1589

Ellis, L. 2021. The Great Disillusionment: College Workers Are Burning Out Just When They'll Be Needed Most. *The Chronicle of Higher Education,* August 25. Accessed November 30, 2023. https://www.chronicle.com/article/the-great-disillusionment

Fink, L. 2016. A Forum for the Contingent Teacher [blog post]. National Council of Teachers of English. Accessed November 30, 2023. https://ncte.org/blog/2016/09/forum-contingent-teacher

Gerwin, V. 2022. A 'Great Resignation' Wave among Academics Has Many Researchers Stepping Off the Tenure Track. *Nature* 606: 211–213.

Hartocollis, A. 2023. Can Humanities Survive the Budget Cuts? *New York Times,* November 3. Accessed November 30, 2023. https://www.nytimes.com/2023/11/03/us/liberal-arts-college-degree-humanities.html

Heller, N. 2023. The End of the English Major. *The New Yorker,* February 27. Accessed November 30, 2023. https://www.newyorker.com/magazine/2023/03/06/the-end-of-the-english-major

Hewett, S. 2019. *The Sponsorship Effect: How to Be a Better Leader by Investing in Others.* Cambridge, MA: Harvard Business Press.

Hubler, S. 2020. Colleges Slash Budgets in the Pandemic, with 'Nothing Off-Limits'. *New York Times,* October 26. Accessed November 30, 2023. https://www.nytimes.com/2020/10/26/us/colleges-coronavirus-budget-cuts.html

Kelsky, K. (2019). “Bad” TT vs. “Good” NTT: A Conversation (Part I). The Professor is In Blog. Accessed September 17, 2024. https://theprofessorisin.com/2019/08/02/bad-tt-vs-good-ntt-a-conversation-part-i/

Kezar, A., T. DePaoloa, and D. Scott. 2019. *The Gig Academy: Mapping Labor in the Neo-liberal University.* Baltimore, MD: Johns Hopkins University Press.

Kinman, G., and F. Jones. 2008. A Life beyond Work? Job Demands, Work-Life Balance, and Wellbeing in UK Academics. *Journal of Human Behavior in the Social Environment* 17 (1–2): 41–60.

Kolata, G. 2016. So Many Research Scientists, so Few Openings as Professors. *New York Times,* July 14. Accessed November 30, 2023. https://www.nytimes.com/2016/07/14/upshot/so-many-research-scientists-so-few-openings-as-professors.html

McClure, K. 2021. Colleges Are Hiring, But Do People Want to Work There? *The Chronicle of Higher Education,* November 22. Accessed November 30, 2023. https://www.chronicle.com/article/colleges-are-hiring-but-do-people-want-to-work-there

McMurtrie, B. 2023. Americans Value Good Teaching. Do Colleges? *The Chronicle of Higher Education,* September 20. Accessed November 30, 2023. https://www.chronicle.com/article/americans-value-good-teaching-do-colleges

McNaughtan, J., H. A. García, and K. Nehls. 2017. Understanding the Growth of Contingent Faculty. *New Directions for Institutional Research* 2017 (176), 9–26.

Mitchell, M., M. Leachman, K. Masterson, and S. Waxman. 2018. Unkept Promises: State Cuts to Higher Education Threaten Access and Equity. Center for Budget and Policy Priorities. Accessed January 29, 2023. https://www.cbpp.org/research/state-budget-and-tax/unkept-promises-state-cuts-to-higher-education-threaten-access-and

Peters, K., and J. Turner. 2014. Fixed-term and Temporary: Teaching Fellows, Tactics, and the Negotiation of Contingent Labour in the UK Higher Education System. *Environment and Planning A* 46 (10): 2317–2331.

Pew Trusts. 2019. Two Decades of Change in Federal and State Higher Education Funding. Accessed November 30, 2023. https://www.pewtrusts.org/en/research-and-analysis/issue-briefs/2019/10/two-decades-of-change-in-federal-and-state-higher-education-funding

Purcell, M. 2007. “Skilled, Cheap, and Desperate”: Non-tenure-track Faculty and the Delusion of Meritocracy. *Antipode* 39 (1), 121–143.

Schuster, J. H., and M. J. Finkelstein. 2006. *The American Faculty: The Restructuring of Academic Work and Careers.* Baltimore, MD: JHU Press.

Sennett, R. 1998. *The Corrosion of Character: The Personal Consequences of Work in the New Capitalism.* New York: WW Norton & Company.

Simula, B., and C. Elue. 2023. Navigating Grief in Higher Ed Career Transitions [blog post]. *Inside Higher Ed,* November 16. Accessed November 30, 2023. https://www.insidehighered.com/opinion/blogs/university-venus/2023/10/16/navigating-grief-higher-ed-career-transitions.

Smyth, J. (2017). *Toxic university: Zombie Leadership, Academic Rock Stars and Neoliberal Ideology.* London: Palgrave Macmillan.

Trower, C. 2021. Negotiating the Non-Tenure Track. *Chronicle of Higher Education,* July 6. Accessed September 17, 2024. https://www.chronicle.com/article/negotiating-the-non-tenure-track/?sra=true

Tuchman, G. 2009. *Wannabe U: Inside the Corporate University*. Chicago: University of Chicago Press.

Washburn, J. 2006. University Inc: The Corporate Corruption of Higher Education. London: Basic Books.

Zahneis, M. 2023. Katalin Kariko Has Thoughts on How to Support Shunned Scholars Like Her. *Chronicle of Higher Education,* October 6. Accessed November 30, 2023. https://www.chronicle.com/article/katalin-kariko-has-thoughts-on-how-to-support-shunned-scholars-like-her

5

TIME MANAGEMENT

Patricia Price

Introduction

Not long after I was appointed as interim provost of Queens College, I was invited to a staff barbeque hosted by our largest division, Arts and Humanities, at the start of semester. This is always a busy time, and I could have declined since the barbeque was a long walk from my office and I had other events on my calendar. But I went, and, as we sat around our plastic tables, one person remarked on how impressed they were that I was there and asked why I had decided to show up. It only took me a moment to genuinely respond, "this is a priority for me."

The barbeque presented a golden opportunity for me, as a brand-new member of the Queens College community, to meet a large number of individuals with whom my staff and I would be working. Indeed, as I began this position, I made a conscious decision to prioritize time for building relationships, by scheduling one-on-one meetings with department chairs and by accepting as many invitations as possible. I shared with this person my conviction – most notably advanced by former US President Dwight Eisenhower – that we may consciously choose to allocate our time to activities of varying degrees of urgency and importance. Particularly when beginning a new position – whether, like me, you are advanced in your career, or perhaps even more so when you are starting out – everything can seem to be both urgent and important. Particularly for those of you just now embarking on academic careers, the main areas of allocation of effort – teaching, scholarship, and service – can seem like three separate, full-time jobs simultaneously held, each with its own set of urgent and important priorities.

While our default tendency is to pay the most attention to – and spend the most time on – those activities that are both urgent and important, it is vital

DOI: 10.4324/9781003342953-6

that we identify those activities that are important but not urgent, and make time for those. That barbeque was one of those important activities that would allow me to build personal relationships with the people who run the college. But there was no real urgency to it. I could have just as easily chosen to attend another, competing event, one that clamored more loudly for my attention.

As early career faculty, you might reasonably suppose that as a person rises through the academic ranks – certainly someone at the level of a department chair, a dean, or a provost – they surely will have mastered the art of time management. Let me reassure you that this is not the case. Time management is a continuous challenge. Ironically, tackling time management may be one of those important but not urgent activities that we somehow fail to make time for. Part of its trickiness is its tendency to shift as our lives, and its attendant demands and priorities, unfold. Time management is better conceptualized as an aspirational horizon that eludes our sustained grasp: once you achieve a place of relative equilibrium, it seems given that something will happen that knocks it awry. In addition, and as I will discuss in more detail later, time management is considerably more challenging for some of us than for others because of the social locations we occupy.

In this chapter, we will explore time management from the perspective of early career faculty – though, as I noted earlier, the advice shared here is mostly applicable across the arc of your career. We will begin by noting that mastery of one's time has a foundation in repeated practices. There is really nothing new under the Sun here – so much of success involves small positive habits that are repeated to cumulatively large impact – though technological advances present both new strategies for managing time, as well as posing novel distractions. The trick really is to identify a handful of tactics that work for you, and engage them steadily. We will then turn to the reality that in order to successfully manage your own time, you must also strategically manage others' time. Structural considerations are also afoot in the time management landscape, and we will consider how such factors as gender, race, age, and other aspects of our social location can shape an individual's time management calculus. We then return to tactics, considering how tasks that take up your time – whether routine domestic work or highly specialized chores such as indexing – may best be outsourced to professionals (or bots), thus freeing you up for higher-value activities. In conclusion, I wish to reinforce the importance of cultivating a mindset of equanimity, small wins, and cumulative habits that will stand you in good stead over the arc of your career by setting you up for success, whether you are a newly minted assistant professor or the president of a large university.

Ideas for managing your time

You do not rise to the level of your goals. You fall to the level of your systems.

James Clear, Atomic Habits

Goal-setting and regular planning

The most readily available and well-known published work on time management is in the popular press. However, for those among you who demand evidence-based scholarship, rest assured that there is sound academic research focusing on time management (Aeon, Faber, and Panaccio 2021). That scholarship is primarily centered in the disciplines of psychology and management. Studies explore, for example, the relationship between time management efficacy and student outcomes such as grades, anxiety and stress levels, and resiliency (e.g., Wolters and Hussain 2015), the relationship between tasks and the perception of time (e.g., Tonietto, Malkoc, and Nowlis 2018), and other avenues of scholarly inquiry. Given that most of you will not have an academic background in these fields, you are likely to be largely unfamiliar with this work. Though, as academics, we are far more likely to be persuaded by research than by the popular press, these two genres intersect and often come to the same conclusions.

There's no shortage of pithy advice when it comes to time management. "You can't make more time, but you can manage the time you have." "Plan your work and work your plan." "How you spend your days is how you choose to spend your life." "What gets measured gets managed." The opening quote for this section is one of my favorites. It highlights the importance of systems to effective human functioning, and because I identify as a systems thinker, it resonates with me. In addition, it articulates enduring wisdom, drawing as it does from the Greek lyric poet Archilochus (680–645 BCE).

Both the academic and the popular press agree on an important observation about time management: it is but a tool (or more accurately, a set of tools), a means to an end rather than an end itself. Effective time management serves higher-order aspirations. In other words, at the end of the day, time management is simply a set of repeated practices that, if not aligned with goals, leads nowhere. So, effective time management starts with setting goals. Devoting quality time and care to setting appropriate, achievable, ambitious goals is, therefore, an extremely useful starting point.

It is worth setting aside some time to reflect on your professional goals. For many of you, a primary goal is achieving tenure (or the equivalent labor permanency at your institution). What else? Having a discernible impact on your field of study? Becoming known by your students and your colleagues for innovative teaching? Engaging undergraduate students in authentic research? Becoming a role model? Making headway on an intellectual challenge in your discipline? Establishing a well-functioning lab? Contributing positively to a community you care about? Garnering significant external funding? While your list will likely vary from this one, I include these examples to underscore that there are many valuable and compelling professional goals beyond simply achieving the milestone of tenure, and it is useful to enumerate them both to distribute your proverbial eggs among several baskets, and to articulate sources of self-worth and identity that do not entirely depend on the evaluation of others.

One comforting thing to note is that, for academics at the early career stage, your options tend to be well-defined and governed by a time-limited tenure clock. But once you've achieved tenure (if that is an option at your current institution), your array of options expands significantly – you can become a master teacher, or write another book, or engage in high-level mentorship, or become an administrator, among others – such that mid-career time management is in some ways more challenging than time management for new faculty members.

While high level goals should be long-term in nature, we cannot expect them to last forever. Think back to your earlier self – as an undergraduate, or when you began graduate studies – and some of the important professional (and personal) goals you held at that time. You will likely agree that, for the most part, those goals are not the same ones driving you today. Being conscious and intentional about setting and revising goals, and setting aside time to periodically evaluate your higher-order goals, is both a good place to begin your time management journey and a good practice to regularly revisit. It is helpful to make goal setting and revision a habit that is pegged to the calendar in a regularly recurring and obvious way. Many people make time for this at the New Year, but there's nothing particularly special about that day other than the ready nudge afforded by the cultural connection between the date and the practice. You could just as well make goal setting and reevaluation an annual habit coming directly after commencement exercises in the spring, or prior to your annual performance assessment activity (if your institution requires this). And, as with all time management practices, it doesn't really matter what date you choose – it matters that you identify a date and stick with it.

A template for the periodic setting and revision of goals is provided by Kerry Ann Rockquemore in "The Sunday Meeting" (2010). Rockquemore advises setting aside a portion of every Sunday (or a day of the week that is relatively quiet for you) as a time when you will commit to aligning your to-dos with your calendar. The first step is essentially a brain dump: put onto paper (or into an electronic document) everything that you wish to accomplish in the upcoming week, large and small. The next step is to schedule time for making progress on these to-dos. Rockquemore notes that, when you attempt do this, it becomes readily apparent that your to-do list far exceeds the time available to you, and you will then – as a third step of the Sunday Meeting – need to do several things: prioritize, balance, triage, unpack, and align. Which activities will move you closer to your professional goals? Are you ensuring that you're making time to allocate your efforts such that those important but not urgent tasks are given space alongside the urgent tasks? What activities can you minimize or triage altogether from your list? Are there ways you can break down complex projects into smaller, more-achievable tasks that are then scaffolded toward larger objectives? Finally, Rockquemore encourages the nesting of these weekly meetings into periodic, bigger-picture semester-by-semester and annual planning, with which the weekly activities should be aligned.

As the Covid-19 pandemic underscored so abruptly for us all, sometimes the best thought-out plans can be utterly derailed by circumstances entirely outside our control. And that can feel very defeating on many levels. "Why plan at all?" you may be asking yourself. Yet while the enormity and profundity of Covid-19's impact is out of the ordinary, it does serve to underscore the importance of having a plan. For while elements of your plan – its cadence, the substance of the milestones, and the means by which you achieve the long-term goals – may be drastically altered through no fault of your own, long-term planning provides a framework for action and a mindset of planfulness that will stand you in good stead over the long term. Indeed, many of us found our planning practice to be a source of comfort and structure that was sorely needed during the darkest days of the pandemic. As I repeatedly tell my students, it is always better to have a plan, even if that plan goes awry, than to not have a plan at all. It is advantageous to operate from a position of proactivity, as opposed to reactivity, because you are in the proverbial driver's seat.

That being noted, establishing a habit of regular, recurring, and nested (weekly, semesterly, annually) planning can seem daunting. In addition, it can take some time to see results from a practice intended to span years of your professional life. Let us now turn to a variety of time management tactics that are easier to implement and quicker to result in discernible progress. They are not a substitute for regular long-term planning, but instead offer an array of tools for ensuring that your day-to-day allocation of effort is not only progressing you toward your goals but also providing much-needed motivation along the way.

Time management techniques

The Internet abounds with techniques for managing time. Many are permutations on a theme. Here are some of them:

1 Make a list of the three things you will do today. For many of us, crossing off a list of to-dos provides immense satisfaction. Keeping a smaller list forces us to prioritize. In addition, it ensures an achievable daily goal that can leave us with a feeling of accomplishment, providing the motivation to tackle the next day's list.
2 Use a calendar. If you don't already use a calendar – whether online or old school paper – seriously consider using one. Calendars are visual reminders of our boundaries, and boundaries help to keep us focused and to protect our time. They also provide a record of past activities that we can look to, for instance when scaffolding small tasks onto larger projects, or

when writing annual progress reports. For those of you transitioning from graduate school, a calendar will help remind you of events, appointments, and tasks you must attend to – perhaps a significant shift from your days as a student that consisted mostly of unstructured time (itself a challenge). Online calendars can be shared with others at varying levels of detail, to help identify mutual blocks of time to collaborate or hold meetings. One caution about calendars: resist the temptation to schedule every moment of every day. Though some claim to find a full calendar to be inherently satisfying, perhaps as a signifier of importance, in practice it results in scattered energy and ultimately exhaustion (Jensen 2021).

3 Plan certain types of work during the times of day that you are typically most alert. This requires some self-awareness on your part. Are you a morning person or a night owl? Consider scheduling tasks that require a lot of energy, focus, and creativity during those times of the day when you are most naturally alert, and reserving other tasks (e.g., copyediting, answering email, and reviewing lab notes) for times when you are normally less energetic. Do you have small children? Experiment with using the quiet time in the morning, before the rest of the family awakens, to do focused writing and thinking – or, if it suits you better, do that at night when the others have gone to bed.

4 Write in structured bursts punctuated with brief periods of rest. One well-known technique is called the Pomodoro method (Cirillo 2018), after the old-school ticking kitchen timers shaped like tomatoes ("pomodoro" is tomato in Italian). The gist of the method is to write in uninterrupted 25-minute blocks, during which all distractions – social media pings, email, phone calls, pets – are ignored. When the timer dings, that's your signal to get up, make a cup of tea, stare out the window, pet the dog, stretch – any non-writing activity you like, for five minutes – then back at the desk for another 25-minute round. If you can only fit in one or two of these blocks, you will at least get some writing done every day, ensuring slow but steady progress toward a writing goal. And, if you've got a large writing project, Cirillo suggests stacking four of these 25-minute blocks together in the morning, taking an hour for lunch, then completing another block of four intervals in the afternoon. I can attest to the power of this approach, which I used over a six-month period to knock out an entire book manuscript. It took a lot of discipline but got the result– an accepted book manuscript – that I needed to earn tenure.

5 Join a group. Another way to provide the structure necessary to complete large, complex writing tasks is to join a group of others in the same boat. There are online groups, usually for a fee, as well as face-to-face groups you can join. You may engage in a "writing challenge" – say, writing at the same time for a defined period – or check in daily to set and report writing goals. Your institution may already have established in-person groups, or an

institutional subscription to an online service such as the National Center for Faculty Development and Diversity that provides access to writing groups and coaching. One benefit of this technique is engaging with and learning from others, breaking up the solitariness that can creep into the life of the mind.

These are just five of many possible techniques you might adopt. No one of them is inherently better than any other. I suggest you try them out and see which ones align with your needs and proclivities. Keep in mind that such techniques are only as effective as your use of them: merely thinking about adopting them does you no good whatsoever. Choose one or several and put them in regular rotation. You will be amazed at their cumulative impact on your professional progress.

A word about technology

Regardless of your personal style, it is important for all aspiring academics to start strong by engaging in good time management hygiene. Adopting and regularly employing techniques like the ones discussed that work for you is part of that process. In addition, we all need a strategy for managing email and other electronic forms of communication. Left unchecked, reading and replying to email, responding to "pings" from Teams, and posting on social media can really add up timewise, insidiously gobbling up large swathes of precious time and sapping our focus and creativity.

It is also vital that you lay a foundation of sound recordkeeping practices, for doing so will save you a wealth of downstream time, reduce anxiety, and ensure that your many professional achievements are not overlooked when it comes time to report on them. Keep a running record of all that you accomplish – it can be as simple as a Word document or an online notetaking program. When you achieve a milestone in teaching, research, or service that "rises to the level," note it immediately. Some institutions subscribe to online platforms to track productivity. If you're lucky enough to be at such an institution, make regular use of it, as the software will allow you to not only gather a running repository of information but also to make short work of compiling annual reports, third-year review materials, and tenure packets, provided you've been entering your productivity information all along. And don't forget to create and maintain backups. Accidents happen that destroy paper (flood, fires, roof leaks, pets, spills, etc.), online platforms cease to be supported or get hacked, hard drives die, and so on. Resign yourself to some housekeeping and stay on top of it.

Enlisting your supporters

No one manages their time in a social vacuum. We all interact with other people – friends, family, colleagues, bosses – some of whom claim a great deal of our time because they live with us, and others who share common time management challenges. In other words, your time management necessarily involves others' time management.

Because this is a book geared toward early career academic professionals, let's start out with workplace time management. Fortunately, being a faculty member is, relatively speaking, a career option that normally carries with it a fair measure of autonomy with respect to time utilization. As you advance through the academic ranks, however, the requests, expectations, and demands of others will begin to segment your daily life in a way you probably never imagined as a graduate student. Managing these requests appropriately, while prioritizing your professional momentum, becomes more and more important.

As a faculty member, you will have a supervisor, most likely a department chair or head. They, in turn, have a supervisor – the dean – and so on up the hierarchy, through the provost, the president, perhaps a system chancellor, and ultimately a governing board. You have peers – fellow faculty members in your department as well as colleagues outside of it, and in your discipline at large. Still others will be looking up the ladder at you – students, office staff – and you may formally supervise others, as well. "Managing up, down, and sideways" is such a vital skill for any professional that I put it in quotes, because it appears in countless titles and catchphrases. You can easily find a great deal of advice about sharpening these skills through role-playing and real-world practice.

Ensuring that you can turn to not one, but several, mentors and advocates is vital to thriving as an early career academic. A mentor inside your department can let you know early on what is non-negotiable with respect to your time, and what requests are best left until later in your career. Service on the college's curriculum committee, for instance, may sound innocuous enough, but at many institutions is one of the most politically fraught committees around. A trusted mentor at your institution will know that and steer you to another, lower-risk service activity. A mentor at another institution can open your eyes to different ways of approaching your work, and can provide a "safe space" to process your thoughts. A caring mentor can help you say "no" to requests in a way that doesn't burn bridges, leading to such service once you are tenured. Such people can also serve as advocates, by championing your successes, nominating you for visible opportunities that will help you to grow, or connecting you with others. Cultivating a handful of mentors and advocates early on in your career, and tending to those relationships over the course of your professional life, will go a long way toward setting you up for success in time management and many other aspects of career navigation.

Friends and family

Most of us are incredibly invested in our identities as academics. We have spent years working our way through graduate school, may have added several years as a post-doctoral student, and have made innumerable sacrifices to obtain that first academic position. And there is still more work ahead, as we learn the ropes of a new role and find our way around a new institution. Yet, no matter how identified you may be with your identity as a faculty member, it is important to bear in mind that it's just a job. And no matter how good of a job it may be, with its attendant creative freedom and relatively flexible time parameters, still ... it's just a job. It is not your family. It is not your health. If we're brutally honest, institutions and their leadership will always prioritize what is best for the institution, not necessarily what is best for you as an individual.

I say this to remind myself as much as anyone else not to neglect those things that bring joy to our lives. Whatever those things may be – togetherness with children, friends, partners, parents, and the friendship of non-work colleagues; engaging in hobbies that have nothing to do with our academic work; being in nature; religious observance – do not sacrifice them on the altar of more and more work. Make commitments to yourself or to others to honor these activities, and keep those commitments. Set firm boundaries around when you will pack up for the day and leave the office. Keep one day per week work-free and fully immerse yourself in non-work activities and relationships. Yes, there may be times when a late night or an early morning cannot be avoided, but do not make a habit of it. There are limits past which you will cease to be as effective, productive, and creative: for me, it's about 60 hours a week. After that my performance plummets, I become chronically tired but find it hard to sleep, my anxiety spirals, and my tendency toward catastrophizing kicks in. Be honest with yourself about what your limits are and keep inside of them.

Last but certainly not least, it is important to recall that you are not alone in this journey. Others – parents, partners, children, friends, relatives, mentors – have provided support and guidance to you along the way. That won't end just because your professional role has shifted. If you find yourself becoming overwhelmed, ask for help: perhaps your teenager can come up with dinner once a week, or your relatives would be willing to watch your young children overnight so that you can spend some quiet time with a partner. There are as many strategies as there are challenges, my point being that by articulating your needs and how these may have shifted lately, and enlisting the support of others, you are likely to find yourself reminded that time management is not an individual challenge but rather, in many ways, a collective endeavor.

Bringing in the professionals

No matter what society or certain individuals may say, you absolutely do not need to do everything yourself. Outsource those activities and areas of your life that will benefit from the attention of a specialist, that you despise doing yourself, or that are simply not priorities. Hire a cleaner instead of spending your weekends scrubbing. Get take-out rather than having dinner at 9:00

PM. Contract an editor to review your tenure statement or a professional indexer to index your book. You may benefit from employing one or more of these individuals at different junctures in your career – or, especially, at those crossroads of career and life when you find yourself assuming multiple, conflicting roles. For many of us, such crossroads are demographically preordained. Being sandwiched between teenage children and aging parents, for instance, is a common predicament for many people in their 40s, and often coincides with the intense period of preparation for going up for tenure and promotion. Outsourcing and relying on professional help can truly be a lifesaver in such situations.

Finally, it is hard to overstate the benefit of working with a professional coach or therapist. Though these professionals offer different services, and some may combine both professional coaching and more-traditional therapy, each provides a trained professional who is in your corner but also has at their disposal tried and true tools for helping you to advance along your life path, maintain your equanimity, and enjoy the journey (Whitaker 2022; Wisdom 2022). Granted, such specialized professionals can command a healthy price tag, and rarely does one's workplace offer high-quality coaching or therapy to employees as a benefit. So you may well be footing the bill for coaching and/or therapy yourself. However, this is an investment that will pay handsome cumulative dividends, not just professionally but for your quality of life over the long term, as you identify the reasons for recurring challenges, learn how to set and manage boundaries, understand why you react in predictable but perhaps less-than-optimal ways in certain situations, process trauma, and so forth.

Inequities

Before bringing this chapter to conclusion, it is worth spending a moment to recognize that the oft-touted notion of work–life "balance" is a fiction. At best, progress in work and in life is typically made in an unbalanced manner: work may need to be foregrounded for a time while non-work activities and relationships receive less attention, whereas at other times non-work relationships and activities may require more of our focus and attention while work-related activities and relationships drop into the background. Under these conditions, "work–life balance" may best be understood as an aspirational horizon of sorts, ever-receding as we move forward.

For many of us, however, "work–life balance" is a particularly pernicious fiction, for work, life, and indeed time itself are not the same for everyone, and our navigation of them is often inflected by our racial, gendered, socioeconomic, and other positions (O'Meara et al. 2022). In other words, there are asymmetries underlying time management that are structural in nature. Saying "no" isn't the same if you've routinely been told, through the actions and words of individuals and by society more generally, that your time and effort are communal property, that your boundaries will not be respected, and that your voice does not carry the same weight as the voices of others. "Leaning in," as Cheryl Sandberg has admonished women to do, is often far more fraught in practice and carries with it real risks when done by someone who has traditionally not enjoyed a seat at the table. And the more I read through the time management literature, the more that much of it appears to have been authored by privileged self-styled geniuses proposing "hacks" and "tricks" to shortcut their way to the top through, say, a four-hour workweek and other mostly ridiculous propositions.

Despite – or, more accurately, because of these asymmetries – it is vital to always bear in mind that you deserve to have time away from work-related activities and work colleagues. It is absolutely necessary to set your boundaries, to communicate those boundaries diplomatically and clearly, and to stick firmly to them. In fact, you will find that others for the most part will respect you for doing so. You must, in other words, look out for your own best interests when it comes to your time management, for the playing field is not a level one and to pretend otherwise is to be dangerously disingenuous.

Conclusion

Focusing on time management allows us to articulate our priorities, allocate our precious time to them above other options, and exert positive control over who and how we are constantly becoming. The opportunities and the challenges of living in a world where our attention, capacity, and drive are limited but the options are nearly limitless forces us – whether by commission or omission – to make choices. These choices have cumulative impacts. We become what we do repeatedly.

Successful time management is an ever-shifting puzzle that accompanies us throughout our professional lives. The substance of your time management – the specific mix of choices, challenges, and opportunities confronting you – will shift as your roles in life change, as well as in response to circumstances outside of your immediate control. The important thing is to be intentional and planful about your time, so that you engage in activities that further your objectives, that support you as a whole person, and that position you in a proactive rather than a reactive manner to the challenges and opportunities that arise along your path.

References

Aeon, B., A. Faber, A. Panaccio. 2021. Does time management work? A meta-analysis. *PLoS One* 16 (1): e0245066. doi: 10.1371/journal.pone.0245066

Cirillo, F. 2018. *The pomodoro technique: The acclaimed time-management system that has transformed how we work.* New York: Currency/Crown/Penguin Random House.

Clear, J. 2018. *Atomic habits: An easy and proven way to build good habits and break bad ones*. New York: Avery/Penguin Random House.

Jensen, D. 2021. Time management won't save you. *Harvard Business Review*, June 23. Accessed March 9, 2024. https://hbr.org/2021/06/time-management-wont-save-you

O'Meara, K., D. Culpepper, J. Misra, and A. Jaeger. 2022. Equity-minded faculty workloads: What we can and should do now. *American Council on Education*, online at: https://www.acenet.edu/Documents/Equity-Minded-Faculty-Workloads.pdf

Rockquemore, K. 2010, The Sunday meeting. *Inside Higher Ed*, September 19. Accessed March 9, 2024. https://www.insidehighered.com/advice/2010/09/20/sunday-meeting

Tonietto, G., S. Malkoc, and S. Nowlis. 2018. When an hour feels shorter: Future boundary tasks alter consumption by contracting time. *Journal of Consumer Research* 45 (5): 1085–1102.

Whitaker, M. 2022. What I gained from a year of professional coaching. *Chronicle of Higher Education*, July 8. Accessed March 9, 2024. https://www.chronicle.com/article/what-i-gained-from-a-year-of-professional-coaching

Wisdom, M. 2022. Beyond productivity hacks and time-management tricks: How coaching can help professors to "flourish in place." *Chronicle of Higher Education*, March 1. Accessed March 9, 2024. https://www.chronicle.com/article/beyond-productivity-hacks-and-time-management-tricks

Wolters, C. and M. Hussain. 2015. Investigating grit and its relations with college students' self-regulated learning and academic achievement. *Metacognition and Learning* 10 (3): 293–311. https://doi.org/10.1007/s11409-014-9128-9

6

CREATING COMMUNITIES OF CARE AND CARING FOR COMMUNITY

Building community through work–life integration

Dan Trudeau and Dydia DeLyser

Geography can be a caring discipline, and it is up to us to make it so. Indeed, Skop et al. (2021) invite us together to uphold an "ethos of care" in academia more broadly. Prioritizing care is crucial in a time when the business emphasis of many higher-education institutions has prompted academics to focus instead on competition and hyper-productivity. Simultaneously, the real and perceived pressures we labor under can foment feelings of loneliness and isolation and undermine practices of reciprocity, mutual support, and inclusion – they can hinder efforts to make progress on broader goals of equity and inclusion and to foster a collaborative workplace and discipline. Though these pressures are manifold and seemingly unrelenting, Skop et al. (2021) urge us to consider how, when, and with whom we can work to create academia otherwise. We can begin this work today by acting intentionally to extend care to others, including the nonhuman world, and build relationships of mutuality and solidarity in the various aspects of our professional practices. When we embrace and then extend practices of care, we can transform ourselves as well as our discipline and create geography as we wish it to be.

These calls for a reorientation of higher education toward care build from the voices of others who urge us academics to act where we are to create a more caring academy, especially one that is not fixated on scholarly productivity and bedeviled by extraction and the incessant circulation of anxiety (e.g., Mountz et al. 2015; Hawkins 2019; Puawai Collective 2019). This shared vision reminds us of the opportunities and the urgency to act in ways that affirm our well-being and ability to thrive, aligning our labor with values, and participating in (re)making geography into a caring discipline.

In this chapter, we frame the very moment you're reading this essay as an opportunity to respond to these calls to action. As people reproducing the

DOI: 10.4324/9781003342953-7

profession and its institutions, it's up to *us* to place care at the forefront of what we do. Centering care can inform what we do in our professional practices as well as how we proceed with realizing our goals, engaging with others, and building a collective future. In this chapter, we review helpful advice for emerging academics on how to begin with self-care in a challenging and busy work environment. While some academics will hold the different aspects of their work and lives apart, we detail how one way to center care is to consider integrating the different facets of our work in the profession into a rewarding career and life through community-engaged scholarship and community geography. Doing so may also create opportunities to make valuable contributions to our relationships with students, with scholarship, and communities wherever we are.

Building communities of care

Our academic lives exist in and operate through a series of communities: communities of practice, intellectual communities in the discipline, learning communities, the community where we live, the community of our family and friends, and so on. In this sense, communities do not exist as fully formed entities but emerge and evolve through activity, interaction, and engagement. Seen this way, though we may sometimes feel we work alone, our work and our lives are always with, and integrally connected to, other people. Adopting a mindset that centers these connections – and cares for them – in what we do can help our communities *and ourselves* to thrive.

We all give and receive care. We all need care. Even if we ignore it, "care is embedded in all of our encounters and interactions" (Lawson 2007, 3). It is therefore worth thinking critically and strategically about care. Considering care from a feminist perspective encourages us to extend and circulate care to others in our world in ways that intentionally foster relationships of mutual benefit and trust. Doing so can be empowering, for it takes us beyond feeling alone in our work and our worlds. It involves a shift from a perspective of "I" to one of "we" (de la Bellacasa 2017; Dombrowski et al. 2018). It invites us to consider how our own journeys are coupled with those of others to whom we are connected through community relationships. Suddenly we are no longer just laboring alone. Thinking and acting from a perspective of "we" can create an environment where care can flow in and through, back and forth, linking our own lives and those of others, creating an abundance – of possibilities, support, and opportunities – that exceeds what we can manage alone.[1]

As we center and circulate care, it's vital to consider that this begins, but cannot end, with caring for ourselves. Centering care in our own lives means taking time regularly to ensure that we, our loved ones, and our communities endure (Lorde 1988). Indeed, we care for ourselves in part by building communities that matter to us. Doing this, in turn, enables us to matter to others.

In this understanding of self-care, we move beyond the outdated self-help model of academia where emerging faculty were forced to sink or swim without support – such strategies only exacerbate inequalities and inequities. Instead, we can situate self-care in the context of communities of care in thoughtful efforts to move beyond and away from the feelings of isolation common among emerging faculty (Foote 2010). We can begin with the self, and move outward embedding our (new) careers and work expectations in a model of work–life *integration*.

We advocate for an embrace of work–life integration over that of work–life balance because "balance" suggests a difficult-to-achieve teetering point where work and life must hold exactly equal weight, or worse, a trade-off in a zero-sum game where adding to work must come out of life and adding to life must come out of work. Instead, *integrating* our work into our lives opens avenues for circulating care in ways that might create what brown (2017, 3) calls "critical connections" wherein new relationships between what may once have seemed different objectives can be formed – seeking ways to intentionally connect activities related to our scholarship, teaching, service, and perhaps even our lives. The business focus of neoliberalism in academic institutions that requires us to engage in scholarship, teaching, and service may not expect that we build synergies across these endeavors. Nevertheless, the "[t]aken-for-granted distinctions between teaching, service, research and outreach or between professional and personal/family life need to be interrogated" (Foote 2010, 16). When we can do so, and when we can move beyond those false boundaries, many may benefit.

We do not mean that our work or vocation should or must infuse all aspects of our lives (though for some it might). Rather, following Boice (2000), we urge consideration of how to coordinate different facets of academic labor so that these may become mutually beneficial – if not mutually constitutive – and aligned with our interests and values.

Recognizing the differences between short-term and long-term goals, between immediate needs/requirements and more overall needs/requirements, can help (Solem et al. 2009). It's easy to allow apparently urgent matters to take priority over what is really most important: without our health, and without our family and friends, we cannot thrive. While pulling all-nighters may have been a part of our lives since we were undergraduates, we must set that into the context of long-term careers, and long-term family lives. For academics, there is always more work to be done: more time we can spend on research and writing, as well as more time we can spend with our students, on our teaching, and on improving our classes.

Yes, there's always more work to be done, but maybe *right now* is the time to stop for the day – if we don't set limits, no one else will do it for us. Understanding this won't give us more hours in the day, but protecting ourselves against burnout is critical (Hawkins 2019): watching for early signs of frustration, depletion, and overcommitment can help keep burnout at bay. Fatigue is ancient and not the province merely of academics (though it does seem to track with privilege) (Vigarello 2023). Setting sane limits, even if they are small ones, *before* a crisis strikes us can help us to flourish even amid the pressures we work under. Following that, we can address those feelings of stress and strain and strive to create otherwise conditions for ourselves and others (Puawai Collective 2019; Mountz et al. 2015), as a footing to ensure we and our communities survive – and thrive.

The networks we began to establish in graduate school will now serve us well, and they can be strengthened. Even if you felt little connection with fellow students back then, you may find a greater connection now the stakes are higher. New connections can build our strengths too: AAG Specialty Groups are a great way to meet others with shared research interests and often so much else as well. Their "Business Meetings" are often sparsely attended, and such small groups can be an easy way to build a network and get involved. Another way to get to know geographers who share your interests is to organize paper or panel sessions at the Annual Meeting. Panel sessions are typically easiest for people to commit to (particularly when asked well before the deadline) and can be a great place to build connections among scholars who like to think together. These people will also share some of our work experiences – they may themselves be fighting to keep burnout at bay. We are not alone.

Even so, our lives must be built from more than just work. Most of us in new academic positions will also find ourselves in an unfamiliar city, town, and/or state, distanced from family and friends. While that's another experience you will be sharing with new and old geography friends, it can be challenging to even think of making time to explore this new world when job pressures feel so high. When the challenge feels greatest, start small. Think of one thing you enjoy doing, and find a way to do that locally, then keep your eyes open for local others with shared interests. Do you like hiking, dog training, or crocheting? While it might seem obvious that you can meet other hikers on the trail and other dog trainers in the dog park, think creatively about your interests and how they might intersect with others: if you love crocheting, is there a boutique yarn store in a town near you or a club of yarn bombers you can join? When Dydia moved to Louisiana, she became fascinated by the rural Cajun and Zydeco dance scenes, and so took dance classes both to meet local people and to better understand and participate in this intriguing world.

Engaging with community: just service?

As we consider how we circulate care as academics, some may be tempted to frame community engagement as a form of "service" – placing our community-care work in the traditionally least valued category of academic labor. But community engagement and caring for community are more than "just" service. Instead, we can seek ways to highlight how community engagement can contribute across categories to scholarship, teaching, and/or our lives. For many of us, this is essential.

The traditionally understood category of "service" often seems to be a way to count labor that is not easily identifiable as either research or teaching, but which is nevertheless related to our positions – derided as the "boring" committee work that others don't want to do. Worse still, service is often perceived as an *overload* of work – we do this in addition to what is required, and engage in more of it at our academic peril. Even so, many who identify as women or with underrepresented communities are sought out at disproportionately high rates precisely for service: because committees seek broad representation, and they need a woman and/or a person of color; because students who share our backgrounds – as women, as first-generation college students, as BIPOC, as LGBTQ+, as people living with disabilities, and so on – may see themselves in us and seek us out for (extra/additional) advising, mentorship, and/or advice on how to manage and thrive in the university system (Peña 2022).

This means that, for some academics, such engagement – *community engagement* – is not elective! It's a critically important part of who we are, of what we do, and of doors we open and *hold open* in our discipline. Such advising and mentorship, whether or not it is formalized, can be foundational to our role as educators and community members. Seen that way, "service" is integral to creating the kind of geography, the kind of academia, and the kind of world we seek (Trauger and Fluri 2014; Rees et al. 2020).

In this light, we can rethink service and connections to community as essentials in our lives, efforts that often represent multiple categories of academic labor. And, rather than dismissing this labor as "just service," we can elevate it as community-engaged scholarship (CES), or extend it in "community geography."

Beyond service: community-engaged scholarship and community geography

One way of building synergies that connect and even inspire different aspects of our work and our lives flourishes in practices of CES. CES refers to scholarly activities that engage community stakeholders in the process of knowledge production that serves the public good (O'Meara 2018). Across its different disciplines and iterations, researchers apply their expertise in working with communities to co-create knowledge as well as specific and tangible products to apply that knowledge, such as a training program, a museum exhibit, or a map representing community assets.

Rather than pigeonholing our activities into just one form of academic labor, an integrative view of our work recognizes that CES can contribute simultaneously to *all* of the categories – research, teaching, service, and life. The benefits extend beyond the livelihood of individual practitioners: they can enrich educational experiences when students are involved, and they can benefit our communities as well as ourselves.

CES is now being recognized as an integral part of geography. Recognition of CES may seem recent, but there is a long tradition of it in geography. While not the first instance, William Bunge (1971) and his collaborators' work in the Detroit Geographical Expedition and Institute (DGEI) serves as an early landmark achievement in CES in geography. Scholars and neighborhood residents and activists collaborated to document and visualize how Black residents in Detroit were experiencing systemic social inequity and what could be done to create a world otherwise. In the decades since the DGEI, the discipline of geography has grown to embrace many forms of community-engaged and participatory scholarship. So much so that Brown and Kyttä (2014) reported in 2014 that there were 3,500 instances of one form of PES, "public-participation GIS," alone.

As an example, the opportunities and contributions of community mapping are evident in Jin-Kyu Jung's work. Jung's expertise in participatory mapping and creative visualization has led him to work with a number of communities. His scholarship in this area began by working with children, supporting an effort to document their perception of the built environment and communicate their perspectives with broader publics. He has expanded this approach to connect with community organizations in Seattle, Washington, where he lives and works, engaging people in mapping their neighborhood in order to facilitate their learning about place and advance their involvement in government planning and decision-making processes that affect their neighborhood (Jung 2018). These experiences have likewise fueled Jung's reflections about his own pedagogical approaches to learning about geographical visualization, which has also shaped his teaching practice (Jung 2020).

In a different example of integrating research, service, and life, when Dydia finished her dissertation, she felt she wanted to give back to the community at Bodie State Historic Park, the California ghost town where her research was situated. She began, in the 1990s, by volunteering. Today, she still volunteers, and for the past decade has served also as Secretary of their non-profit Board. Later, as she sought to make her work community engaged from the outset, she began her research on how neon signs help shape American communities by building connections with communities of neon-sign makers, afficionados, and preservationists. That led to more Board service – for the American Sign Museum and the Museum of Neon Art. Academics end up with a lot of expertise in how (and how not!) to run meetings, job searches, and performance evaluations. These skills can be valuable assets to small non-profits, and many welcome new Board members with diverse backgrounds and experiences.

For Dydia, it has also been possible to use the skills she's gained as a writer in her public-facing engaged scholarship: co-authoring the Introduction for *Saving Neon: A Best Practices Guide* (Barna and Homan 2018), and collaborating with that book's author-designers to produce *Neon: A Light History* (DeLyser and Greenstein 2021).

CES can also translate well into teaching. For example, Dan centers his Qualitative Research Methods course on a community-defined research project as a way to teach research design, data collection strategies, and analysis. Working in this way also presents opportunities to practice ethical conduct, reciprocal exchange, and writing for public audiences. In each instance of the class, Dan works with a local community organization to identify and develop a suitable inquiry for a semester-length undergraduate qualitative research course, and then the students and he work with the organization to prepare a response that is relevant to its work while applying scholarly approaches to the production of knowledge.

These stories about individuals reveal personal commitments and motivations, and it's important to know that support for CES more broadly is building within the discipline, often under the frame of "community geography." We can define community geography as a praxis that focuses on "generating new practices, knowledge, and social identities rooted in particular times and places" that is guided by a focus on "place and place-based concerns; diverse positionalities, committed and reciprocal community partnerships; flexible epistemologies and methods; and open research practices and public scholarship" (Shannon et al. 2021, 1149–1151). Practitioners in community geography engage in community partnerships that produce knowledge with and for community partners and frequently involve undergraduate students in the process (Fischer et al. 2022). Reflections on such work have been gathered in a focus issue of *The Professional Geographer* (Barcus and Trudeau 2018), and Shannon et al. (2021) have likewise laid important groundwork for establishing community geography as a particular practice of geography and identified exemplars in the approach. As a parallel effort, the Community Geographies Collaborative (https://cgcollaborative.org/) has emerged to offer support and resources to practitioners and formed as a Specialty Group of the AAG. Today, there is much support for CES and community geography!

Indeed, across the US, many colleges and universities are taking up the call. For example, look to Michigan State University (2023), where they formally identify ways in which community-engaged activities can be counted as scholarship, teaching, and service *at the same time*. And MSU is not alone: they are part of a movement across higher education in the US to recognize and value this work (Box 6.1). These developments represent an expansion of how academic labor is conceptualized (HuMetrics Team 2022).

Even if our institution is not (yet!) formally a participant in the broad-based-and-growing CES movement, there are steps we can take to acknowledge and account for the multiple ways community-based efforts contribute to our work in the academy, and you may find others on your campus who are already participating.

Some of us may have heard the outmoded-and-misguided advice that engaging in "too much service" or doing "too much" community-based work could jeopardize our tenure cases (O'Meara and Terosky 2010). For all of us, there are resources. KerryAnn O'Meara (2014, 2018) has written extensively about the ways to support CES and has identified a number of practices that enable this work to count within existing frameworks of tracking academic labor. These include documenting the impact of CES activities and their importance for addressing matters of public concern as well as the activities' perceived merit among members of the community and/or funding agencies. O'Meara also stresses the importance of fostering peer-review (and tenure-review) processes that include the perspectives of community partners engaged in CES activities, as well as tracking the multiple ways in which knowledge is produced and disseminated.

Community-engaged scholarship and community geography as practices of care

Significantly, for many of us, circulating care through CES and community geography can help us integrate academic work with our lives: catalyzing work, life, and community can be revitalizing. Scholars participate in CES for a variety of reasons – from an intentional effort to connect scholarly expertise with communities that matter to us, to a bottom-line commitment to mentoring others that may involve unplanned labor at unexpected times. Regardless of the reason, engaging in CES can feel personally satisfying and professionally rewarding (Trudeau and Kruse 2014; Hawthorne and Jarret 2018). And this means that there are good reasons for higher-education institutions to find ways to recognize and support CES as there are longer-term benefits for creating conditions in which individuals feel "more satisfied and committed" (O'Meara and Terosky 2010, 48).

Of course, participating in CES and integrating our work with our communities may not be appropriate for everyone, or for every research project, class, or aspect of our lives. Not all research programs fit with this format. Not all aspects of our lives can be or should be linked to our work. And not all academic institutions have review processes configured in a way to recognize CES as the valuable form of academic labor that it is. It is therefore imperative to be strategic about whether and when we engage in CES. Moreover, it is probably the case that there are no requirements or expectations for us to participate in CES. Accordingly, it is prudent to examine how work in community geography or CES more broadly will be valued by institutions (Trudeau et al. 2018) and to pursue arrangements to specify how this type of work counts as academic labor (Robinson and Hawthorne 2018). These qualifications notwithstanding, CES, however we practice it, offers opportunities for work–life integration and the chance to circulate care to ourselves and others.

On that point, CES and community geography provide a way to demonstrate higher education's relevance to evolving societal issues and concerns of our communities. These also engender high-impact learning experiences (AAC&U 2023) that model service to society as a part of a comprehensive education. Valuing CES can likewise aid the recruitment and retention of diverse faculty by recognizing that faculty from historically marginalized groups in particular may participate in academia through a motivation and commitment to produce scholarship connected to and benefiting specific communities. Most academic institutions include in their mission statement or core values some version of pursuing academic excellence, preparing students to be engaged citizens and leaders, and fostering involvement in service to society. It is clear that when CES and community geography are done well, they can resonate with all of these and even serve as models for how to combine them. These, then, are important practices of care.

Conclusion

Building a stimulating and challenging career that is integrated into one's life, and seeking to find connections between scholarship, teaching, and community can be challenging. But it *is* rewarding. And it can be synergistic – serving to catalyze strengths and draw personal interest and social/community values into academic labor. And we can start small.

Consider your own community values and priorities. How might those intersect with organizations in the area where you find yourself living and working? Consider how you might integrate one project for just one of your classes with an organization you value or a need in your community. CES and community geography take time. Time to build connections to communities, time for community members to identify and understand how an academic geographer can help. Time for projects to develop, mature, and expand. Time for care to circulate. This time is the time of our lives. The time to care for ourselves and our communities as we build thriving academic careers. When we do so, we can help make geography a caring discipline.

BOX 6.1 SUPPORT RESOURCES FOR COMMUNITY-ENGAGED SCHOLARSHIP AT COLLEGES AND UNIVERSITIES

- The Academy of Community Engagement Scholarship – https://academyofces.org/
- Center for Community Learning, University of California, Los Angeles – https://communityengagement.ucla.edu/wp-content/uploads/2020/11/TP-for-Publicly-Engaged-Scholars.pdf

- Campus Compact – https://compact.org/resources/tenure-and-promotion-for-engaged-scholarship-a-repository?f%255B0%255D%3Dresource_tag=698
- Michigan State University – https://engage.msu.edu/about/overview/common-types-of-community-engaged-scholarship-reported-by-faculty
- Swearer Center, Brown University – https://www.brown.edu/academics/college/swearer/engaged-research
- TRUCEN Engaged Scholarship Resource collection – https://compact.org/resources/trucen-engaged-scholarship-resource-collection?f%5B0%5D=practice_area%3A198&f%255B0%255D%3Dpractice_area=198
- University of Minnesota – https://engagement.umn.edu/sites/ope.umn.edu/files/umn_pes_criteria_11.08.18_1%20Final.pdf
- University of North Carolina, Chapel Hill – https://academicpersonnel.unc.edu/wp-content/uploads/sites/1069/2020/05/PT-Final-Report-092016.pdf

Note

1 Caring is complex. Extending care to others may not always feel virtuous. As Samuels and Freeman (2021) explain, we may not always experience care from others as ultimately supportive. Moreover, intending to care without attentive listening or solidarity can have no impact, or even the opposite effect (Bartos 2021). Yet, when tethered to collaboration and reciprocity, care can orient relationships toward mutuality (Rees et al. 2020).

References

American Association of Colleges and Universities. 2023. High-impact practices. https://www.aacu.org/trending-topics/high-impact

Barcus, H., and D. Trudeau. 2018. Out in the world: Geography's complex relationship with civic engagement. *The Professional Geographer* 70(2): 270–276.

Barna, A., and R. A. Homan. 2018. *Saving Neon: A best practices guide.* San Francisco, CA: Giant Orange Press.

Bartos, A. 2021. Troubling false care: towards a more revolutionary 'care revolution' in the university. *ACME: An International Geography for Critical Geographies* 20(3): 312–321.

Boice, R. 2000. *Advice for new faculty members.* Boston, MA: Alyn and Bacon.

brown, a. 2017. *Emergent strategy.* Chico, CA: AK Press.

Brown, G., and M. Kyttä. 2014. Key issues and research priorities for public participation GIS (PPGIS): A synthesis based on empirical research. *Applied Geography* 46: 122–136.

Bunge W. 1971. *Fitzgerald: Geography of a revolution.* Cambridge, MA: Schenkman.

DeLyser, D., and P. Greenstein. 2021. *Neon: A light history* San Francisco, CA: Giant Orange Press.

Dombrowski, K., A. Watkins, H. Fitt, J. Frater, K. Banwell, K. Mackenzie, L. Mutambo, K. Hawke, F. Persendt, J. Turković S. Ko, D. Hart. 2018. Journeying from "I" to

"we": Assembling hybrid care collectives of geography doctoral scholars. *Journal of Geography in Higher Education* 42(1): 80–93.
Fischer, H., D. Block, A. Bosse, T. Hawthorne, J. Jung, H. Pearsall, A. Rees, and J. Shannon. 2022. Doing community geography. *GeoJournal* 87: 293–306.
Foote, K. 2010. Creating a community of support for graduate students and early career academics. *Journal of Geography in Higher Education* 34(1): 7–19.
Hawkins, H. 2019. Creating care-full academic spaces? The dilemmas of caring in the 'anxiety machine.' *ACME: An International Geography for Critical Geographies* 18(4): 816–834.
Hawthorne, T., and O. Jarret. 2018. Developing the next generation of community-based scholars. *The Professional Geographer* 70(2): 291–297.
HuMetrics Team. 2022. *Walking the talk: toward a values-aligned academy.* HuMetricsHSS. https://hcommons.org/deposits/objects/hc:44634/datastreams/CONTENT/content
Jung, J. 2018. Mapping communities: geographic and interdisciplinary community-based learning and research. *The Professional Geographer* 70(2): 311–318.
Jung, J. 2020. Teaching creative geovisialization: Imagining the creative in/of GIS. *The Canadian Geographer* 64(4): 512–528.
Lawson, V. 2007. Presidential address: Geographies of care and responsibility. *Annals of the Association of American Geographers* 97(1): 1–11.
Lorde, A. 1988. *A burst of light: Essays.* London: Sheba Feminist Publishers.
Michigan State University. 2023. Common types of Community engaged scholarship reported by faculty. https://engage.msu.edu/about/overview/common-types-of-community-engaged-scholarship-reported-by-faculty
Mountz, A., A. Bonds, B. Mansfield, J. Loyd, J. Hyndman, M. Walton-Roberts, R. Basu, R. Whitson, R. Hawkins, T. Hamilton, W. Curran. 2015. For slow scholarship: A feminist politics of resistance through collective action in the neoliberal university. ACME: *An International Journal for Critical Geographies* 14(4): 1235–1259.
O'Meara K., and A. Terosky. 2010. Engendering faculty professional growth. Change: *The Magazine of Higher Learning* 42(6): 44–51.
O'Meara, K. 2014. Legitimacy, agency, and inequality: Organizational practices for full participation of community-engaged faculty. In *Publicly engaged scholars: Next generation engagement and the future of higher education*, ed. M. A. Post, E. Ward, N. V. Longo, and J. Saltmarsh, 96–109. Sterling, VA: Stylus.
O'Meara, K. 2018. Accurately assessing engaged scholarship. *Inside Higher Education,* August 21. https://www.insidehighered.com/advice/2018/08/22/academe-needs-better-evaluate-engaged-scholarship-opinion
Peña L. 2022. *Community as rebellion: A syllabus for surviving academia as a woman of color.* Chicago, IL: Haymarket Books.
Puig de la Bellacasa, M. 2017. *Matters of care: Speculative ethics in more than human worlds.* Minneapolis: University of Minnesota Press.
Puawai Collective. 2019. Assembling disruptive practices in the neoliberal university: An ethics of care. *Geografisk Annaler: Series B, Human Geography* 101(1): 33–43.
Rees, A., T. Hawthorne, D. Scott, P. Solís, and E. Spears. 2020. Toward a community geography pedagogy: A focus on reciprocal relationships and reflection. *Journal of Geography* 120(1): 36–47.
Robinson, J., and T. Hawthorne. 2018. Making space for community-engaged scholarship in geography. *Professional Geographer* 70(2): 277–283.

Samuels, E., and E. Freeman. 2021. Introduction: crip temporalities. *South Atlantic Quarterly* 120(2): 245–254.

Shannon, J., K. Hankins, T. Shelton, A. Bosse, D. Scott, D. Block, H. Fischer, L. Eaves, K. Jung, J. Robinson, P. Solís, H. Pearsall, A. Rees, and A. Nicolas. 2021. Community geography: Toward a disciplinary framework. *Progress in Human Geography* 45(5): 1147–1168.

Skop, E., M. Caretta, C. Faria, and J. Smith. 2021. An ethos of care. *Inside Higher Education*, October 14th. Available online: https://www.insidehighered.com/advice/2021/10/15/pledge-foster-more-equitable-research-collaborations-opinion

Solem, M., K. Foote, and J. Monk. 2009. *Aspiring academics: A resource book for graduate students and early career faculty.* Upper Saddle River, NJ: Pearson Prentice Hall.

Trauger A., and J. Fluri. 2014. Getting beyond the 'God trick': Toward service research. *The Professional Geographer* 66(1): 32–40.

Trudeau, D., and T. Kruse. 2014. Creating significant learning experiences through civic engagement: Practical strategies for community-engaged pedagogy. *Journal of Public Scholarship in Higher Education* 4:12–30.

Trudeau, D., H. Barcus, L. Smith. 2018. Coda: Making geography relevant. *The Professional Geographer* 70(2): 333–337

Vigarello, G. 2023. *A history of fatigue: From the middle ages to the present.* New York: Polity Press.

7

SLOWING DOWN AND SAVING OUR HUMANITY

Shannon O'Lear and Kenneth Foote

Dear Reader,

Take a moment to set down this book and take a few deep, nourishing breaths. Go ahead. We'll wait…

…Better?

If you are feeling a little overwhelmed as you flip through the chapters of this book, this one is a good place to spend a little extra time.

You have chosen a career path that isn't easy, but is one that can be rewarding and (yes) fun. We hope this chapter will help you keep your humanity and humor intact as you prepare for the next steps in your professional adventure.

As you take on new responsibilities, a new role, or even a new identity, remember that it is also ok to leave behind any "baggage" that is no longer needed on this leg of your new journey.

Safe travels,

The Editors

Academics undergo rigorous intellectual training that provides a foundation for research, for teaching in our discipline, and that fosters an interest in service. It is not required in most academic programs to learn how to balance all of those activities within a life well-lived.

For many of us, this aspect of life as an academic was not an explicit part of our training or preparation, and it is not guaranteed that an academic career will include learning valuable lessons in self-care. Practices of care

DOI: 10.4324/9781003342953-8

received unprecedented attention during the Covid pandemic when there was mainstream coverage on challenges in working from home and dealing with social isolation. At the same time, more and more institutions such as the American Association of Geographers (2021) also facilitated conversations about care as part of their diversity, equity, and inclusion plans. Yet practices of care and living in an overall balanced way have not been routinely embedded or even recognized as part of professional development in many graduate programs or academic institutions. Fortunately, however, there is a growing appreciation of these practices that may be integrated, modeled, and adapted throughout the spectrum of academic research, teaching, and service and into our personal lives as well.

These are important points to address because studies dating back at least to the 1980s have documented high levels of stress among people in new academic appointments (Fink 1988, Fisher 1994; Gmelch 1993; Seldin 1987; Sorcinelli 1992). This sense of stress seems to be felt across the full range of early-career appointments, from temporary, short-term post-docs and teaching positions to appointments leading to tenure and long-term employment.

We can see these concerns playing out over a decade in a longitudinal study that was started by a group of graduate students who were interested in the concept of work–life balance in the lives of geographers and other scholars in the making (Manzi, Ojeda, and Hawkins 2024). This study tracked a group of research participants through graduate school into their early and mid-careers with a focus on care work and life–work trajectories. In ten-year follow-up interviews, they observed:

> At first glance, things were going well for participants but, as we examined the interview transcripts more carefully, we noticed that feelings of despair, overwhelm, exclusion, and isolation dominated people's work-lives at these stages in their careers – including the participants who explicitly claimed that 'things were getting better.' More freedom and flexibility led to more work. More leadership roles led to more experiences of discrimination. More responsibilities often led to burnout. More long-term connections to students, colleagues, and communities led to more emotional labor. We realized that the participants' sense of whether or not things were getting better in their work-lives was directly related to their possibility of creating and maintaining spaces for collective and self-care.
>
> *(Manzi, Ojeda, and Hawkins 2024, 2)*

After tracking this group of emerging and now established scholars for ten years, these authors observe that "The flexibility of academia can lead to overwork. The idea of academia as a vocation can led to further exploitation" (Manzi, Ojeda, and Hawkins 2024, 8). The authors also note these issues are often amplified for groups marginalized by gender, race, ethnicity, and LGBTQ identities. Similar results were reported in earlier studies in geography, as well as the effects of nationality or immigration status on stress (Fink 1984; Foote et al. 2008). In one of these earlier studies involving interviews with 40 early career faculty in the United States, the researchers found that stress arose from a wide range of sources as varied as time management; challenges of balancing work, personal, and family roles; lack of clarity over roles and expectations; the absence of mentoring; and many other factors (Solem and Foote 2004, 894–895).

In large part, these factors revolve around a few key issues. One of the most important is that early career academics often feel tremendous pressure to prove their abilities in a relatively short period, that "the clock is always ticking" even when they try to relax. Another key source of pressure is that new faculty are often asked to excel across a wider range of tasks than they were asked to master while in graduate school, particularly the teaching and service roles they are now expected to fulfill. An additional problem is that help and mentoring by colleagues may be haphazard or missing, particularly the sorts of implicit knowledge needed for success, but which is privileged by racial, ethnic, national, gender, and sexual identities. Finally, many early career faculty may feel a need to succeed on their own, and therefore be hesitant to ask for help for fear that any such request will show they are unprepared for their positions.

For starters, consider your physical well-being

People understand that taking care of our bodies supports mental health and intellectual capacity. The Covid pandemic brought these connections into stark light. One report from the World Health Organization noted a worldwide 25% increase in anxiety and depression stemming from the conditions of the pandemic just in 2021, and many of us are familiar with the unexpected, simultaneous challenges of the loss of in-person interactions, adapting to online working conditions, and the discomfort of tensions around what was known or not understood about the virus. Even as we move away from the deepest throes of Covid, many people seem to be getting the message that our physical well-being is crucial for supporting our mental well-being. It is not uncommon on college and university campuses to see people carrying around yoga mats, and reusable water bottles are a ubiquitous feature in student backpacks.

Beyond any visible signals of physical well-being, it is helpful to consider what kinds of practices help you to feel healthy, rested, and recharged. We know that some form of movement and regular exercise, nourishing food and hydration, and sleep are all important even if there are different views on the precise details of each of those categories. Most of us know what kind of activities and social interaction provide the relaxation or stimulation that we need:

group activities with a lot of people, with a few people, with family or friends, or activities by ourselves or with animal companions. Most of us also have had opportunities to learn what kinds of activities to step back from to avoid feeling depleted or overwhelmed (Looking at you, Doom scrolling!).

Next, consider your mental and emotional health

Beyond paying attention to things that we do (or avoid) to feel better; it matters how we think and talk about what we do. Robert Boice (1991, 113–114) set out to understand why some new hires seem to transition smoothly into faculty roles. He noted a number of characteristics shared by these "quick starters," as he called them, and four of those characteristics are particularly noteworthy:

- They verbalized (to me) uncritical, accepting, and optimistic attitudes about the undergraduate students on their campuses.
- They displayed low levels of complaining and cynicism about their campuses and their colleagues in terms of supportiveness and competence.
- They showed a marked disposition to seek advice about teaching, from colleagues, via reading and observing, and from faculty development programs. Specifically, they spent an average of four hours per week in social contacts with colleagues that included discussions about teaching.
- They displayed high energy, broad interests (for example, singing in choirs), concern with self-presentation, and a sense of humor.

What is interesting about Boice's observations is that they are related not just to things we do, but to how we approach our work and roles as well as the ways we create personal and professional relationships that support our well-being. These are points that are also highlighted in Patricia Price's chapter on time management and Dan Trudeau and Dydia DeLyser's chapter on creating communities of care and caring for community. As academics, we do have to demonstrate our own track record and independent contributions to scholarly work, but our contributions will be stronger and probably more interesting if they are developed through collaboration. Collaboration goes beyond co-authorship; it is about building intellectual networks that can support us in success as well as at times when we face challenging situations or receive negative feedback on manuscripts, proposals, or evaluations.

If getting a graduate degree were easy, everyone would do it. Pursuing an academic career is just as difficult. It involves multiple, steep learning curves: extensive reading to understand landscapes of academic thinking; writing in new ways and learning how to take critical feedback; designing a research project; and all the while steeping in a different kind of work environment where we may sometimes experience the imposter syndrome, feeling as though our success is an illusion that our colleagues uncover.

These feelings may relate to how we are sometimes socialized into academic life. As Foote (2010, 10–11) has noted, the early career period is frequently viewed as a "sink or swim" experience: either new faculty succeed in meeting job expectations on their own, or they fail and leave, and "this attitude is deeply embedded in academic life and stems from a view of the early career period as a rite of passage," a journey that must be taken alone, even to the point that "helping an early career faculty member is not only unnecessary but counterproductive, even detrimental" because they can only succeed in an academic career if they do it on their own. Graduate students and early career faculty are sometimes encouraged to see academic life in terms of solitary competition and struggle. What this view ignores is that many of us choose academic life because we enjoy being part of a community that values the sharing of ideas and learning from others. Adopting a sink or swim mindset also overlooks the value of building and sustaining our own communities because they are keys to self-care, well-being, and creativity.

Take inventory and give yourself credit

One of the most important things we can do as we step into an academic career is to acknowledge our accomplishments and see how they have prepared us for the next steps. It is worth taking the time to write out a list of accomplishments and recognize that even a few years ago some of these items might have seemed impossible – submitting a manuscript for publication, successfully completing fieldwork, successfully developing and offering a completely new course, and so forth.

In most tenure track positions and other types of entry-level academic positions, department chairs will seek to limit the expectations for service-related tasks. It is in the department's interest that new faculty members are able to devote time to research and teaching so that they clear the hurdles for promotion and remain as members in the department. Having clear and regular conversations with supervisors (and mentors) can be helpful in identifying the type and number of tasks that should be prioritized to retain one's position. The priorities for eventual promotion are akin to a 30,000-foot view of self-care and preparation that will help you to stay focused on the tasks that will aid you to get promoted (or retained) and to de-prioritize tasks that will not be as valued. Having the 30,000-feet, "big picture" understanding of how to spend your time and energy will support your mental wellness by aligning your longer-term goals with shorter-term decisions.

Beyond the "big picture," there are many decisions that can either leave you feeling overwhelmed and even panicked or, preferably, that can support your mental well-being and a sense of meaningful engagement. Several chapters in this volume offer useful and actionable suggestions to guide a healthy navigation of academic life. Hilda Kurtz's chapter on mentoring offers perspective on finding support for various aspects of an academic career. Another chapter in this volume, Patricia Price's "Time Management," offers constructive thoughts on how to prioritize tasks and activities to make the most meaningful use of our time. The chapter on "Creating communities of care and caring for community" by Daniel Trudeau and Dydia Delyser offers a wealth of support for caring for ourselves and others in an academic setting and why it matters.

Another helpful insight comes from Annmarie Caño's (2021) work on being a healing leader. Her comments are aimed particularly at people in leadership positions, but her suggestions for observing and validating people, checking in regularly, building inclusive teams, and looking for holistic solutions are useful workplace strategies for everyone. She also encourages us to be aware of our own burnout and seek our own healing in whatever form or means it is available. It is difficult, if not impossible, to contribute ongoing and meaningful work if we do not recognize and honor our own needs. Caño (2021) writes,

> Recalling what is important and meaningful to us, and connecting with other people who share those values, are essential for sustaining the energy and imagination required to engage in healing leadership.

For emerging scholars and new faculty members, leadership in this case might not refer to leadership of an academic department. It could certainly refer to leadership of one's own research project, evolving research team, or even committee work focused on service. Leadership at this stage of one's career can simply mean taking responsibility for conducting oneself and navigating this phase of work with purpose and intention.

What next?

Given the pressure in academia to do more, produce more, and achieve more in the same 24-hour days and 7-day week that everyone lives, perhaps a more radical rethinking of how we work is overdue. One approach that is catching attention is the slow scholarship approach popularized by Berg and Seeber in their book *The Slow Professor* (2016) and by a group of geographers led by Allison Mountz (Mountz et al. 2015, 1237) who recognize that:

> Good scholarship requires time: time to think, write, read, research, analyze, edit, and collaborate. High quality instruction and service also require time: time to engage, innovate, experiment, organize, evaluate, and inspire. This kind of slow work both defies and is threatened by the myriad demands on our time as academic laborers.

They consider how universities and other institutions of higher education assess faculty work through particular metrics and values, with a strong tendency to value or make visible products of research such as publications or profits from research grants. Less visible are the process of research and the time devoted to ongoing, creative, intellectual work. Mountz and her colleagues consider the trend to put the responsibility for finding work–life balance on to the individual, and they urge us to push back on this isolating strategy in favor of collective action that demands altered and meaningfully improved work conditions for everyone. They "recognize the need to move beyond the individual strategy of 'just say no' to institutional changes so that they 'just don't ask' for unreasonable levels of service" (Mountz et al. 2015, 1251).

Slow scholarship, for Mountz et al. (2015), can take several forms of shared response, a few of which are discussed here. At the top of their list is **talking about slow strategies**:

> Part of the stress of the neoliberal university is the notion that everyone else is working harder and longer than you – or is more successful at managing time...The relentless acceleration of work will continue until we say 'no" to wildly outsized expectations of productivity...We seek slowness not only for ourselves, but as an attempt to change the academic cultures of our discipline and workplaces.
>
> *(p. 1250)*

They describe the importance of **organizing**:

> Creating spaces for new modes of scholarship and intentional communities helps us to move from individually focused solutions to solutions with potential to create institutional and structural changes that nourish and support slow scholarship.
>
> *(p. 1250)*

Take care:

> We must take care of ourselves before we can take care of others. *But we must take care of others.* Find concrete ways to support and find support in someone else who might be struggling to move a project forward or just wants to talk through ideas...Do not shy away from talking about life and how intertwined life and work are...Ensuring that our scholarship is taking care of others may also help us to engage in different ways of experiencing and valuing time.
>
> *(p. 1251)*

Write fewer emails and find times when you can **turn off email**: "Writing less email may, in fact, allow us more time to think, read, and write more clearly and carefully" (p. 1251). They point to the email charter (https://www.emailcharter.info/) which provides a practical approach to spending less time on emails.

Mountz et al. also describe approaches to **make time to think and make time to write (differently)**. They urge us to **reach for the minimum** (i.e., good enough is the new perfect) so that we can shift focus from the number of publications or the number of grant dollars to focus on quality work that will allow us to clear hurdles or benchmarks for promotion. They offer the following advice:

> Be unwilling to be undermined or belittled for not conforming to hegemonic agendas that are devoid of the responsibilities and joys of life beyond the ivory tower.
>
> *(p. 1253)*

The "ethos of care" advocated by Skop, Caretta, Faria, and Smith (2021) is another approach that also prioritizes self-care, but in a broader vision of care for broader communities and worlds. It is a vision that acknowledges the "long-standing inequities and injustices of academe" with the goal of pursuing antiracist, feminist approaches to knowledge production and social justice advocacy. Key elements of their pledge relate to our concern for self-care:

- Make space to hear and learn from uncomfortable, innovative, and transgressive ideas.
- Create transparency and fairness by setting, communicating, and respecting clear boundaries.
- Take time each year to reflect and revise these boundaries in recognition that, over the course of our personal and professional lives, those boundaries adapt and change.
- Protect our mental, emotional, and physical well-being and growth in the research process by supporting each other's professional and personal aspirations.
- Share and rotate the labor of intellectually joyful and tedious tasks.
- Mentor up, down, and across professional and personal life-course stages to unsettle hierarchical relationships and promote an ethos of care.
- Disrupt perfection: Share insights and experiences overcoming challenges, failures, and rejections as well as motivations, successes, and ambitions.
- Humanize our work by valuing the intellectual and ethical centrality of friendship, connection, and responsibility.

Conclusion

These writings by Mountz, Skop, and their colleagues support the value of positioning our lives and work within broader personal and academic communities to "Mentor up, down and across professional and personal life-course stages to unsettle hierarchical relationships and promote an ethos of care" at a pace that pushes back against the often-false urgency of academic life. This approach also makes clear that we are not alone but rather part of communities that can provide support for our physical, mental, and intellectual well-being. This is not to say that we sometimes face personal and professional setbacks and tragedies like those that Jennifer Ho reflects on in her chapter "It's OK to not be OK." Sometimes we benefit from reaching out to skilled professionals such as a campus ombudsperson, advocate, or peer counselor who can help tremendously in some situations. But building networks can help us understand the situations we are in as well as share compassion and support for others. A sense of caring and community is perhaps the most enjoyable element of academic life, and we all play roles in sustaining and promoting these values in our work.

References

American Association of Geographers (AAG). 2021. Three-year justice, equity, diversity & inclusion strategic plan. October 28. https://www.aag.org/aag-3-year-jedi-strategic-plan/

Berg, M., and B. Seeber. 2016. *The slow professor: Challenging the culture of speed in the academy*. Toronto: University of Toronto Press.

Boice, R. 1991. Quick starters: faculty who succeed. In *Effective practices for improving teaching*, ed. M. Theall and J. Franklin, 111–121. San Francisco, CA: Jossey-Bass.

Caño, A. 2021. Today's academic leaders must be healers. *Inside Higher Ed*, 11 August, Accessed March 4, 2024. https://www.insidehighered.com/advice/2021/08/12/how-be-healing-leader-during-these-difficult-times-opinion

Fink, L. D. 1984. *The first year of college teaching*. San Francisco, CA: Jossey-Bass.

Fink, L. D. 1988. The trauma of the first year. *Journal of Geography in Higher Education* 12 (2): 214–216.

Fisher, S. 1994. *Stress in academic life: The mental assembly line*. Bristol, PA: Open University Press.

Foote, K. 2010. JGHE Annual Lecture 2008: Creating a community of support for early career academics. *Journal of Geography in Higher Education* 34 (1):7–19.

Foote, K., W. Li, J. Monk and R. Theobald. 2008. Foreign-born scholars in U.S. Universities: issues, concerns, and strategies. *Journal of Geography in Higher Education* 32 (2): 167–178.

Gmelch, W. H. 1993. *Coping with faculty stress*. Newbury Park, CA: Sage.

Manzi, M., D. Ojeda, and R. Hawkins. 2024. Creating "wiggle room": Spaces of care and possibility within the neoliberal academy. *Geoforum* 149: 1–10. https://doi.org/10.1016/j.geoforum.2024.103962

Mountz, A., A. Bonds, B. Mansfield, J. Loyd, J. Hyndman, M. Walton-Roberts, R. Basu, R. Whitson, R. Hawkins, T. Hamilton, and W. Curran. 2015. For slow scholarship: A feminist politics of resistance through collective action in the neoliberal university. *ACME: An International E-Journal for Critical Geographies* 14 (4): 1235–1259.

Skop, E., M. Careta, C. Faria, and J. Smith. 2021. An ethos of care: Pledging to make research collaborations more equitable. *Inside Higher Education*. October 14. https://www.insidehighered.com/advice/2021/10/15/pledge-foster-more-equitable-research-collaborations-opinion

Seldin, P., ed. 1987. *Coping with faculty stress.* San Francisco, CA: Jossey-Bass.

Solem, M. and K. Foote. 2004. Concerns, attitudes, and abilities of early career geography faculty. *Annals of the American Association of Geographers* 94 (4):889–912.

Sorcinelli, M. 1992. New and junior faculty stress. In *Developing new and junior faculty*, ed. M. Sorcinelli and A. Austin, 27–37. San Francisco, CA: Jossey-Bass.

SECTION II

Becoming a curriculum leader

LaToya E. Eaves and Jongwon Lee

Navigating the path to becoming a curriculum leader involves more than just understanding the content; it's about shaping the educational experiences that prepare students for the challenges of the modern world. This journey requires a nuanced approach to pedagogy, incorporating diverse methodologies and innovative technologies to meet the evolving needs of students and society. The chapters in this section explore various aspects of curriculum leadership, from decolonizing educational practices to integrating cutting-edge technologies and fostering global perspectives. Each chapter offers unique insights into the transformation of teaching and learning environments, emphasizing the importance of inclusivity, critical thinking, and real-world relevance. By exploring these themes, this section aims to equip early career faculty with the tools and perspectives necessary to lead curricular innovation and make a lasting impact on their students and the broader educational landscape.

Eric Fournier and Meg Gregory's chapter, "Aligning Objectives, Activities, and Assessments: Essentials of Course Planning," places the backward design model at the heart of curriculum development, advocating for a strategic approach that starts with defining clear learning outcomes, and then aligns assessments and learning activities accordingly. Fournier and Gregory emphasize the essential role of each course element in contributing meaningfully to the primary learning goals. They argue that assessments should do more than just gauge student learning; they should also transparently convey success criteria, enhancing the academic confidence and sense of belonging for all students, particularly those from minority backgrounds. Addressing the contemporary challenges in education, including the rise of AI tools like ChatGPT, Fournier and Gregory offer forward-thinking strategies to maintain academic integrity and ensure the authenticity of student work. They advocate

DOI: 10.4324/9781003342953-9

for the creation of comprehensive, inclusive assignments that mirror the complexities of the real world and foster genuine student engagement. Fournier and Gregory underscore the importance of active learning strategies as crucial for stimulating student interaction, critical thinking, and a sense of community within the classroom. They present universal design for learning (UDL) as a foundational strategy for creating accessible and equitable learning environments, enabling every student to participate fully. This chapter is a directive for curriculum leaders to cultivate diversity, innovate in assessment practices, and craft educational experiences that are impactful and relevant to all students.

The chapter, *"Getting Started in the Classroom: First Steps toward Active, Engaged Pedagogy"* by Andrew Jolly-Ballantine and Ken Foote, serves as a valuable resource for early-career faculty tackling the challenges of teaching large lecture classes. It offers practical strategies to help new instructors build confidence and foster a more dynamic, student-centered learning environment. Key points include the importance of collaboration with colleagues and engaging in teaching workshops to create a supportive network that enhances instructional skills. The authors recommend leveraging essential resources, including foundational books like *McKeachie's Teaching Tips and Understanding by Design*, to help organize and design effective courses. A central theme is avoiding over-preparation by focusing on a few core learning objectives, allowing for student interaction and reflection on key concepts. Active learning techniques such as 'Minute Papers,' 'Think-Pair-Share,' and 'Role-Playing' are highlighted as ways to encourage deeper engagement and experiential learning, fostering meaningful connections with the material. The chapter also discusses the effective use of technology—such as Clickers, Google Earth, and digital whiteboards—to create interactive, data-driven lessons, especially in large classes where individual participation may be more challenging. Finally, the authors emphasize continuous improvement through self-reflection and feedback from students and peers, encouraging educators to adapt and enhance their teaching approaches over time. This chapter ultimately provides new instructors with a solid foundation to reduce initial anxieties and build a fulfilling, student-engaged classroom experience.

In "Decolonizing the Curriculum," Zuhri James and Sarah Radcliffe engage in a comprehensive discussion on the necessity of restructuring geography education to move away from Eurocentric perspectives toward a more inclusive understanding of global geographies. Through personal narratives and reflective analyses, the authors explore the transformative potential of decolonizing pedagogies, highlighting innovative approaches that incorporate non-traditional and marginalized sources, such as music videos and social media content, to challenge students' preconceived notions and foster a critical understanding of coloniality and its ongoing impacts. The chapter explores the practicalities and philosophical underpinnings of decolonizing efforts within the geography discipline, advocating for an educational shift that questions the roots, sources, and limitations of the prevailing Eurocentric curriculum. It addresses the challenges and barriers faced in implementing decolonial practices, from institutional resistance to the professional risks for educators advocating for change. The chapter concludes with an optimistic yet realistic outlook on the potential of decoloniality to foster hope and envision alternative futures beyond the confines of colonial legacies. Through a discussion, this chapter presents decoloniality as an ongoing process of interrogation, adjustment, and engagement with the geographies of hope, urging educators and students alike to actively participate in reimagining and reshaping the landscape of geography education.

In "Creating a Socially Just Classroom," LaToya Eaves and Miguel Hernández center the transformative impact of the college classroom. Given how crucial the college classroom is for a student's - and a teacher's - development, they discuss the importance of creating a socially just classroom. They find this particularly important to consider due to the inherent diversity of today's college student body. Put differently, they contend that our college campuses are populated by students representing a range of racial, ethnic, gender, sexual, and socioeconomic identities, generations, (dis)abilities, and who are forthcoming about their mental health. The reality of today's student body composition requires reflection on and adaptation of traditional teaching practices. The chapter provides key definitions and a conceptualization of what it means to foster an inclusive and equitable learning environment for today's college students. The definition of social justice is characterized by core values, including respect, belonging, and empowerment. The authors emphasize the necessity for educators to practice reflexivity and strive to understand positionality as the first step in transforming teaching practice as a way for teachers to understand how their experiences as students, their life backgrounds, and their fears about implementing socially just teaching practices. The authors suggest that a social just classroom calls for an equity-minded approach, which involves examining curriculum choices and implementing differentiated instruction. The chapter argues for setting the foundation for our classrooms as brave spaces, which they differentiate from the more commonly used term "safe spaces." Both concepts are aimed at cultivating inclusive spaces but diverge in key areas, particularly around assumption of risk and ideas of safety. They call for using a brave space framework to develop classroom guidelines in collaboration with students. The chapter concludes by acknowledging that creating a socially just classroom is both a goal and an ongoing process and that the practice offers opportunities for personal and professional growth.

In "Teaching in the Field," Douglas Gamble, Peter Haproff, and Chen Wu offer an insightful exploration into the unique challenges and rewards of field instruction within the geosciences. Through the lens of both physical geography and geology, the chapter illustrates the varied scales and scopes of field instruction, from short-term trips that complement classroom learning to intensive, capstone field courses that synthesize students' accumulated knowledge and skills. The authors share successful strategies for engaging students in real research, utilizing cutting-edge technologies, and overcoming common pitfalls of field assignments. They highlight the transformative power of field instruction in enhancing student engagement, knowledge retention, and the development of essential observational and analytical skills through direct interaction with natural landscapes. This chapter serves as a comprehensive guide for early career researchers, blending a review of pertinent literature with personal insights on creating impactful field experiences in physical geography and geology.

The chapter, "Integrating Web Mapping and Geospatial Technologies," authored by Joseph Kerski, is a resourceful guide for early career faculty members aiming to enrich their curriculum with the dynamic and versatile field of geospatial technologies. The chapter demonstrates the relevance of geospatial tools, such as Web mapping, Geographic Information Systems (GIS), remote sensing, and Global Navigation Satellite Systems (GNSS), across various disciplines by emphasizing their ability to enhance students' understanding of complex spatial phenomena and real-world problems. He argues for the importance of geospatial technologies in aligning with educational goals at multiple levels. Instructionally, it equips students with hands-on experience using professional tools, fostering deep engagement with course content through active problem-solving and critical thinking. Institutionally, it contributes to producing graduates who are equipped with marketable skills for the workforce, particularly in high-demand areas like data science, geodesign, and sustainability. The chapter outlines a pragmatic framework for teaching with geospatial technologies, advocating for a scaffolded integration of tools into the curriculum. It identifies ten core skills akin to "teaching how to fish," which empower students to apply their knowledge to diverse issues and locations. Moreover, it presents ten strategies to make geospatial technology instruction effective, including making learning holistic, focused, multiscale, varied, and relevant, while encouraging field-based activities and multi-level engagement.

In "Using Technology to Promote Critical Thinking," Ken Foote and Chris McMorran explore the critical role of technology in enhancing critical thinking and analytical skills within university education, particularly through the lens of geography and spatial data analysis. The authors explore the dual nature of technology as both an enabler and a disruptor in academic research, highlighting the challenges and opportunities presented by the digital landscape, including the vast availability of online information, the emergence of AI systems like ChatGPT, and the complexities of navigating online mapping tools and search engine algorithms. Foote and McMorran advocate for a balanced approach to incorporating technology into educational practices, emphasizing the need for educators to guide students in navigating the complexities of the digital age thoughtfully. They propose practical classroom activities designed to enhance students' digital literacy and critical thinking skills, covering topics such as the evaluation of search engines, understanding content standards on the web, dealing with misinformation and bias, critically analyzing online data, and considering the ethical implications surrounding digital information. The chapter highlights the importance of fostering an environment where students are encouraged to question the reliability of online information and the implications of their findings, thus preparing them for a future irrespective of technological advancements. Through a combination of practical strategies and thoughtful reflection on the role of technology in education, authors offer a roadmap for preparing students to navigate the complexities of the digital world effectively and ethically.

In "Teaching Online: Strategies to Promote Engagement and Learning," Brandi Gaertner, Adrienne Goldsberry, and Beth Fletcher King offer an insightful exploration into the transformative power of online education, with a special focus on asynchronous courses within the Penn State Online Geospatial Education (OGE) Programs. Central to the chapter is the advocacy for project-based learning (PjBL) and its critical components, which are presented as the scaffolding for effective online learning. Through detailed examples from Penn State's OGE programs, authors demonstrate how projects, ranging from short-term to semester-long endeavors, can significantly enhance critical thinking, collaboration, and practical skills among students. Further, the chapter explores the importance of learning through research, adopting a student-centered approach that mirrors the scientific method. This segment outlines an eight-step process that encourages self-driven learning and reflection, offering a roadmap for conducting meaningful research in an online setting. Professional experience is also identified as a valuable asset in the online learning environment. This chapter discusses integrating students' work experiences into the curriculum, enhancing the relevance and applicability of coursework. Additionally, it suggests leveraging faculty and industry expertise to further enrich the learning experience, providing students with insights into real-world applications of their studies. The authors provide practical advice on designing effective discussions, emphasizing the need for inclusivity, relevance, and engagement.

In "Internationalizing the Curriculum," Gil Latz, Robin Matross Helms, and Keiko Ikeda present a comprehensive guide to embedding global perspectives within higher education. The chapter intricately weaves together the necessity of curriculum internationalization, offering a strategic framework that spans from theoretical underpinnings to practical implementation, particularly emphasizing the innovative use of virtual exchange (VE) and collaborative online international learning (COIL) at Japan's Kansai University as a model for global learning. It provides early career faculty with actionable insights for seamlessly integrating international themes into their courses, advocating for using VE/COIL to democratize global learning experiences across student demographics. Moreover, the chapter highlights the critical support role of faculty disciplinary associations and professional organizations in fostering curriculum internationalization, offering field-specific guidance and resources to enhance faculty efforts toward global engagement. It also explores the complexities of how faculty's international experiences are recognized within tenure and promotion frameworks, encouraging a deeper understanding of institutional policies and cultures. This multifaceted approach not only highlights practical examples of internationalization at work but also underscores the importance of faculty initiative and institutional support in preparing students with the skills, knowledge, and attitudes necessary for success in a globalized society. Through this lens, the chapter acts as a roadmap for faculty to navigate the challenges and opportunities of curriculum internationalization, ensuring their contributions significantly impact student learning, career development, and the broader academic community.

8

ALIGNING OBJECTIVES, ACTIVITIES, AND ASSESSMENTS

Essentials of course planning

Eric Fournier and Meg Gregory

Introduction

At its core, course alignment is about ensuring that what we teach (objectives), how we teach it (activities), and how we measure learning (assessments) are connected. This connection helps ensure that every moment in the classroom, whether physical or virtual, serves a purpose and propels students toward achieving desired learning outcomes. Through a blend of theoretical foundations, research-based practices, and practical examples, this chapter will help early career faculty design and deliver courses that are not only coherent and logical but also inspiring and potentially transformative. In course design and teaching, assessment and course activities reveal how well students are learning what we intend for them to learn. Developing strategies to articulate and enact the connections between activities and assessments to intended objectives reinforces the ways teachers can impact student learning and motivation.

Often when instructors are developing courses, they start by thinking about a reading list or a list of topics to lecture on. This is a forward-thinking process of designing a course. By contrast, Wiggins and McTighe (2005) recommend a backward approach that encourages you to consider your outcomes (goals) for student learners first. The following are the three steps of backward design:

1 Identify learning goals for your class
2 Design assessments that will allow you to measure how well students have achieved those goals
3 Create activities and an inclusive environment that supports students in achieving those goals

DOI: 10.4324/9781003342953-10

Once your goals for the course are clear, the next step is to think about assessments or how you want students to demonstrate their learning in relation to the outcomes that you've set for them. The learning goals or outcomes ideally inform your assessments as well as the learning activities that students will engage in to build their knowledge and grow their skills in your discipline of expertise. The benefit of following backward design is the creation of a student-centered course that ensures clarity, transparency, and alignment between all aspects of your teaching (Weimer 2013). You can be confident that students who succeed in the course will achieve the goals that you set for them at the beginning.

This chapter will follow a backward design model, where we first consider developing learning goals for your students, then we examine how to assess student learning, and offer suggestions for ways to engage students with your course content. We will conclude by discussing some additional considerations that are essential for creating an equitable, accessible learning environment where all students have the potential to succeed.

Consider situational factors

In his backward design-inflected model, Dee Fink argues that considering a variety of situational factors is a key first step in course design. According to Fink, situational factors are instructional priorities that will inform any learning goals, activities, and assessments that you design for a course. These might include obvious things like the level of student, the size of a class, and characteristics of the assigned room; as well as less-obvious factors like an instructor's learning philosophy. Once you identify the factors that are critical to student success in your course, you'll use these answers to inform many of the choices that you'll make when planning. According to Fink, situational factors in student learning include considerations like those mentioned in Table 8.1.

Identify learning goals

Now that you've taken some time to reflect on the situational factors involved in your course, we can consider how to apply what you know about the course and your students to the process of identifying and refining specific learning goals. A learning goal is a statement of what students should know or be able to do as a result of successfully completing your course.

Ideally, learning goals for a course are developed by considering contextual factors and the kinds of knowledge production activities (e.g., synthesis, analysis, comparison) you want your students to be comfortable performing. Starting from situational factors and considering types of learning on a macro-level should make it easier to identify specific course-level learning goals for your students.

TABLE 8.1 Key Situational Factors in Course Design

Situational Factor	*Questions to Consider*
The specific context of the learning situation	• How many students are in the class? • What modality will the course be taught in? • What physical elements of the learning environment might affect the class? • How often does the class meet, and for how long?
How the course fits into the wider university curriculum	• What learning expectations are placed on this course by the university? • Is the course part of a sequence? • Are there any prerequisites for the course? • Are there professional accreditations that must be considered?
The nature of the course content	• What level is the course material? • Is the course primarily theoretical, practical, or a combination? • Are there important changes or controversies occurring in the field?
The characteristics of the learner	• Who are your students? • What are their goals and expectations for taking the course? • What prior knowledge or experience do students bring to this course?
The characteristics of the instructor	• What beliefs and values do you have about teaching and learning? • What is your attitude about the course and your students? • What prior knowledge or experiences do you bring to this course?
Other concerns that could significantly impact student learning	• What were the circumstances of their learning in the past? • What special challenges or other commitments might students be tackling this semester?

Source: Fink (2013, pp. 76–77).

You might be wondering at this point: What's the big deal about learning goals? Why are they so important that this is where we are starting instead of a daily outline of activities or a plan for how we are going to assess students? You might even be annoyed if you see learning goals as simply an output of the corporatization of higher education. The truth is, however, that even if you haven't used the words "learning goals" before to describe your classes, instructors always have in mind what it is they want their students to get out of a course. And, the best, most meaningful classes for students tend to be those in which that foundational set of goals drives every other decision that is made about the course: What assignments should I ask my students to complete?

What should they read or watch? What should we do in class? How should they interact with each other? In short, learning goals can be our compass and can keep us from veering offtrack in ways that don't support our students' learning.

In a time when new technologies and other considerations seem overwhelming, learning goals are all the more critical. They help us remember what is most central and most important to our courses. They can also help us triage our own learning and preparation: What part of the learning management system do I need to learn? What educational technologies should I adapt for my class? How will I navigate communicating with different groups of students? If we are willing to start from our learning goals, the noise of possibilities will begin to die down, and everything that is truly essential for us to know in order to support our students' learning will become clearer.

Characteristics of Effective Course Goals

- **Describe What Students Will Learn and Be Able to Do** – Because the focus is on student learning, course goals are written in terms of what students will have learned and will be able to do at the end of the course.
- **Are Actionable, Visible, and Measurable** – Goals state what students *do* to demonstrate their learning. *Each goal statement should include an action verb.* The goal uses an action verb to describe what the learner will be *doing* when demonstrating the achievement of the goal.
- **Are Clear and Understandable to Students, as Well as Instructors** – Goals should be phrased in fairly simple terms so that they are clear to both students and instructors.
- **Have an Appropriate Level of Generality** – One of the most difficult tasks in writing a course goal is selecting the *appropriate level of generality.* It is important to craft a goal that encompasses a broad view of the subject without too much specific detail.
- **Require High Levels of Thinking and Learning** – Good course goals are generally at the higher levels of learning. This does not mean that fundamental knowledge or skills are unimportant. Indeed, your high-level goals are likely to be supported by lower-level course goals. However, course goals are more than just a collection of low-level goals.
- **Are Developmentally Appropriate** – Course goals should be challenging, yet achievable, for a student at your institution, and match the educational levels identified in situational factors.
- **Lead to Authentic/Motivating Tasks** – Authentic tasks are those in which students face challenges similar to those faced by professionals in the field. Students find authentic tasks highly motivating, as it is easy to see the relevance of completing them.

As you develop and refine your learning goals for students, you'll want to make sure they are specific and measurable. It's critical that the goals that you

TABLE 8.2 Bloom's Revised Cognitive Taxonomy and Related Verbs

	Remembering	*Understanding*	*Applying*	*Analyzing*	*Evaluating*	*Creating*
Bloom's definition	Exhibit memory of previously learned material by recalling facts, terms, basic concepts, and answers	Demonstrate understanding of facts and ideas by organizing, comparing, translating, interpreting, giving descriptions, and stating main ideas	Solve problems for new situations by applying acquired knowledge, facts, techniques, and rules in a different way	Examine and break information into parts by identifying motives or causes. Make inferences and find evidence to support generalizations	Present and defend opinions by making judgments about information, validity of ideas, or quality of work based on a set of criteria	Compile information together in a different way by combining elements in new patterns or proposing alternative solutions
Verbs	Choose Define Describe Find Label List Match Name Omit Recall Relate Select Show Spell	Classify Compare Contrast Demonstrate Explain Extend Illustrate Infer Interpret Outline Relate Rephrase Show Summarize Translate	Apply Build Choose Construct Develop Identify Interview Model Organize Plan Select Solve Utilize	Analyze Assume Categorize Classify Compare Contrast Discover Dissect Distinguish Divide Examine Inspect Simplify Survey Test for	Appraise Assess Award Choose Compare Conclude Criticize Decide Defend Determine Disprove Estimate Evaluate Judge Justify Prove Rate Recommend Select Support	Adapt Build Change Combine Compile Compose Construct Create Design Develop Formulate Imagine Improve Invent Modify Plan Propose Solve

Source: Anderson and Krathwohl (2001).

TABLE 8.3 Student-focused Learning Goals Using Bloom's Taxonomy

Example Course	*Bloom's Level*	*Course Goal*
Introduction to Environmental Studies	Remember	**Describe** the key elements of the carbon cycle in a given ecosystem.
Introductory Urban Studies	Understand	**Identify** the social, economic, and political forces that have shaped urban growth and decline in the North American city
Research Methods in Social Sciences	Apply	**Select** and **implement** appropriate statistical tools to explore complex data sets
Macroeconomics	Analyze	**Compare** and **contrast** the way tax changes and government spending changes impact employment rates
Contemporary Issues in Public Policy	Evaluate	**Evaluate** the potential social and economic impact of two competing redevelopment proposals
Senior Capstone in Geography	Create	**Design** and **produce** an original GIS-based research project

choose are ones that can be measured -- that is, it would be possible for you to assess how well students have been able to accomplish this goal in your class. A good way to start drafting a specific learning goal is to identify what you want students to actually do with the knowledge that you hope they will gain in your course. Examining a list of verbs can be helpful for identifying the specific things that you'd like students to be able to do with knowledge acquired (Table 8.2). One common way to break down these cognitive activities (what students are "doing") is Bloom's (revised) taxonomy, a hierarchical framework for constructing and classifying learning goals (first developed by Benjamin Bloom (1956) and later revised by Anderson and Krathwohl (2001)).

The revised taxonomy includes the following levels of cognitive engagement: remember, understand, apply, analyze, evaluate, and create. This taxonomy suggests that one isn't ready to do more complex cognitive tasks (e.g., application, analysis) until one has a firm grasp on the lower levels (remember, understand). Traditionally, learning goals are written from the student's point of view. The examples in Table 8.3 provide an illustration of student-focused goals that are linked to the different levels of Bloom's taxonomy.

Assessment

Ideally, assessment activities are aligned with learning objectives; that is, assessments test students on their progress toward achieving the learning goals that you set out for them. Further, a transparent connection between goals and course assessments is essential for grounding students' understanding of the deep structure of your course. Clearly connecting your learning goals with what you are asking students to do on assessments or assignments is a crucial component of course design.

Summative assessments: overview and strategies

Summative assessments are assignments which evaluate student learning at the end of an instructional unit by comparing it against some standard or benchmark. Summative assessments are generally time-intensive, comprehensive, and high-stakes (they have a relatively large effect on the student's overall course grade). Traditional summative assessments, like closed-book unit exams and research papers, can be effective tools for demonstrating student learning in a course. However, as we've already noted, it is critical to carefully question the alignment between modes of assessment and learning goals during the course design process. Therefore, it is important to reflect on the learning objectives for your course in order to determine if the summative assessments you have been using in your courses are truly the best tools for measuring your students' learning.

Exam design

Mid-semester and end-of-term exams are the hallmark of many traditional college courses. Some key aspects of designing effective exams include developing strong questions and question stems, carefully aligning questions with the learning goals that you've set out to test, and putting structures in place that minimize exam anxiety and increase the validity of the results.

Traditional closed-book, in-person exams can work well for measuring learning, but there are other options for exam design that may be used. For example, you might consider an open-note or open-book format. Open-note exams work best when they test higher-order learning rather than recall or definitions that can be directly looked up. Consider testing learning objectives that ask students to "apply," "discuss," or "compare." This leverages the fact that students have access to notes, textbooks, research articles, and other readings and encourages them to demonstrate their ability to engage with resources from your course. There may be a concern that open-note exams will lead to higher grades from students or a lack of long-term retention of content. However, a recent study conducted in a psychology classroom indicated that while students may initially score higher when using notes, the long-term retention was statistically the same between an open-note exam and a closed book exam (Gharib, Phillips, and Mathew 2012). This study suggests that a well-designed, open-note exam can be an effective assessment and testing tool for students.

You may also wish to consider the format of the exam: in-person versus online via LMS; timed versus not timed; multiple choice versus short answer or essay-type questions. Ideally, each aspect of assignment design should align with the goals that you've set for your students. For example, if a learning goal for the students is to synthesize key theoretical texts, then questions that allow students to articulate their understanding of the relationship between the texts (perhaps an essay-type question) is key. If being able to solve a problem quickly and efficiently is a learning goal in your class, then a timed exam would help to illustrate students' learning.

Paper assignment design

Research papers are another staple of traditional summative assessment in college courses that help students practice research and disciplinary writing skills. Instructors wishing to re-imagine a traditional research paper might consider some of the following adaptations. If conducting the research is less important than using it effectively, instructors might provide students with a curated bank of sources linked in the LMS. This might include scanned primary and secondary sources that are relevant and maintain the appropriate level of scholarly rigor to be cited in their work. If independent research is an important component, then instructors may want to grade not just the product, but some process aspect of the research. This might include finding and summarizing quality sources. No matter what, particularly in lower-level courses where students are new to writing in the discipline, instructors will want to build structures into their course to help students become aware of what university and library resources are available to them.

Generative AI and challenges of assessment

Since it was unveiled in November 2022, ChatGPT (and other artificial intelligence powered composition tools) has been the topic of much concern among the academic community. Many of the concerns involve issues of academic integrity; however, detection tools have proven unreliable and sometimes biased against non-native English speakers (Liang et al. 2023). In general, we advocate for transparency, open communication with students, and clarity in expectations. Some specific suggestions include the following:

- **Learn how ChatGPT would respond to your class assignments**: Try out *ChatGPT and other tools* yourself to understand its capabilities and limitations.
- **Focus on higher-order thinking tasks in Bloom's taxonomy (e.g., analysis over recall)**: ChatGPT is best at summarizing from its training data and less good at applying knowledge to new areas. Focus student tasks on

the higher levels of Bloom's taxonomy such as creating, evaluating, and analyzing.

- **Make assignments more authentic and robust**: Develop authentic assessments that ask students to complete complex tasks similar to those they would in a non-academic setting.
- **Add non-traditional ways of assessing learning (podcasts, oral exams, projects)**: Consider other ways of letting students demonstrate their knowledge beyond the traditional college essay.
- **Use local or class-based prompts**: One limitation of ChatGPT (and similar tools) is that it does not know what happens in your class. Consider designing prompts related specifically to your course. For example, one could ask students to expand upon a particular class discussion or answer questions requiring knowledge from a lecture slide.

Group projects and assignment design

Group assignments can be a great opportunity to build community among your students who may be feeling isolated in their learning. However, there are also many complaints about group assignments: one student does all the work, students split up work rather than truly collaborating, students can't find time to meet or communicate effectively outside of class, interpersonal issues arise, and so on. Some key considerations for effectively assigning group projects include the following:

- Create interdependence (if collaboration is a key goal of this assignment, then make an assignment where they must collaborate, and not divide and conquer).
- Assign group roles (if everyone has a role, it's less likely that someone will "check out").
- Devote time in class to teamwork skills.
- Build in accountability and check points.
- Build a space or framework for students to communicate with each other.
- Form teams carefully and intentionally.

Formative assessments: overview and strategies

There's another type of assessment though that can also positively influence student learning: **formative assessment**. These kinds of assessments are often low-stakes, informal activities that do not have a large impact on students' grades, but that can be used by instructors to provide students with key information about how to improve their skills or understanding of specific course concepts. Formative feedback benefits both students and instructors: students receive ongoing feedback that will help improve their learning, and instructors receive feedback that allows them to adjust and improve their teaching. More specifically, formative assessments help students identify their strengths and weaknesses and target areas that need work. They also help faculty recognize where students are struggling and allow them to address problems immediately.

Examples of common formative assessments include asking students to:

- Write a one- or two-sentence summary of the main points of a lecture.
- Describe the "muddiest point" or concept they are least clear about after completing a learning activity.
- Provide a draft version of a paper or assignment.
- Submit a proposal that outlines plans for a larger paper or assignment.
- Answer a polling question about a concept that's just been covered or a scenario that applies something learned.
- Briefly reflect on a concept or course text.
- Complete a short quiz after (or in the midst of) a learning activity.

Transparency

An important consideration in designing assessments is how to be transparent with students about what you are asking them to do, how they should approach the task, and what criteria you will use to measure their success and achievements. Transparency in assessment significantly boosts student success in terms of academic confidence and sense of belonging (Winkelmes et al. 2016). In particular, transparency showed statistically significant gains for minority students' success when TILT (Transparency in Teaching and Learning) guidelines were followed during design. In a related study, faculty reported increases in student motivation, a higher-level of class discussion, more on-time completion of assignments, and fewer grade disputes as positive outcomes in following this design framework (Winkelmes et al. 2015).

The framework includes three key elements of transparency: **Purpose**, which helps students understand why they are being asked to complete a task and what skills or knowledge they'll need to be able to do it; **Task**, which explains what students will do and how they should accomplish the task; and **Criteria for Success**, which describes what excellence will look like and how it will be assessed.

Learning activities

Now that you have a set of learning goals for your students, and you've identified some ways that you can assess how well your students have achieved those learning goals, we turn toward thinking more about how your students will engage with the content in your course.

We will explore the role that active learning and student engagement can play in effective teaching and learning. Building on the core idea of backward design, your learning goals will guide you in choosing what in-class engagement activities and tools to integrate into your teaching.

One of the main factors contributing to student satisfaction is the extent to which students feel connected to each other and to their instructor. Students value regular opportunities for positive, constructive interaction where they can learn together and share ideas. As you design your course, consider how you will create these opportunities for your students.

Active learning "engages students in the process of learning through activities and/or discussion in class, as opposed to passively listening to an expert. It emphasizes higher-order thinking and often involves group work" (Freeman et al. 2014). "Active learning" is an umbrella term that covers traditional classroom activities such as discussion and debate, small group work, problem-solving, brainstorming, peer review/critique, and reflection. Research on learning consistently demonstrates that students who are asked to actively engage with course content learn better and more deeply than those who engage with material only passively (e.g., listening to someone else speak) (Barkley and Major, 2018)

Another key step in course design is to consider which of these activities might be appropriate and adaptable for your current teaching context. The following sections provide some examples of common collaborative activities and some strategies for using them successfully in your teaching. As you read, consider which activities might work well for you.

Designing large group activities

Large group discussions or question/answer sessions tend to be the traditional default modes of student engagement in many courses. Documented benefits of utilizing discussion, even in asynchronous online forms, include that "discussion can build students' critical thinking, argument, and self-reflection skills, promote engagement among introverted students, encourage the thoughtful integration of content, [and] build community" (Herman and Nilson 2018, p. 49).

Here are some suggestions for engaging students in large group discussion:

- Establish a clear trajectory so that students understand the goals of the discussion.
- Provide clear, specific directions, how to respond, and (in the case of asynchronous discussions) when engagement should happen.
- Be transparent about what students should accomplish in discussion and why they are being asked to participate.

- Assess students' contributions in some way against well-defined criteria that incentivizes both listening and contributing.
- Set ground rules for interactions with peers.

Designing small-group activities

Small-group activities are a great way to facilitate the development of social presence in your classroom. As Josh Eyler (2018) notes in *How Humans Learn*,

> at a very fundamental level, we need others in order to live functioning lives. As an extension of this, our sociality drives our learning in many essential ways. Sure, we can learn things on our own, but we rarely learn them as deeply, because so much of our learning derives from our social nature and our visceral need to communicate with other people.
>
> (*pp. 66–67*)

According to Barkley et al. (2014), some key suggestions for effective small-group activities include:

- Make purpose of the group work clear and transparent to encourage student buy-in.
- Provide a clear structure by which students accomplish various tasks.
- Set expectations for some kind of end-product or result based on work done by the group.
- Consider how best to structure groups and whether it makes sense to assign roles to different students.
- Continue to facilitate and circulate during the time that students are working with each other.
- Provide meaningful closure.

Designing polling activities

Another common active learning strategy is polling students. Polling can be used for a number of different pedagogical purposes at various points during the class. Instructors can ask students to recall, classify, explain, reflect, predict, and more. Well-designed polls help students self-reflect while providing instructors with valuable feedback on students' perspectives or understanding of the content. Further, polling is very adaptable because it can be high-tech (Poll Everywhere) or low-tech (hand/finger raising).

Universal design for learning (UDL)

Now that we've considered how to create course content and activities, it's time to consider what you need to do to make your course accessible and engaging for all students. A key feature is the idea that an accessible classroom is an equitable classroom. By taking steps to lower barriers to learning for all students, you are creating a more welcoming and equitable learning space.

A traditional approach to making content accessible to students is to follow an accommodation framework. An accommodation framework provides students with a disability (physical, mental, learning) with the support they need in order to have equal access to course material and equal opportunities to learn. A major drawback of a course design with this framework is that it requires that students self-identify and request accommodation. The required self-advocacy involved in the accommodations model is a lot to ask of our students and can lead many to choose not to disclose a disability for fear of judgment from peers or instructors.

Universal Design for Learning (UDL) takes a different approach to course design. UDL is defined by Center for Applied Special Technologies (CAST), a leading non-profit educational organization, as a "framework to improve and optimize teaching and learning for all people based on scientific insights into how humans learn." In their book *Reach Everyone, Teach Everyone: Universal Design for Learning in Higher Education*, Thomas Tobin and Kristen Behling (2018) explain, "UDL is a way of thinking about creating the interactions we have with our learners so that they do not have to ask for special treatment, regardless of the types of barriers they may face – time, connectivity, or disability" (p. 130). In other words, a UDL framework acknowledges the diversity of the students in your class and designs to reach diverse learners rather than waiting for a student to request accommodation in order to adapt your course content. Through the use of this framework, instructors create a more equitable learning environment through the use of explicit expectations and varied modes of content delivery, knowledge building, and assessment.

UDL is centered on providing multiple means of engagement, representation, and expression. **Engagement** is often described as the "why" of learning. Considering how students engage with your content will help you garner student interest, build persistence, and help students self-regulate their learning. **Representation** is considered the "what" of learning. Through multiple means of representation, you offer students avenues to read, listen, or watch content to best suit their ability and time constraints while also helping build comprehension of the content. **Expression** is considered the "how" of learning. In this area, the focus is on allowing students many ways to demonstrate that they have mastered the course material and achieved their learning goals.

The following is a list of a number of high yield, small changes you can make to your course design that will help you incorporate the principles of UDL into your course (adapted from Burgstahler 2013).

- Include a statement on your syllabus that invites students to meet with you to discuss learning needs.

- Provide multiple ways for students to engage with content (text, videos, podcasts, images).
- Use examples that appeal to students with a variety of experiences, interests, and identities.
- Scaffold assignments to allow students to get feedback from the instructor before the final version is due.
- Provide class outlines and notes in an accessible format.
- Clearly connect material to course learning goals.
- Assign a note taker for your lectures or in-class sessions and make these notes available to all students.
- Assess students' learning through multiple formats (quizzes, writing assignments, presentations, creative formats).
- Include alt-text for any graphic images you share with students.
- Caption videos that you share with students.

Inclusive course design

Inclusive teaching can be defined as teaching that creates "a culture in which all learners feel welcome, valued, and safe., and it requires intentional and deliberate strategies"(Hogan and Sathy, 2022, pp. 5). In their recent book *What inclusive instructors do: Principles and practices for excellence in college teaching*, Addy et al. (2021) add that inclusive teaching "dismantles the historic notion that cultural and other differences among students make them in some way deficient as learners. Instead, inclusive teaching embraces and responds to such diversity to create effective teaching and learning environments" (pp. 4–5).

Inclusive teaching practices are essential for fostering and sustaining a classroom environment in which all students can engage and be successful. The UDL framework we've explored already is a major component of creating an inclusive and more equitable classroom. We will now explore other ways that inclusive teaching can infuse your course design.

One way to make your course more inclusive is to include diverse, relevant perspectives in course text or media with which students engage. This may involve assigning texts or media produced by underrepresented groups in your field, providing accounts or examples from a variety of perspectives, supplementing textbook readings, or providing in-class examples that demonstrate that your field is perhaps more diverse than students might expect. As Addy et al. (2021) note, this approach supports the social belonging of all students in that it both has the potential to "help students who are historically underrepresented see that they have a place in the field" while also showing students "who are used to seeing and learning from people like themselves that other perspectives have value" (p. 53). Further, incorporating perspectives that resonate or modes of media that are familiar to students can "incentivize students to engage more fully" (54). In other words, this action is not only more welcoming for a variety of students but is also likely to increase student motivation to learn (beyond simply achieving a high grade).

Another aspect of inclusive course design includes developing a course structure that models a "growth" rather than a "fixed" mindset toward learning. Carol Dweck (2006) and her colleagues describe a "growth mindset" as one which conveys the idea that intelligence is not a reflection of fixed, natural abilities (indicative of statements like *I'm just bad at math*) but can change and grow over time. In further studies, helping students adopt a growth mindset has shown a positive effect on student GPA and persistence through difficulty (Yeager et al. 2016). Course structures that exhibit a growth mindset are as follows:

- Offer students the chance to practice core skills in a low-stakes environment before high-stakes assessments.
- Provide opportunities for productive failure without major consequences (e.g., allow students to earn back points for seeking help on an exam, allow for revision and resubmission, weigh grades toward student performance following growth or revision); and
- Reward students for growth throughout the semester (e.g., offer the chance to drop the lowest score on a quiz or exam, provide a reflection opportunity that asks students to reflect on how they've achieved goals they set for themselves over time).

Syllabus

Faculty often think crafting a syllabus is synonymous with course design, but you might have been surprised that this chapter did not start by asking you to craft your syllabus. Instead, using backward design, we identified learning goals, assessment strategies, and student engagement strategies that acting in alignment together will build a strong foundation for student learning. Once you have done all the hard work of course design, constructing a syllabus should be the final step.

Many students and instructors recognize the syllabus as **a reference guide** for a particular course. It provides a compendium of information that can be consulted throughout the course including logistical information, prerequisites, the instructor's contact information, learning goals, a course description, course policies, due dates and requirements, a list of resources, and grading criteria. In short, effective syllabi outline clearly what a student must do to be successful in the course.

As Ken Bain noted in his *What the Best College Teachers Do*, the most effective syllabi not only act as a reference guide for students but also function as **an invitation to learning** (p. 75). Syllabi set the tone for the course by informing students about what they can expect, why they should take the course, and what they'll learn and learn to do while engaging in it. In this way, the syllabus acts as a "promise" rather than a contract. We invite you to think about how your syllabus can act as an *invitation* to joining your community of learners (Table 8.4).

TABLE 8.4 Common Syllabus Elements

Syllabus Element	*Context*
Instructor and TA Contact Information & Office Hours	You may decide to offer meeting or office hours either in person, remotely via Zoom, or some combination of both. As you plan for office hours, consider if it makes sense to meet in person or via Zoom.
Course Description	Provide some detailed information about the course that will help students feel more "invited" into the learning experience
Required Texts, Materials, Supplies	It's now common to assign more multimedia texts (or texts viewed in digital form such as a .pdf). Students may require instructions for how to access or borrow these materials. You may also consider a statement indicating free or reduced-cost options that exist for obtaining course materials.
Classroom Climate	There are many steps you can take to foster an inclusive classroom climate. Steps that can be taken on the syllabus include using "we" language, adding a classroom civility statement, describing ground rules for appropriate classroom interactions, and providing information for students on what to do if they have concerns about classroom dynamics and/or classroom climate
Guidelines for Academic Integrity	Students often take courses across departments and come from a myriad of cultural contexts. What constitutes plagiarism, appropriate collaboration, and appropriate sharing may not be obvious to all students.
Guidelines for Attendance and Participation	Our experiences with Covid-19 have obviously presented challenges for establishing traditional attendance, participation, and late work policies. Even without the Covid-19 complications, university commonplaces like "participation" may mean different things in different settings
Campus-Wide Resources	Students may not know how to contact campus resources which have been established to support them in their learning. They may not know about additional options available if they are seeking help. Keep in mind that how you frame these resources on the syllabus and in your discussion of them may dictate students' willingness to seek out help when they need it

A special area of emphasis in syllabus design is in how you craft your course policies. If you are determined to construct an inclusive course, this is one place where many of the tools and suggestions we've already discussed can come in handy. Students may face any number of complex challenges that make it difficult for them to stay on track in any given semester. The development of flexible, student-centered policies can ensure that your class is ready to support students' learning whatever the situation. It's important to note here that this call for flexibility does not mean we are advocating for "anything goes." You are still responsible for the course, and you must balance your own needs with those of your students. Yet, striking a balance and building some flexibility into your course design through refining class policies will create an improved learning environment.

As you think about your course policies, you'll want to determine what boundaries are the most comfortable for you. While you should communicate that you care for student well-being and success, setting reasonable boundaries is also critical for setting you and your students up for success (there's a fine line sometimes between supporting and enabling). A non-exhaustive list of some boundaries that you might think about setting include when and in what time frame you'll respond to student emails, when and in what circumstances you will accept late work, and when and in what circumstances students will be allowed to make up missed participation.

Conclusion

In conclusion, this chapter highlights the importance of aligning course objectives, activities, and assessments in effective course planning. This alignment is not merely an academic exercise but a vital strategy in enhancing the quality and effectiveness of educational experiences. The backward design model, as detailed in this chapter, serves as a robust framework, guiding educators to begin with clear learning goals and then to tailor assessments and learning activities to these goals. Such an approach ensures that every component of the course contributes meaningfully toward the desired learning outcomes.

We stress the significance of setting specific, measurable learning goals. These goals, informed by an understanding of both the subject matter and the students' contexts, act as the foundation upon which the entire course is structured. By employing tools like Bloom's taxonomy, educators can create a hierarchy of learning objectives that progressively build students' knowledge and skills.

The alignment of assessment strategies with learning goals is crucial. This alignment not only provides clarity and direction for students but also ensures that assessments are meaningful and genuinely reflective of students' understanding and mastery of the course content. The implementation of both formative and summative assessments, designed with transparency and inclusivity in mind, fosters an environment where students are continually engaged and supported in their learning journey.

In summary, this chapter serves as a guide for new faculty, equipping them with the tools and insights needed to design courses that are not only academically rigorous but also deeply engaging and inclusive. By aligning objectives, activities, and assessments, educators can create learning experiences that are transformative, relevant, and memorable.

References

Addy, T. et al. 2021. *What inclusive instructors do: Principles and practices for excellence in college teaching*. Sterling, VA: Stylus Publishing.

Anderson, L. W., and Krathwohl, D. R. 2001. *A taxonomy for learning, teaching, and assessing*, Abridged Edition. Boston, MA: Allyn and Bacon.

Bain, K. 2004. *What the best college teachers do*. Cambridge, MA: Harvard University Press.

Barkely, E. F., Cross, K. P., and Major, C. H. 2014. *Collaborative learning techniques: A handbook for college faculty*. Hoboken, NJ: John Wiley and Sons.

Barkley, E., and Howell Major, C. 2018. *Interactive lecturing: A handbook for college faculty*. San Francisco: Jossey-Bass.

Bloom, B. S. 1956. *Taxonomy of educational objectives, handbook I. The cognitive domain*. New York: David McKay Co Inc.

Burgstahler, S. 2013. "Introduction to universal design in higher education. In S. Burgstahler (Ed.). *Universal design in higher education: Promising practices*. Seattle: DO-IT, University of Washington.

Dweck, C. 2006. *Mindset: The new psychology of success*. New York: Ballantine Press.

Eyler, J. 2018. *How humans learn: The science and stories behind effective college teaching*. Morgantown: West Virginia Press.

Fink, L. Dee. 2013. *Creating significant learning experiences: An integrated approach to designing college courses*. San Francisco, CA: Jossey-Bass.

Freeman, S. et al. 2014. "Active learning increases student performance in science, engineering, and mathematics Download Active learning increases student performance in science, engineering, and mathematics." *Proceedings from the National Academy of Sciences of the United States of America* 111, pp. 8410–8415.

Gharib, A., Phillips, W., and Mathew, N. 2012. "Cheat sheet or open-book? A comparison of the effects of exam types on performance, retention, and anxiety." *Psychology Research* 2(8), pp. 469–478.

Herman, J., and Nilson, L. 2018. *Creating engaging discussions: Strategies for 'avoiding crickets' in any size classroom and online*. Sterling, VA: Stylus Publishing.

Hogan, K.A. and Sathy, V. 2022. Inclusive Teaching: Strategies for Promoting Equity in the College Classroom. Morgantown: West Virginia University Press.

Liang, W., Yuksekgonul, M., Yining, M., Wu, E., and Zou, J. 2023. "GPT detectors are biased against non-native English writers." *Patterns* 4(7), pp. 100779. https://doi.org/10.48550/arXiv.2304.02819

Tobin, T. J., and Behling, K. T. 2018. *Reach everyone teach everyone. "UDL in the Next Twenty Minutes Download UDL in the Next Twenty Minutes"*. Morgantown: West Virginia University Press.

Weimer, M. 2013. *Learner-centered Teaching: Five Key Changes to Practice*, 2nd Edition. San Francisco, CA: Jossey-Bass.
Wiggins, G., and McTighe, J. 2005. Understanding by design. Alexandria, VA: Assn. for Supervision and Curriculum Development.
Winkelmes, M. et al. 2015. "Benefits (some unexpected) of transparently designed assignments." *National Teaching and Learning Forum* 24(4), pp. 4–6.
Winkelmes, M. et al. 2016. "A teaching intervention that increases underserved college students' success." *Peer Review* 18(1/2), pp. 31–36.
Yeager, D. S., et al. 2016. "Teaching a lay theory before college narrows achievement gaps at scale." *Proceedings of the National Academy of Sciences* 113(24), pp. E3341–E3348.

9

GETTING STARTED IN THE CLASSROOM

First steps toward active, engaged pedagogy

Andrew Jolly-Ballantine and Kenneth Foote

Teaching can be one of the most challenging aspects of starting an academic career. No matter how well we plan, or how much experience we've gained in graduate school, leading classes on our own for the first time can sometimes be stressful. Eric Fournier's chapter on the essentials of course planning will be of tremendous help, but here we provide insight on teaching new classes, tools for creating a more engaged classroom, and recommendations for getting feedback and reflecting on your class to improve it in the future. We've picked tips from among the many others suggested in the books in our reference list because they are particularly good first steps for organizing and teaching your classes, with particular stress on bringing active pedagogy into your classroom. All are intended to help you get your feet on the ground in the classroom, while also improving student learning and engagement.

We also hope these suggestions will help you develop a sense of comfort with your materials that will reduce the stress and anxiety you may feel as your step to the front of a classroom for the first time. The importance of finding comfort with our classes (and with our students, colleagues, and campuses) comes from the writings of Robert Boice (1991, 114), a researcher who focused much of his research on understanding and improving the work of early career academics (Boice 1992, 2000). His point is not that we can dispel every anxiety and stress, but there are strategies that can help us achieve success (and comfort) in the classroom and our careers.

Talk with, share, and learn from others

Teaching is best seen as a team effort, involving not only our students but also our colleagues inside and outside of our programs. Issues of effective teaching,

DOI: 10.4324/9781003342953-11

course planning, and curriculum development are usually not stressed in most graduate programs. Academic culture often leads to the sense that our courses are our own domain, so getting started in the classroom can be a lonely process, and therefore a source of anxiety rather than enjoyment. We recommend that you cast off the mantle of the solitary academic and don't hesitate to discuss with others the strategies, expectations, and materials, and share ideas and know-how. Ask if you can visit classes taught by colleagues, and invite them to attend, participate, or co-teach with you from time to time. Don't think that you need to know everything when you walk through the door to a new department.

Reach out to people and resources beyond your own program, too. Most colleges, universities, and professional associations now invest in resources to help faculty get started, improve their teaching, or learn new techniques and technologies, like the one on our campus (University of Connecticut 2024). Workshops, coaching, access to sample syllabi and learning materials, and colloquia are often among the mix of resources provided by these centers for teaching and learning. Taking advantage of these resources can help us reach our classroom comfort zone sooner, but it also helps us build a network of friends and colleagues on campus who can answer other questions that might come up as you get started. Try also to document your efforts to improve your teaching, since an upward trajectory of improvement in the classroom can be useful to point out during promotion reviews.

Build a library of useful books and online resources

A host of excellent materials have been published, or are available online, and some are worth having close at hand as you prepare learning materials for your classes. Some of these books are comprehensive guides mentioned elsewhere in this volume, like those by Dee Fink's *Creating significant learning experiences* (2013) or Grant Wiggins's *Learning by design* (Wiggins and McTighe 2005).

Perhaps more important are books that provide short, readable summaries of key topics and strategies. Perhaps the most widely used is *McKeachie's teaching tips*, now in its 14th edition (Svinicki and McKeachie 2013). But a few other excellent sources include Gooblar's *The missing course: Everything they never taught you about college teaching* (2019); Nilson's *Teaching at its best*, now in its 5th edition (Zakrajsek and Nilson 2023); Brown, Roediger, and McDaniel's *Make it stick: The science of successful learning* (2014); Gilmore and Hatcher's *Preparing for college and university teaching* (2021) and the *Handbook for teaching and learning in geography* (Walkington, Hill, and Dyer 2021).

Several volumes provide overviews of dozens of active-learning strategies that can be used to promote student learning such as *Classroom assessment techniques* (Angelo and Cross 1993) and Barkley and Major's *Learning assessment techniques* (2016). Other authors, such as Bligh (2000a, 2000b), provide excellent tips on organizing and getting the most out of discussions and lectures when we use them in our classes.

Be aware that a number of journals publish articles on pedagogy in geography, environmental studies, and related fields. The *Journal of Geography*, *Journal of Geography in Higher Education*, and *Journal of Environmental Education* publish research articles as well as innovative approaches to pedagogy.

Don't try to do everything at once

Being well prepared for classes is a good thing, but it can sometimes seem overwhelming, particularly at the start of an academic career. We may have to prepare to teach several courses simultaneously, while at the same time trying to make progress on all the other projects we'd like to get started. The good news is that these stresses begin to dissipate quite rapidly as new hires get their feet on the ground (Boice 1989). Several tips may help to reach this comfort zone more quickly.

Be careful about overpreparing for classes. As Boice (1991, 112) notes, "By their own admission. new faculty typically went to class overprepared; that is, they prepared so much to say that they had to rush to say it all. In so doing, they inadvertently discouraged students from active participation in classes." Pausing and saving time for questions, comments, discussion, or structuring such activities into a class period can help students understand and apply new concepts and ideas. Rather than seeing class as a 50- or 75-minute solo performance, it might be better to think in terms of Wiggins's backward design: what are the two or three ideas you would like students to carry forward from a given class, then organize the class to focus on these takeaways, with time for some follow-up between each takeaway point.

This approach also implies giving students opportunities to make decisions about activities and topics addressed in a class. This could involve giving students options for assignments; looking at the list of majors enrolled in a survey class to draw examples from those fields; or taking notes of the most popular topics mentioned by students in their assignments and addressing those in subsequent classes.

Finally, keep a record of your plans and activities, as discussed at the end of this chapter. Be sure to include after-class notes, focusing on those approaches that worked, those that didn't, and the sorts of changes or improvements you might make in your plan the next time you teach a particular topic. And don't give up on a good idea; activities that may not work the first time, may work well the second time around.

Be clear about expectations and provide tips about how to reach them

Use your syllabus and each class session as a way of moving toward your learning goals. If discussions are important to your teaching, build toward that goal by highlighting its importance in your syllabus and reinforce the ideas by incorporating discussions regularly in your schedule – each one offering a bit more challenge. If teamwork is needed for group projects, set expectations early and use some low-stakes, team-building activities that encourage student participation and interaction both inside and outside the classroom. If your classes raise controversial topics, it is useful to set, discuss, and gain student input on community standards before the controversial topics are introduced. We can also set expectations by modeling our own behavior. Sometimes, we too have trouble getting started on writing projects, learning new software, or making mistakes in our work. We can hide these from students or, instead, use these issues as a way to see how we try to resolve these sorts of problems in our own work and teaching.

Another reason for setting expectations is that teaching doesn't always go as planned. We may find ourselves working with students who are struggling academically or facing family, health, or emotional issues that impact their studies. At other times, we may face situations in which students cause classroom management issues for you or other students in a class. Universities have offices, policies, and training to help address many of these issues as well as staff who can answer questions and provide advice. Some universities require that links to relevant campus policies and offices be listed in syllabi. Here at our universities, these links include those to the student code and policies: for people with disabilities; against discrimination, harassment, and interpersonal violence; and on academic and professional integrity and misconduct, among others (University of Connecticut 2024). Students may view these policies as "syllabus bloat" and skip over them when reading the syllabus, so it is good practice to point these out, provide a short quiz on them, and discuss them in class when they are relevant to particular assignments or classwork. Don't hesitate to reach out and meet some of the staff who are in charge of these policies, such as our Institutional Equity and Title IX offices here on our campus.

Discussing expectations can also be important in addressing what are termed classroom incivilities – student behaviors that negatively affect a classroom or learning environment. There are a wide range of behaviors that can be disruptive in any given context, and the impact of these disruptions can vary greatly depending on the setting and individuals involved. Students may be more disruptive, for instance, in courses they are required to take or in large classes. Unfortunately, incivilities can also stem from political and religious viewpoints as well as sexism, misogyny, racism, xenophobia, ableism, ageism, and other animosities. This means that incivilities are typically experienced more frequently by women, people of color, and international faculty (Alberts et al. 2010, Gutiérrez y Muhs et al. 2012; Niemann, Gutiérrez y Muhs, and Gonzalez 2020; Vargas 2002). Fortunately, research indicates a number of strategies that can help reduce or eliminate these incivilities (Boice 2000, 81–98; Sorcinelli 1994, 2002; Svinicki and McKeachie 2013, 172–188; Zakrajsek and Nilson 2023, 137–150).

Weave active learning into your classes

Gaining some background in educational psychology, particularly constructivist learning theories, can be a great help in developing more effective learning experiences for our students. Extensive research over many decades indicates learning is best viewed as a process of knowledge construction in which learners are always using their existing knowledge, preconceptions, and interests to understand new concepts and experiences. Constructivism positions the teacher as a facilitator of learning, as a guide who develops activities or realistic problems that promote student exploration and discovery in ways that build upon their existing expertise, but may also sometimes push against misconceptions. Or as Kolb writes, "'Learning is the process whereby knowledge is created through the transformation of experience" (Kolb 1984, 38).

Two particularly good summaries of key issues in this area are, *How people learn: Brain, mind, experience, and school* (Bransford, Brown, and Cocking 2000) and *How learning works: Eight research-based principles for smart teaching* (Lovett et al. 2023). We find Kolb's work particularly useful because he stresses the value of experiential learning – the ways learners gain understanding through experience with new concepts and phenomena (Kolb 1984; Healey and Jenkins 2000). His insight was learning as a four-step cycle involving concrete experience, reflective observation, abstract conceptualization, and active experimentation (Figure 9.1).

Lectures and discussions can certainly play roles in this four-step cycle, but Kolb also suggests the value of active pedagogies that allow students to experience, observe, conceptualize, and experiment with the topics they are studying. It also indicates that learning (and teaching) may be a nonlinear process in which we may need to return to concepts that can be difficult to grasp, or occasionally have students take on new experiences to gain practice transferring their knowledge to new domains. In the following sections, we provide some suggestions for active and experiential learning techniques that can be used to realize Kolb's ideals.

Build a repertoire of active learning strategies for your classes

Large lecture classes often are given a mandate to present an immense amount of content to the students. It is easy to get caught up in the sense that more

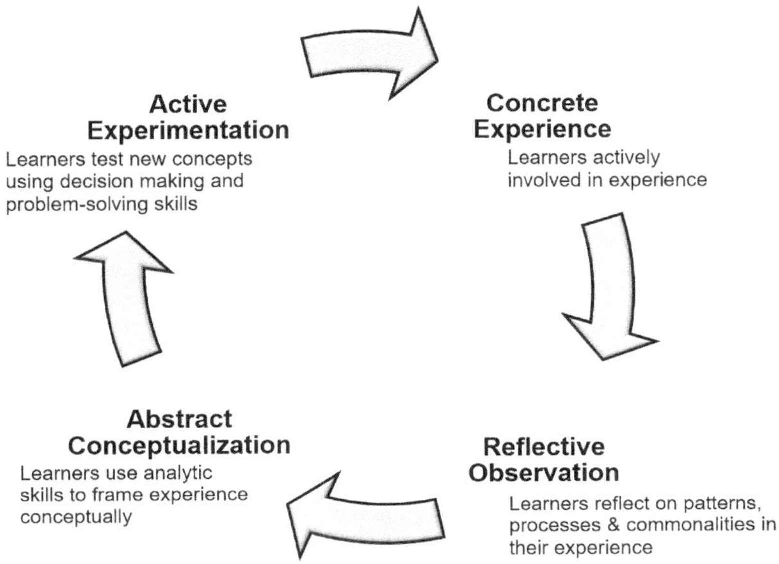

FIGURE 9.1 Kolb's (1984) experiential learning cycle. Kolb argues that mastery of a concept involves all four activities, though not necessarily in a set order (Foote 2012).

material is equivalent to a more worthwhile class. However, numerous studies indicate that "less is more" in class, when the material is addressed in an engaging way with a mix of active learning interspersed with the expertise of the professor (Zakrajsek and Nilson 2023). The typical attention span for adults is about 10–15 minutes. Using this knowledge, we can organize class periods as three or four shorter interlinked episodes, all focused on a particular learning goal, but using different types of activities in each. The downside of this approach is that it takes planning and may sometimes seem to slow the pace of classes by reducing the amount of material that can be covered in a conventional lecture. What is gained is greater student comprehension, and solutions like "flipping the classroom" can set expectations for students to cover more of the basic material at home, prior to class.

When identifying activities to engage students and provide novel ways to introduce or reinforce content, be creative. An abundance of techniques and tools can be used in classes of all sizes, and Table 9.1 is a good starting point. Listed are ideas from many sources and, as mentioned above, many colleges and universities have centers for teaching and learning that provide links to many other strategies. While some of these tools (e.g., classroom response systems) lend themselves to regular usage through a semester, some can get stale if overused. Be thoughtful in mixing these different techniques as you match class topics to active learning strategies throughout a semester.

TABLE 9.1 Examples of Activities for In-Class Learning

Individual Activities

Minute Paper: Each student takes a minute to reflect on content they have just received

Journal: Students write regular journal entries expressing their thoughts about class material and/or how it relates to their lives

Quiz: Having quizzes on a regular basis can test the students' knowledge about the material in a relatively low-stakes way

Presentation Response: Students watching others' presentations and providing a response

In-Class Research: Students use devices to look up materials related to the topic at hand to answer a few, open-ended questions

Video Responses: Students answer questions about a video, both factual to make sure they are watching, and opinion-driven to get them thinking

Paired Activities

Think-(write)-pair-share: Students think about a problem, write their thoughts down (it can be submitted as a simple participation activity), pair up with their neighbor to compare thoughts, then open the discussion to sharing with the whole class

Note Comparison: Students swap notes to compare what each sees as important

Peer Review: Students review one another's work in class for immediate feedback

Small-Group Activities

Cooperative Groups In-Class Share: Student groups tackle a problem and share with class

Active Review Sessions: Students review a study guide and then actively pose potential questions for an exam for the class to review (professor gets new questions too!)

Work at a Common Surface: Use black/white board, large sheets of paper, online whiteboards, and so forth as a space for students to share work and thoughts

Concept Mapping: Student groups build a conceptual map of a problem and solutions

Visual Lists: Students come up with a list of factors (e.g., pros and cons) as a group

Jigsaw Group Projects: Students work on a topic in a group and then bring the perspective of the first group to a second group for further discussion (see Box 9.1)

Role Playing: Students act out roles in a situation to demonstrate how a negotiation or awkward discussion might work in a low-risk environment

Panel Discussions: Students research a perspective on a problem and bring their combined knowledge or opinions to the class through a panel discussion

Debates: Students research a perspective on a contentious topic and debate with students from an opposing perspective, using debate rules to keep things civil

Games: There are numerous ways to gamify content through modifying "classics" like Jeopardy or Monopoly, or through having students create a game around the content

Whole-Class Activities

Whole-Class Discussion: Taking a moment to ask for volunteers to provide their thoughts on a class topic is a great way to break up lecture and identify misunderstandings

Concept Mapping: Create a concept map on screen or on the board using ideas from the class

You'll see in Table 9.1 that active learning strategies can be grouped into several broad categories, based on the number of students involved: individuals, pairs, small groups, or full class. Individual activities are important for continually assessing how each member of the class is keeping up and contributing. Group work can be frustrating for students when some don't carry their weight, so individual activities help to reward those who are staying on top of everything and identify those who may not be contributing as much.

Conversation and one-on-one paired interaction are some of the best ways to switch gears and engage students. Treating these activities as ways to get students talking with the person next to them helps to build classroom community in a quick way without the sometimes-cumbersome elements of group discussion. Sometimes, direct feedback from a peer is helpful to see how others view the material.

The diversity of viewpoints available in a small group, along with manageable size, can make for terrific, engaged discussion. However, it is good to establish ground rules for respectful discussion for the whole class that students can refer to when topics get contentious. There are a tremendous number of options here, and these activities are a good way to mix things up on a regular basis.

Many of the aforementioned activities help students gain insight into the material. In most of those cases, you can ask for a sampling of responses to share with the whole class to spur a discussion. Creating a list of solutions from everyone's ideas values what they were doing and shows the whole class various perspectives on a problem.

BOX 9.1 TRYING A JIGSAW WITH 320 STUDENTS IN A WORLD REGIONAL GEOGRAPHY CLASS

Teaching very large classes can be a challenge, especially engaging so many students. Professor Jolly-Ballantine has been honing a modified jigsaw activity with up to 320 students in his World Regional Geography class for several years. In this activity, students are divided into groups of six, with each group representing an African culture and each member of each group given a "topic" representing a "universal" attribute of cultures (family structures, arts, food, etc.). After meeting with the cultural group, each student goes home to research their aspect of the culture and write a short paper. A couple of classes later, the groups come back together to create a "cultural profile" of their culture. After that, the students are swapped into "interest groups" where a group of six students studying a topic (e.g. governance) would get together to compare their respective cultures. At first, the professor and TAs assigned students to cultures and topics, using an online spreadsheet, and told the students where to sit, leading to utter chaos in such a large class (times two since it happened again when switching to topical groups). Rather than discard the activity for being too chaotic, the spreadsheet was made open, so students could fill themselves into the appropriate spots once they had formed a group of six neighbors in class. Students tend to sit in similar spots, so returning to their group for discussion of research was generally easy. Rather than do a "proper" jigsaw where the topic groups met in a second round, the discussion was placed in the online discussion tool of the Course Management System, where each group posted their cultural profile and then each individual had to review a number of cultures and compare with their own. Being flexible and trying new methods to adjust from year to year can dramatically improve an activity.

Activities with technology

Many of these activities can be aided by technology, such as having students use laptop computers to do in-class research. Sometimes, this involves demonstrating or using lab or field equipment in class, but digital, networked technologies are the focus here. Clickers, smartphones, tablets, and laptop computers are opening up a whole world of additional activities that can help students reach learning objectives as well as gain a critical understanding of the strengths and limitations of these technologies.

Classroom response systems (CRSs), such iClickers, Kahoot, Socrative, and Top Hat, can be very effective in large classrooms. These systems allow students to use their phones, tablets, and computers to respond to questions posed during a lecture. Once set up, these systems can yield a quick check on student comprehension without interrupting the flow of a lecture. Most CRS apps are able to generate response data on the fly, so the distribution of responses can be displayed immediately and discussed in class. Tools like slido (https://www.slido.com/) can also be used to quickly create engaging displays for class data collected in real time. This means you can generate sample exam questions, see how well students grasp the concepts, and provide follow-up, all while giving the students a sense of the types of questions to expect in the exam. These tools can also be used for both short- and open-ended questions that will generate word clouds or response frequency charts for discussion (Figure 9.2).

Systems like Google Drive and Microsoft SharePoint also allow us to create shared virtual spaces for individual and group activities, even in large classes. Students can respond to questions individually or hold a "silent" discussion in groups by typing their posts and responses into a shared document. They can brainstorm ideas using the whiteboard app in Zoom, or an app like Miro, or present their key findings to the full class using shared slides.

Google Earth, ArcGIS Online, and other online mapping tools are simple, often free ways to help students master geographic concepts, both inside and

FIGURE 9.2 Word Cloud from a CRS prompt to a World Regional Geography class of "What comes to mind when you hear the word safari" during a section on sub-Saharan Africa. This can be used to stimulate discussion around many students' conception of safari as an exotic vacation to see animals or environments as compared with the colonial legacy or eco-tourism potential of safaris.

outside the classroom. Students can venture into a Google Street View, for example, to discuss what they are seeing at ground level. Some data sets allow delving deeper into critical issues such as the FEMA flood mapping viewer (https://hazards-fema.maps.arcgis.com/home/index.html), where a map viewer can allow students to see how different amounts of climate-driven sea level rise or river flooding from extreme events can threaten neighborhoods and sites that are familiar to them. Similarly, commercial products, such as Zillow's real-estate viewer, can be used to see how property values vary across a community in response to variables like access to amenities, crime data, and access to transit. Joseph Kerski's chapter delves intoe geospatial tools in more detail.

Other new apps are appearing regularly, so take notes on the ones you hear about and don't hesitate to try them to gauge their suitability for your classes.

Continuous improvement to make things better

Improvement is a continuous process. Not every activity or set of lecture materials is going to work the first time around, but each one provides practice and a chance to gain feedback to make improvements. Such feedback, whether from students or colleagues, gives you a chance to reflect on your own teaching and make improvements from week to week and year to year.

Most American universities use standardized evaluations of classes and instructors. These surveys can provide useful feedback that can point to improvements we might make in our classes and teaching strategies, but they are only a starting point for the sorts of feedback you might gather. We advocate gathering a broader range of feedback. Sometimes, this might be formative feedback on the value of particular assignments or activities that will help improve a class while in progress. Sometimes, we can gather detailed summative data from students at the end of a term. We can gain input from colleagues in our own program, or from peers in other departments or universities. Information from all of these sources can help us reflect on and improve our teaching. A few ideas we've found useful include the following:

1 Place a question or two at the end of an exam or assignment, asking, "How could this assignment be improved?"
2 After introducing a difficult topic, give each student a sheet of paper or link to an anonymous web form to ask them a question, like "What is the most confusing or 'muddiest' point we covered today?".
3 Provide incentives for students to complete the standard course evaluations administered by your university so that you receive as broad a range of responses as is possible. Some universities forbid extra credit for these, but treats or other emoluments are possible.
4 Save time once or twice during the semester to gather feedback on the pacing of the course and on any lingering questions. This feedback might be gathered anonymously in writing or via a Google doc form.
5 Save time in the final week to ask everyone to answer two questions: 1) What was your favorite element of our course, that is what one topic, assignment, or activity would you keep; and 2) What element of our course would you change, add, remove, or improve?
6 Take time after class to make note of whatever did or didn't work well that day and how to make improvements the next time you use the activity. If you can, fix or make changes to your materials right then, while the ideas are fresh and before you forget.
7 Invite colleagues to visit your classes occasionally and ask them for suggestions for improvement. Ask for invitations to visit classes taught by experienced colleagues and talk with them about their approaches to organizing their classes.

Using information of these sorts for self-reflection at the end of each term is very valuable. Spending some time at the end of the semester is a way of "paying it forward" to yourself by taking time to write some reflections about your class over the whole semester. These needn't be well-structured or follow a formula. It might be easiest to see it as a sort of end-of-class diary entry or as a letter you're sending to yourself the next time you teach the class.

With this feedback in mind, set up a four-part to-do list before putting your class to bed. In the first part, note the essential work that needs to be done before you can teach next time, such as updating the syllabus with new dates, TAs, and office hours. Second, make a list of "essential" tasks that you could complete with a couple of days of work. Third, prepare your "in an ideal world" category – things you'd like to do, like redo that discussion prompt or build a new assignment – but the class will be fine if these can't be completed. Fourth, prepare your wish list for when you have time, money, or help to make substantial changes to class over a summer, year, or longer. Maybe you want to reframe your class around a new research theory, use a flipped classroom technique, or build a set of new labs based on new technology. These all take time and effort to develop. Add to your wish list and carry them forward from year-to-year so when that funding opportunity or eager graduate student does come along, you will be ready with a plan.

Summary

Teaching is at the core of what we do as faculty, and yet a source of anxiety. We hope these suggestions will help you reduce any anxiety and help you reach your comfort zone as soon as possible. As you build your classes, it is important to network with your colleagues, inside and outside of your program, to develop a community that helps support your teaching and in which you can support others. There is an abundance of resources including books, journals, online resources, your network, and center for teaching and learning from which you can seek help and feedback.

Using backward design, as Fournier advocates in his chapter, will help you organize your materials more effectively and find ways to actively engage students. There is a world of active learning that allows students to take charge of what they are learning, enhances mastery and retention, and opens up the process to us learning from teaching as much as our students do. Being sure to invite feedback and engage in self-reflection to improve your teaching each time you step into a classroom.

References

Alberts, H., H. Hazen, and R. Theobald. 2010. Classroom incivilities: The challenge of interactions between college students and instructors in the US. *Journal of Geography in Higher Education* 34(3): 439–462. doi: 10.1080/03098260903502679

Angelo, T., and K. Cross. 1993. *Classroom assessment techniques: A handbook for college teachers.* 2nd ed. San Francisco, CA: Jossey-Bass.

Barkley, E. and C. Major 2016. *Learning assessment techniques: A handbook for college faculty.* San Francisco, CA: Jossey-Bass.

Bligh, D. 2000a. *What's the point in discussions?* Portland, OR: Intellect.

Bligh, D. 2000b. *What's the use of lectures?* San Francisco, CA: Jossey-Bass.

Boice, R. 1989. Procrastination, busyness and bingeing. *Behaviour Research and Therapy* 27(6): 605–611.

Boice, R. 1991. Quick starters: Faculty who succeed. In Effective practices for improving teaching, ed. M. Theall and J. Franklin, 111–121. San Francisco, CA: Jossey-Bass.

Boice, R. 1992. *The new faculty member: Supporting and fostering professional development.* San Francisco, CA: Jossey-Bass.

Boice, R. 2000. *Advice for new faculty members.* Needham Heights, MA: Allyn and Bacon.

Bransford, J., A. Brown, and R. Cocking, eds. 2000. *How people learn: Brain, mind, experience, and school.* Washington, DC: National Academy Press. Free download: https://www.nap.edu/download/9853

Brown, P., H. Roediger III, and M. McDaniel. 2014. Make it stick: The science of successful learning. Cambridge, MA: Belknap Press of Harvard University Press.

Fink, L. 2013. *Creating significant learning experiences: An integrated approach to designing college courses.* Rev. ed. San Francisco, CA: Jossey-Bass.

Foote, K. 2012. Scope and sequence in GIS&T curricula: Learning theory, learning cycles, and spiral curricula. In *Teaching GIScience and Technology in Higher Education*, eds. D. Unwin, K. Foote, N. Tate, and D. DiBiase, 81–95. Chichester: Wiley.

Gilmore, J. and M. Hatcher, eds. 2021: *Preparing for college and university teaching: Competencies for graduate and professional students.* Sterling, VA: Stylus Publishing.

Gooblar, D. 2019. *The missing course: Everythig they never taught you about college teaching.* Cambridge, MA: Harvard University Press.

Gutiérrez y Muhs, G., Y. Niemann, C. Gonzalez, and A. Harris, eds. 2012. *Presumed incompetent: The intersections of race and class for women in academia.* Boulder: University Press of Colorado.

Healey, M., and A. Jenkins. 2000. Kolb's experiential learning theory and its application in geography in higher education. *Journal of Geography* 99: 185–195.

Kolb, D. 1984. *Experiential learning: Experience as the source of learning and development*.Englewood Cliffs, NJ: Prentice-Hall.

Lovett, M., M. Bridges, M. DiPietro, S. Ambrose, and M. Norman. 2023. *How learning works: Eight research-based principles for smart teaching.* 2nd ed. San Francisco, CA: Jossey-Bass.

Niemann, Y., G. Gutiérrez y Muhs, and C. Gonzalez, eds. 2020. *Presumed incompetent II: Race, class, power, and resistance of women in academia.* Louisville: Utah State University Press/University Press of Colorado.

Sorcinelli, M. 1994. Dealing with troublesome behaviors in the classroom. In *Handbook of college teaching: Theory and applications,* eds. K. W. Prichard and R. M. Sawyer, 365–373. Westport, CT: Greenwood Press.

Sorcinelli, M. 2002. Promoting civility in large classes. In *Engaging large classes: Strategies and techniques for college faculty*, eds. C. A. Stanley and M. E. Porter, 44–57. Bolton, MA: Anker Publishing.

Svinicki, M. and W. McKeachie. 2013. *McKeachie's teaching tips: Strategies, research, and theory for college and university teachers.* 14th ed. Belmont, CA: Cengage Learning.

University of Connecticut, Center for Excellence in Teaching and Learning (CETL). 2024. Teaching your course. Accessed February 2024, https://cetl.uconn.edu/resources/teaching-your-course/

University of Connecticut, Office of the Provost. 2024. References for syllabi links. Accessed February 2024, https://provost.uconn.edu/faculty-and-staff-resources/syllabi-references/

Vargas, L., ed. 2002. *Women faculty of color in the white classroom: Narratives on the pedagogical implications of teacher diversity*. New York: P. Lang.

Walkington, H., J. Hill, and S. Dyer, eds. 2021. *Handbook for teaching and learning in geography*. Cheltenham: Edward Elgar Publishing.

Wiggins, G., and J. McTighe. 2005. *Understanding by design*. 2nd ed. Alexandria, VA: Association for Supervision and Curriculum Development.

Zakrajsek, T., and L. Nilson. 2023. *Teaching at its best: A research-based resource for college instructors*. 5th ed. San Francisco, CA: Jossey-Bass.

10

DECOLONIZING THE CURRICULUM

Zuhri James and Sarah A. Radcliffe

Identifying the problem and the aim of decolonizing

In this section, we consider how coloniality is embedded within the Euro-American knowledge systems that shapes the worldview of students entering universities. We consider how decolonization seeks to destabilize this worldview through processes of un/re-learning, often from overlooked sources outside of the academy.

Sarah (academic staff member and teacher): In my experience, geography students come to university with a particular worldview, which reflects the implicit yet powerful legacy of histories of colonialism and imperialism in the ways we live in – and seek to understand – the world. Decolonizing geographers analyze the processes that embed these legacies in relations and institutions (with detrimental consequences) and call for an openness to ways of living in and knowing the world that owes less to colonialism-imperialism. In the discipline of geography then, the process of decolonizing decenters "western" ways of writing about the world, and combines that with a critique of racializing, exploitative, and marginalizing practices associated with colonial-modern power. While all higher education aims to transform students' knowledge and understanding, decolonizing approaches encourage learners to question the roots, sources, and limitations of

DOI: 10.4324/9781003342953-12

their worldview. For postcolonial critic Gayatri Spivak, this involves an active process of *un*-learning what one has learned previously in order to *re*-learn the role of imperialism and colonialism in structuring the world socially, economically, politically, as well as the knowledges we live by today. Geographers who work on these themes and analysis may use the term coloniality to label these processes.

In terms of a geography curriculum, decolonizing entails providing learners with the tools to understand how we arrived at the situation where Anglophone geography reflects coloniality, and the skills to analyze its consequences for places and spaces, and for their discipline. Decolonizing the curriculum then highlights two key elements: first, a Eurocentric view of the world is not the only way of living in/understanding the world; second, learning about coloniality in the present day contributes to a deeper understanding of geographies at multiple scales. Coloniality frameworks identify the issues, and decolonizing provides some analytical, conceptual, and pragmatic steps toward extending critical accounts. Decolonizing then is not one-step and done, but rather an ongoing critical approach to the geographies around us.

One key issue in decolonizing the curriculum is recognizing that the usual format of delivering content has been through academic written sources. Decolonizing approaches demonstrate that these sources (called the "canon") are profoundly influenced by British, European, and North American historical geographies of imperialism-colonialism, which results in a selective geographical knowledge written by the powerful. One purpose of decolonizing the curriculum is to learn from processes and events that happen outside the classroom.

Zuhri (geography graduate, studying for a master's in geography): Mobilizing pedagogies which abstain from incessantly foregrounding dominant Euro-American sources is something I've been meditating upon for a while. I distinctly recall one of Tami Okamoto's postcolonial geography lectures in my final year which included a music video created by the Northern-Peruvian Amazon group the Wampis, who were utilizing music as a way to creatively express the contested geographies surrounding their ways of living with increasingly threatened territories and waterways. That was a pivotal moment, not least because three years of undergraduate geography had, up until that point, assumed quite a traditional form insulated within an academic bubble – lecture given, journals and books recommended to read, essay task set, and so forth.

Playing one song, and letting it sound out for four minutes, animated all sorts of questions around coloniality and self-determination. This opened up, albeit momentarily, an alternative pedagogical approach predicated upon letting non-academic sources permeate the theoretical frame. Against an academic backdrop pervaded by coloniality-induced institutional racism, what decolonial potentialities lie within this constellational weaving in and out of historically uncharted and overlooked sources?

Sarah: The video of the Wampis youth making music is powerful because the instructor knew the material in depth, having collaborated with them for many years. She could contextualize the video in a way so students would appreciate the short clip's disruptive qualities. Under-resourced institutions are therefore not prevented from engaging with plural geographies as powerful resources are available from diverse knowledge holders.

Zuhri: Absolutely! Some of the most evocative pieces I've read this year came from these kinds of uncanny sources. A blog piece by Sara Ahmed (2014) exploring racism's atmospheric workings, for instance, deeply resonated with some issues I had been grappling with but had not quite found the grammars to articulate. I also recently came across a thought-provoking Twitter thread by the geographer Katherine McKittrick (2023: np), discussing how to mobilize scholarly lenses in ways that diverge from the dominant individual liberalism of "what can I take from this scholar." She advocates an alternative position that considers, "how does this scholar I respect provide a lens, and lessons, that... and allow me/us to [do] our best work?" Critically, as you pointed out with unorthodox sources, it is not necessary to invite people in to speak to these geographies if resource constraints prohibit such a maneuver.

Sarah: This is interesting, as I wonder if you would have been able to look at those pieces in your first year of university geography and appreciate them. Instructors are careful to organize teaching in an inclusive way that encourages learners through analytical steps toward different insights and understandings. Pitching it at the right level is so important!

Zuhri: The retrospective aspect is certainly significant here: would these alternative kinds of pedagogy have been legible to me as an incoming learner? I was being taught about Indigenous self-determination in my first years of study, but that teaching was largely oriented around reading journal articles, book chapters, and so on. I would therefore argue that what alternative sources including music videos, blogs, and other historically overlooked non-academic sources offer is not so much the risk of illegibility to early learners. Rather, it marks a *decolonial potentiality*: to introduce myriad ways for students to think critically with – and through – creative Indigenous geographies, which are grappling with the speculative dynamics of neoliberal capitalism.

Sarah: Daigle and Sundberg's (2017) piece talks about their intervention in the geography 101 classroom with themes of Indigenous peoples, Indigenous resistance, the university built on Musqueam territory, and so on. That vividly brings home to learners/ students that we are located in and inextricably part of colonialism-imperialism *and* subject to Indigenous and other contestations. We also might want to reflect on the fact that decolonizing is not just about Indigenous peoples, as it encompasses the processes that lead to oppressive labor and racial and spatial orders. A decolonizing curriculum in this sense connects these diverse experiences by referring to underlying power relations that gave rise to them, even as we hold on to the very different experiences of being Roma, Indigenous, or Black, and the complex interconnections between positionalities (Radcliffe and Radhuber 2020). In other words, we situate Black and Roma experiences in relation to broader structures of power rather than viewing them as disconnected. (On a side note, Indigenous peoples' experiences are too often represented – mistakenly in my view – as the archetypal dominated subject of colonialism-imperialism.)

Zuhri: That is an important point. Daigle and Ramírez's (2019) conceptualization of decolonial geographies as intrinsically constellational is worth recalling, given their emphasis on a dialogical approach which foregrounds the rhizomorphic networks of coloniality. "We understand constellations in formation as the embodied knowledge of Indigenous peoples coming into dialogue and relationship with those of Black and other dispossessed peoples" (Daigle and Ramírez's 2019: 79). The dialogical entails holding onto the geohistorical specificities of, say, Blackness across North American contexts, and Indigeneity across Latin America, whilst also recognizing that these contexts need not foreclose ostensibly disparate connections that can be made across time and space.

Prompts to action on decolonizing the curriculum

In this section, we share our experiences of becoming aware of decolonizing debates. These experiences shape how we understand decolonization, define its objectives, and seek to further decolonize the curriculum in our context.

Zuhri: My engagements with decolonization cannot be disentangled from my experiences as someone from an ethnic minority background within Cambridge's geography department. Attending in-person teaching after the Covid pandemic immediately exposed a distorted politics of representation: walking into lecture theaters occupied by predominantly white students and hardly seeing any BIPOC staff. I constantly felt like a perturbation from normality – matter out of place. Yet nobody spoke about this. It was only when I became Cambridge University Geographical Society's BIPOC officer and we did a survey which found underrepresented students were grappling with similar issues, that I realized there was a "normalized quiet of unseen power" (Said 2005: 24) circulating across the department. For me the question became: how might we dismantle this pervasive, overwhelming whiteness?

Sarah: Personally, two experiences influenced my engagement with decolonizing the curriculum. The first was the friction between working in a normatively white department, which has been an almost canonical place in this country, working as a white woman, and, on the other hand, working in Latin America/ Abya Yala with racially marginalized groups in conditions of extreme inequality. Becoming involved in Indigenous peoples' discussions and mobilizations over the long-term made me understand the structures of thinking and living (beyond solely the discourses of social movements) embedded in their critiques and practices. It was vital to bring these dimensions into the classroom, while reflecting critically on established geographies. The then geographer Bhaskar Vira and I co-taught a course to bring Latin American and South Asian geographies into conversation that encouraged students to explore connections and differences. It was a great space to think South-South, not addressing decolonizing explicitly yet highlighting contrapuntal thinking from each region. A second step was listening to debates coming from critical Latin American debates about coloniality, racial formations, and decolonizing, especially from Brazilian geographers (including Carlos Walter Gonçalves) and Latin American geography collectives.

Processes of decolonizing

In this section, we explore some potential pathways for decolonizing from both learner and instructor perspectives.

Zuhri: Minelle Mahtani's (2009) work on diversifying pedagogical practices provides a useful starting point to think about processes of decolonizing because she presents to aspiring academics not a rigid to-do list but a series of self-reflective questions, suggestions, heuristics, and recommendations. This approach is oriented around a sensitivity toward the grounds upon which we stand – our positionalities and situated knowledge (Haraway 1988) – rather than an all-applicable set of considerations. So, working with this fluidity, what kinds of avenues might be relevant in dissolving the remarkably persistent whiteness of geography (Pulido, 2002)?

As Glassman (2010: 510) argues an "important space for potentially transformative articulations of racial politics is the classroom." We might recall the music video example discussed earlier, which illustrated an "anti-racist pedagogy" (Saldanha 2011: 27) unfolding in real time. For me, it registered *experientially* as a kind of perturbation. These pockets of disorientation experienced by learners through pedagogies that cognitively challenge academic orthodoxies are particularly generative. After all, the weird is a particular kind of perturbation. As Mark Fisher writes, "it involves a sensation of wrongness: a weird entity or object is so strange that it makes us feel that it should not exist, or at least it should not exist here. Yet if the entity or object is here, then the categories which we have up until now used to make sense of the world cannot be valid" (2016: 15). There is a kind of pressure, then, that experimental pedagogies can exert upon normative academic practices, which connect to sociologist Ramón Grosfoguel's (2013) discussion of dismantling the westernized university in favor of decolonial pluriversities.

Staying with pluralized pedagogical practices that affect how leaners might engage with decoloniality, we might also consider student experiences with reading lists. Sara Ahmed recently noted how being in conversation with someone's work over a sustained period can be *a queer kind of intimacy.* Rather than simply broadening reading lists, instructors might contemplate the ways to encourage students to interact with readings over time. This offers the possibility of eschewing liberal individualistic tendencies in favor of fostering alternative lines of ethical engagement and decolonial orientations.

Sarah: I agree that decolonizing the curriculum has to be open and plural for instructor and learner alike. In a curriculum, this can be done by situating it in dialogue and interaction with the location of learning, a specific university, its place, and historical geography. Situating the curriculum creates a conversation between learning in the classroom and wider spatial processes (at national and transnational scales). This decolonizing does not inflate the length of a reading list (as is sometimes thought) but provides an opportunity to engage meaningful situated issues. By focusing a curriculum on a colonial-modern locale, learners build on their diverse and unequal familiarity and embeddedness in that space. A decolonizing approach offers an opportunity to learn from plural voices and contextualize them in relation to socio-spatial processes. Starting "small" in this way can combine with openness, plurality, and different modes of learning, while engaging with the experiential. Field visits or the visit of knowledge-holders to the classroom is pluralizing and invaluable. It's also a question of leaving time for students to think through what they've been exposed to.

Decolonizing the curriculum also, as noted earlier, is about pluralizing students' understanding of marginalized and unfamiliar world knowledge and practices. The UK public – and most students – are unfamiliar with Latin America/Abya Yala. To highlight the region's plural voices, my reading lists indicate if an author is Black, Indigenous, or Latin American. This schematic, reductive strategy is not ideal, yet makes diverse positionalities and contributions to geography part of the conversation rather than invisible (students often assume authors are exclusively white scholars from the North Atlantic metropole). This practice disrupts the curriculum's normative context, facilitating consideration of less canonized authors. Students thereby un-learn and re-learn. Nevertheless, colonial knowledge production predominates as writers connected to the Euro-American world are more frequently published and translated into English. Finding pathways for students to alternative situated knowledge means turning to blogs, specialist news organizations, and interviews with key people. Videos provide insights into distinctive social, spatial, and epistemological systems in a visually powerful medium. So, a decolonizing curriculum is interested in openness, plurality, the "disruption" of existing expectations of authorship, and contextualization.

Zuhri: This point about how instructors might display and communicate the contents of a reading list to learners is important. We are not talking about a reading list *per se*, but the geographies of – and to – a reading list. In urban studies, for instance, scholars outside of the Anglosphere have not always been translated into English (often by way of exorbitant translation services). The result is scholarly canons which fundamentally coalesce around Eurocentrism. So, processes of decentering (in a way that scholars such as Jenny Robinson and Ananya Roy discuss vis-à-vis urban theory) are critical to decolonial efforts which attune to the epistemically unjust geographies to, and of, that theory.

Sarah: On the political economy of citation, that's correct at the level of academic tenured jobs with the exclusion and marginalization of certain voices. However, within the classroom, an instructor has opportunities to disrupt learning by engaging (say) Indigenous peoples, their scholarship, and lived experience. Geography undergraduates are generally unconcerned with the political economy of citation, so a series of lectures centered on Black geographies or Indigenous geographies will comprise their learning experience. Implicitly, they will experience a disruption in the political economy of citation and a recalibration in how understanding and insight are circulated. Hopefully, instructors have an agency to make the curriculum a space for rebalancing inequality in the academy (which is stubbornly engrained in university tenure systems). Alongside that, an instructor in a supportive institution inhabits a space where setting different parameters is possible by "hacking" the reading lists.

Following that thought, newly appointed academics are often given an introductory-level course to teach. Some departments say, "Well, Geography 101 must deliver the geography canon, Halford Mackinder and so on." By contrast, Michelle Daigle and Juanita Sundberg (2017) had room for maneuver in designing a curriculum. Arguably, whichever level one is teaching, there are possibilities. As I've written (Radcliffe 2022), Anglophone geography is probably less tied to "the canon" in comparison with sociology or political science. In the United States, geography undergraduate programs exist in a context with less pressure to adopt a canon. Decolonizing the geography curriculum might benefit, then, from the relative degree of academic freedom in comparison with other disciplines.

Challenges to decolonizing

In this section, we consider the barriers to a broad decolonizing of the curriculum and discuss how to recognize and address barriers.

Sarah: One challenge can be inserting decolonizing approaches and material in an institution. In this context, student feedback in vocal support of decolonizing content can be important in leveraging institutional change. At the same time, it can alert professors to learners' interests and concerns across the institution, including inappropriate content in "non-decolonizing" classes.

Zuhri: The implementation of decoloniality certainly marks a salient challenge, not least given its protean nature which necessarily requires a unique approach in every case. Here, I think it is useful to return to the fundamental question of *why are we decolonizing?* As the geographer Tariq Jazeel notes, decolonizing our epistemic systems of course includes the potential to learn from, for instance, historically overlooked border epistemologies, but it also suggests the academy "might have something of relevance to say to that world, that it might 'speak to' places, peoples, and communities in politically germane ways. That it might intervene" (Jazeel 2017: 336). This potential to speak to not only marginalized groups within the academy but also to marginalized people facing alienating states of emergency marks what can often feel like the amorphous unclassifiability of decoloniality, "an unknown that does not terrify" (Glissant 1997: 9). After all, as we continue to decolonize our curriculums and institutions, "we will inevitably see more and see differently and feel a new sense of wanting and being and becoming" (Harney and Moten 2013: 6).

Sarah: We might also remember that meso-level challenges arise as the burden of decolonizing often falls on certain people within an institution. Mainstreaming decolonizing by contrast entails distributing the emotional, intellectual, and political work of decolonizing across a wide and mutually supportive group of instructors and administrators. Without that, a "lone" (Tolia-Kelly 2017) decolonizer must undertake that labor alone or with few allies. In this context, an instructor faces the challenge of making a meaningful contribution without burnout or without becoming a lightning rod for opposition. In this context, a lone decolonizer is entirely justified in ensuring their burden is manageable in the long-term. Students will take something away, as decolonizing opens up the curriculum in significant ways.

Another pragmatic challenge is facilitating physical geography instructors' access to material which makes them reappraise established forms of knowledge transmission and content. In *Decolonizing Geography: An introduction*, I found it a challenge to identify a wider literature, although here we can learn from a few singular developments in cognate disciplines (Radcliffe 2022). In terms of case studies, there is a burgeoning range of materials, even if fewer than in human geography. Another strategy in this context is to juxtapose standard physical geography content with an Other knowledge of a physical geography phenomenon. For instance, one colleague switched one of four introductory lectures on glaciology to teach about Indigenous ways of thinking about knowledges of and interactions with glaciers. The challenge for physical geographers, then, is probably greater than for human geographers, in terms of becoming familiar with ways of engaging decolonizing.

Risks of decolonizing

In this section, we discuss risks that decolonizing agendas face.

Sarah: Decolonizing is, as with other critical geographies, liable to be co-opted and de-fanged. Academic decolonizing has rightly been criticized for making minimal difference to the lives and futures of dispossessed and oppressed groups. Decolonizing also fundamentally challenges the structure and purposes of educational establishments in the neoliberal and racial capitalist world, and as such it struggles to transform those institutions (Bhambra et al. 2018; Cupples and Grosfoguel 2018). In settler colonial contexts such as North America and Australasia, Tuck and Yang (2012) identify what they term the moves to innocence by which the decolonizing impetus and vision are slowly but thoroughly emptied of effectivity, materially and epistemologically.

Zuhri: These resistances to decolonial agendas also connect with how some instructors resist giving a positionality statement at the start of a lecture/course (Pirbhai-Illich et al., 2017; also, **Additional Resources**). People who have been in the academy for decades might brush aside decolonial imperatives or deem them irrelevant to their contributions.

Sarah: It's difficult to deal with, especially for a "lone" decolonizer. That type of academic denial represents a risk to decolonizing the curriculum in that colleagues may express interest when you introduce decolonizing themes informally, but they do not consider themselves as part of that process. Arguably a greater risk is professionalization and co-optation; instructors are accountable to learners and colleagues. But from a decolonial perspective, one is also accountable to groups and individuals around the world. How to decolonize in ways that nurture that accountability?

Conclusion

This chapter has discussed the process of decolonizing in relation to fundamental aims, potential action, and processes, alongside inherent challenges and risks; these are all, of course, deeply intertwined. Implementing decoloniality thus requires instructors to reflexively straddle all these dimensions. Whilst this process is fraught with uncertainty – not least because decoloniality is, and never has been, a foregone conclusion – we have highlighted decoloniality as a set of ongoing interrogations, attunements, and (re)adjustments. Decoloniality is thus suffused with the geographies of hope, as we continually reach toward alternative futures beyond the splintering horizons of coloniality.

Additional resources

The Berkeley Black Geographies Project https://theblackgeographic.com/ is committed to "the material study of Black life."

Decolonizing Geography website – https://decolonisegeography.com/ for UK-based geographers. Includes blogs/thematic discussions that link to teaching materials, e.g., why the word "slum" should not be used in classrooms, interviews with academics on racial capitalism, spaces of Black history, and more.

Geogramblings: The Life Geographic blog on Decolonizing geography. https://geogramblings.com/2020/08/01/decolonising-geography/ also includes a YouTube video. Both include US and North American examples. At the end, there's a list of resources on culturally responsive pedagogy.

Morreira, Shannon and Luckett, Kathy: "Questions academics can ask to decolonize their classrooms." *The Conversation,* October 17, 2018. Their "questions [aim] to encourage academics across faculties to unearth some of the norms, assumptions and everyday practices that are taken for granted and which may be entangled in the 'hidden curriculum.'" Some are given here:

1 What principles, norms, values, and worldviews inform your selection of knowledge for your curriculum? (think about absences as well as presences, centers as well as margins)
2 Do you articulate your own social and intellectual position, from which you speak when lecturing?
3 For whom do you design your curriculum? Who is your ideal, imagined student, and what assumptions do you make about their backgrounds, culture, languages, and schooling?
4 Does your curriculum reflect its location in Africa and the Global South? To what extent does it draw on subjugated histories, voices, cultures, and languages?

Positionality statements – For a definition, see https://www.queensu.ca/ctl/resources/equity-diversity-inclusivity/positionality-statement. An introductory video can be found here: https://www.youtube.com/watch?v=GpcIVzGYhVs – also Holmes (2020).

Salaam Geographia – A website with resources from Iram Sammar (who describes herself as anti-racist geography consultant, writer, and speaker): https://salaamgeographia.com/

> In this ever-changing world it is essential to get to know one another for who we are. ***Salaam Geographia***, or 'peace to geography' is a celebration of everyday geographies young learners should experience and learn about regardless of race and ethnicity in the new century. A space for people who want to learn about anti-racist approaches to teaching geography in schools and to help understand decoloniality.

University of British Columbia Reading list of Black geographers (compiled by students, faculty, and staff at UBC Geography, June 4, 2020) "A collection of work by Black geographers, and additional recommended readings on Black geographies." https://geog.ubc.ca/?s=reading+list+of+black+geographers&submit=Search

References

Ahmed, Sara. 2014. Atmospheric walls. https://feministkilljoys.com/2014/09/15/atmospheric-walls/ [Last Accessed 4th October 2023].

Bhambra, Gurminder K., Dalia Gebrial, and Kerem Nişancıoğlu, eds. 2018. *Decolonising the university*. London: Pluto Press.

Cupples, Julie, and Grosfoguel, Ramón, eds. 2019. *Unsettling eurocentrism in the westernized university*. Abington, MA: Routledge.

Daigle, Michelle, and Sundberg, Juanita. 2017. From where we stand: Unsettling geographical knowledges in the classroom. *Transactions of the Institute of British Geographers*, 42(3), pp. 338–341.

Daigle, Michelle, and Ramírez, M. M. 2019. Decolonial geographies. *Keywords in Radical Geography: Antipode at 50*, Antipode, pp. 78–84.

Fisher, Mark. 2016. *The weird and the Eerie*. 1st edn. London: Repeater Books.

Glassman, Jim. 2010. Critical geography II: Articulating race and radical politics. *Progress in Human Geography*, 34(4), pp. 506–512.

Glissant, Édouard. 1997. *Poetics of relation*. Ann Arbor: University of Michigan Press.

Grosfoguel, Ramón. 2013. The structure of knowledge in westernised universities: Epistemic racism/sexism and the four genocides/epistemicides. *Human Architecture: Journal of the Sociology of Self-knowledge*, 1(1), pp. 73–90.

Haraway, Donna. 1988. Situated knowledges: The science question in feminism and the privilege of partial perspective. *Feminist Studies*, 14(3), pp. 575–599.

Harney, Stefano, and Moten, Fred. 2013. *The Undercommons: Fugitive planning and black study*. New York: Minor Compositions.

Holmes, A. G. D. 2020. Researcher positionality: A consideration of its influence and place in qualitative research – a new researcher guide. *Shanlax International Journal of Education*, 8(4), pp. 1–10. https://doi.org/10.34293/ education. v8i4.3232

Jazeel, Tariq. 2017. Mainstreaming geography's decolonial imperative. *Transactions of the Institute of British Geographers*, 42(3), pp. 334–337.

Mahtani, Minelle. 2009. Teaching diverse students: Teaching for inclusion, In M. Solem et al. (eds.) *Aspiring academics*. New Jersey, NJ: Pearson Education, pp. 121–133.
McKittrick Katherine. 2023. Twitter page: demonicground https://twitter.com/demonicground/status/1678628059031691266 [Last Accessed 5th October 2023].
Pirbhai-Illich, Fatima, Pete, Shauneen, and Martin, Fran, eds. 2017. *Culturally responsive pedagogy: Working towards decolonization, indigeneity and interculturalism*. London: Palgrave Macmillan.
Pulido, Laura. 2002. Reflections on a white discipline. *The Professional Geographer*, 54(1), pp. 42–49.
Radcliffe, S. A. 2022 *Decolonizing Geography: An introduction*. Cambridge: Wiley & Sons.
Radcliffe, S. A. and Radhuber, I. M. 2020. The political geographies of D/decolonization: Variegation and decolonial challenges of/in geography. *Political Geography*, 78, p. 102128.
Said, Edward. 2005. The public role of writers and intellectuals. In *Nation, language, and the ethics of translation. STU - student*. Princeton, NJ: Princeton University Press, pp. 15–29.
Saldanha, Arun. 2011. The concept of race. *Geography*, 96(1), pp. 27–33.
Tolia-Kelly, Divya. 2017. A day in the life of a geographer: 'Lone', black, female. *Area*, 49(3), pp. 324–328. https://doi.org/10.1111/area.12373.
Tuck, E and Yang, K. W. 2012. Decolonization is not a metaphor. *Decolonization: Indigeneity, Education & Society*, 1(1), pp. 1–40.

11

CREATING A SOCIALLY JUST CLASSROOM

LaToya E. Eaves and Miguel Á. Hernández

> The work of social justice is complex. To do this work well, individuals must be willing to deeply reflect on themselves and identify ways in which they consciously and unconsciously benefit from and perpetuate oppression.
>
> (Hernández, 2018, p. 35)

Education is fundamentally a site of transformation. College campuses function as small ecosystems comprising multifaceted living and learning communities. Today's classroom asks college teachers to explore differences in both content and approach. It invites an opportunity to engage a more diverse student body, which benefits from the more widely accessible and engaged campus culture than previous generations. Historically, the college classroom has been populated by and privileged a particular kind of student who had economic and social capital to participate in higher education in the United States. Today, the student body is more diverse, representing individuals from every walk of life. Therefore, as teachers, we cannot continue with "business as usual." Instead, college instructors are encouraged to engage in practices that create a socially just classroom.

Central to shifting the classroom environment is not just an analysis of *what* we teach but *how* we teach. Integral to approaching social justice practice is finding common ground with a set of agreements and definitions. For this, we include a set of terms and definitions that provide you with language to get you started (Box 11.1).

DOI: 10.4324/9781003342953-13

BOX 11.1 RELEVANT TERMS AND DEFINITIONS

Bias (Implicit): Also known as implicit social cognition, implicit bias refers to the attitudes or stereotypes that affect one's understanding, actions, and decisions in an unconscious manner. These biases, which reside deep in the subconsciousness and encompass both favorable and unfavorable assessments, are activated involuntarily and without an individual's awareness or intentional control.

Brave Space: A learning environment for social justice dialogue should encourage individuals to explore the ways they've been socialized and how it has impacted their identity development while also examining both how they view themselves and relate to others. Brave spaces challenge participants to move beyond the comfort often associated with ground rules and "safe spaces."

Civic Learning: It involves acquiring the knowledge, the intellectual skills, and the applied competencies that citizens need for informed and effective participation in civic and democratic life; it also means acquiring an understanding of social values that underlie democratic structures and practices. Civic learning can happen through courses, co-curricular activities, and engagement with communities beyond the classroom.

Discrimination: This means treatment of an individual or group based on their actual or perceived membership in a social category, usually used to describe unjust or prejudicial treatment on the grounds of race, age, sex, gender, ability, socioeconomic class, immigration status, national origin, or religion.

Diversity: There are many kinds of diversity, based on race, gender, sexual orientation, class, age, country of origin, education, religion, geography, physical, or cognitive abilities. Valuing diversity means recognizing differences between people, acknowledging that these differences are a valued asset, and striving for diverse representation as a critical step toward equity.

Educational Equity: The condition that is achieved

> when educational policies, practices, interactions, and resources, are representative of, constructed by, and responsive to all people such that each individual has access to, can meaningfully participate, and make progress in high- quality learning experiences that empower them towards self-determination and reduces disparities in outcomes regardless of individual characteristics and cultural identities.

Equity: Equity means fairness and justice and focuses on outcomes that are most appropriate for a given group, recognizing different challenges, needs, and histories. It is distinct from diversity, which can simply mean variety (the presence of individuals with various identities). It is also not equality, or "same

treatment," which doesn't take differing needs or disparate outcomes into account. Systemic equity involves a robust system and dynamic process consciously designed to create, support, and sustain social justice.
Inclusion: This means being included in a group or structure. More than simply diversity and quantitative representation, inclusion involves authentic and empowered participation, with a true sense of belonging and full access to opportunities.

Microaggression: It means verbal, behavioral, or environmental slights that result from an individual's implicit bias. They are often automatic or unintentional and occur on a daily basis.Safe Space: It is an environment in which everyone feels comfortable in expressing themselves and participating fully, without fear of attack, ridicule, or denial of experience.Social Justice: It connotes both a process and a goal. The goal is full and equal participation of all groups in a society that is mutually shaped to meet their needs. The vision of such a society promotes (a) equitable distribution of resources, (b) physical and psychological safety and security, (c) self-determination and independent agency, and (d) a sense of self and responsibility to society as a whole. The vision of a socially just process includes practices and procedures that are democratic and participatory, inclusive, and affirming of human capacities for working collaboratively to create change.

Stereotype: It is the characteristics ascribed to a person or groups of people based on generalization and oversimplification that may result in stigmatization and discrimination.

Definitions excerpted from Racial Justice in Education: Key Terms and Definitions (National Education Association 2021) and *SLP-CLDE KC Anti-Racism Terminology Guide 2020–2021* (Collins et al. 2021).

We enter this conversation as two educators with backgrounds in student affairs practice and teaching. Spanning more than two decades, our relationship has evolved from being involved with national and regional student leadership organizations to being colleagues in university housing at the same institution, then to diverging paths in higher education. Hernández has continued his role in student affairs, with increasing responsibilities in housing and residential education, retention and engagement, leadership development, and multicultural affairs, and now holds the position of Dean of Students. Eaves left student affairs for faculty life/academic affairs. We both have extensive experience as workshop facilitators on a range of social justice topics and college-level faculty serving higher education, educational leadership, and geography units. What we include in this chapter is couched not only in research but also in practices we have incorporated and honed over our careers in higher education. Throughout, we include quotes from conversations with higher education professionals whose work involves promoting diversity, equity, and inclusion on college campuses and in college classrooms.

Reflexivity and positionality: keys to transformative pedagogy

Taking time to think about our experiences and backgrounds as we enter the teaching profession is central to the socially just classroom. The practice of reflection is an active choice, one in which we develop an understanding of our positionality in social categories and the power we have in the exchange of knowledge, otherwise known as reflexivity. Reflexivity is characterized as "the recursive practice of self-examination and self-transformation in the exchange of knowledge" (Kobayashi 2020, p. 213). Reflexivity and understanding the role of positionality are key aspects of taking the first steps to transform pedagogical practices and pursue the creation of a socially just classroom. Initiating these changes signals a willingness to acknowledge that at the most basic, foundational level, institutions have functioned to reproduce socio-structural inequalities in all aspects – in the community and in the classroom.

> ...we have such a disparity in terms of socioeconomic backgrounds. We have such different cultures of immigrant refugee communities, working communities. So it's about, I think recognizing those material inequalities. But I think it's also about recognizing the social, cultural and mental harms of being seen as a problem or seen as a source of violence or seen as a source of contagion. There's all these kind of psychic harms that are directed at people and communities that are seen as other. And so trying to figure out how do we create the conditions where we can collectively thrive, whether it's economically or emotionally and culturally.
>
> ~ *Judy*

Yet, the avoidance of varied experiences is unproductive and, in fact, damaging. As bell hooks asserts, "The unwillingness to approach teaching from a standpoint that includes awareness of race, sex, and class is often rooted in the fear that classrooms will be uncontrollable, that emotions and passions will not be contained" (1994, p. 39). Moving beyond the fear of losing control of the classroom empowers the faculty member to be open and vulnerable to engaging in teaching practices that facilitate learning in transparent and cooperative ways (Figure 11.1).

As Alderman et al. (2021) write, with regard to anti-racist pedagogy, "Anti-racism undermines long-standing and traditionally unquestioned geographic knowledge systems and practices that work to marginalize, misrepresent, and inflict violence upon people of color while also distorting and misinforming the worldview of a dominant White society" (p. 189; see also Kinkaid and Fritzsche 2022; Thayer 2023). These assertions reaffirm bell hooks' reminder that education is not neutral and that it always has wider implications in educational settings.

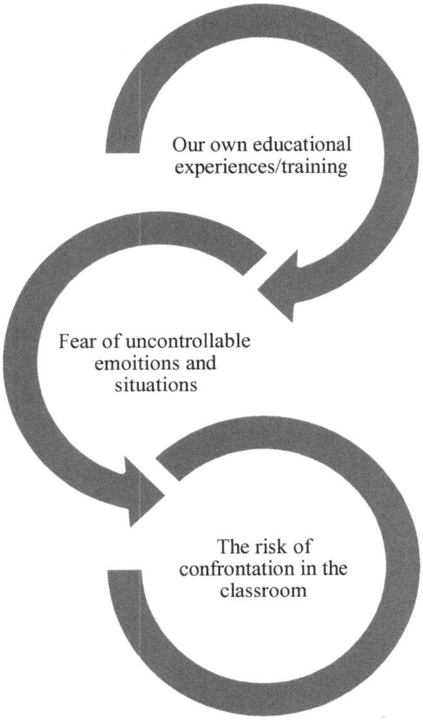

FIGURE 11.1 Understanding the roots of discomfort in creating brave spaces.

Defining a socially just classroom

> *"We don't always control who is in our physical classrooms in terms of people in seats, or in virtual spaces... what we do have more influence over, control over is the content...[to] think about how we are representing different experiences and identities."*
>
> *~Sonja*

Bonnie Thornton Dill and Ruth Enid Zembrana (2009) remind us that, "Inequality and oppression are deeply woven into the tapestry of American life. As a result large disparities exist on measures of income, wealth, education, housing, occupation, and social benefits. These disparities are neither new nor randomly distributed throughout the population, but occur in patterns along such major social divisions as race, gender, class, sexuality, nationality, and physical ability" (p. 1). These societal patterns do not disappear on our campuses. As we stated earlier, the college classroom today is unlike any other in history, representing a student body that comes from every walk of life – gender identity, racial and ethnic difference, socioeconomic status, ability, nation of origin and immigrant status, multilingual, multiethnic, biracial, queer, and trans-identified. Our students are multi-generational, with many adult learners joining "traditional-age" college students on our campuses. In addition, today's students are more open to disclosing their mental health status than ever before. At no other time has higher education been made more accessible to a wide range of people. This reality means that our teaching practices must be equity-oriented, reflecting on what it takes to ensure all students have access to a welcoming and rigorous learning environment.

A socially just classroom is a learning environment that is geared toward inclusion and encourages respect for all students, regardless of their backgrounds, identities, or abilities. Avoiding discriminatory practices is fundamental to the creation of the classroom space. But teachers should go further. A socially just classroom is grounded in principles that include the following:

- Respect and feeling valued: Every student should feel valued and respected in the classroom. They should feel that their experiences and perspectives matter and are appreciated as part of the college learning experience.
- Connection and belonging: Every student should have a classroom experience that offers them a sense of community and where they can show up as their full, authentic selves.
- Empowerment: Every student should feel encouraged to reach their full potential and feel capable of their skills. The classroom space should encourage students to learn and grow.

These principles can be applied in all aspects of the classroom space, from course planning and content creation to lecture design, in-class activities, and assessments. Importantly, the practice of creating a socially just classroom goes beyond mere "multiculturalism" or "diversity" but engages in a wider transformation of the classroom environment and renovation of many pedagogical practices we learned in the past. In the next section, we expand on these principles by discussing what it means to be equity-minded and then offer suggestions for creating a socially just classroom with an equity mindset.

Creating a socially just classroom

> *"I would define social justice as everyone has a fair and just opportunity to learn and engage with the material without consideration of their race, their gender, socioeconomic status, ability, and that they have all of the necessary resources to...equally engage with the material and navigate through whatever program."*
>
> *~Reggie*

Creating a socially just classroom involves creating an environment that is equity-minded, in which instructors take seriously the role of individual differences and/or structural and systemic inequalities that have impacted access across all communities (Adams 2007; Kobayashi 2020). Equity in the classroom requires theory and practice, thought and action. Scholarship in higher education that centers equity calls for multifaceted approaches in the learning environment: "Educators with an equity talk *and* an equity walk critically examine institutional policies, practices, and structures through a lens that questions why inequities exist to change the educational environment to support the success of students – especially students who have been historically and continuously marginalized in our educational systems" (McNair, Bensimon, and Malcom-Piqueux 2020, p. 2). By deciding to have an equity-minded pedagogy, we can review our teaching practices to examine how decisions produce unequal experiences in the classroom. Put differently, we can ask ourselves, "Who is being left out as a result of this approach?" (Sathy and Hogan n.d.). In Box 11.2, we include teaching tips related to providing instruction in an accessible environment for students with disabilities. It should be noted that taking pedagogical approaches that address accessibility needs has a larger benefit and is more inclusive for all students.

BOX 11.2 TEACHING TIPS FOR AN ACCESSIBLE CLASSROOM

Classroom	*Examinations*	*Labs*
• Select course materials early so that students and SDS have enough time to convert it to accessible formats (e.g., audio, large print) as needed. • Make course materials (e.g., syllabus, handouts, assignment sheets) available in electronic format. • Write lecture outlines on the board or incorporate into a PowerPoint presentation. • Write key phrases, technical vocabulary, formulas, etc., on the board or provide as a handout.	• Ensure exams test the essential skills or knowledge needed for the course. • Provide accommodated testing, as appropriate. Some students may require extended time or assistance on exams. Contact SDS regarding options for accommodated testing.	• Provide a lab orientation, take a tour of the lab, and discuss safety concerns. • Provide oral and written lab instructions. • Arrange lab equipment so that it is easily accessible. • Assign group lab projects in which all students contribute according to their abilities.

(*Continued*)

(Continued)

Classroom	*Examinations*	*Labs*
• Do not speak when you have your back turned to the class. • Repeat or rephrase questions or comments from the class before responding	• Consider alternative assessment methods (e.g., portfolio, multimedia presentation).	• Provide adaptive lab equipment.

"Teaching Techniques" from the University of Tennessee, Knoxville's Student Disability Services (n.d.).

We identify several key features to consider when approaching a teaching practice that promotes a socially just classroom. First, we suggest that instructors consider their course(s) curriculum. A socially just classroom includes a curriculum that is inclusive and diverse. This means that the instructional choices in the course material represent multiple perspectives and voices. Additionally, the selected materials address issues of power, privilege, and oppression. One strategy for doing so is including a discussion on a current event in the community or providing case studies outside of the textbook that involve social justice topics.

In addition, curricular choices should be culturally responsive. Research has shown that materials that connect to students' lived experiences and backgrounds affirm their learning experiences (Hackman 2005; Ladson-Billings 2005; Adams 2007; Pappamihiel and Moreno 2011; Cochrane et al. 2017). This is not to be misrepresented as tokenism (hooks 1994). Tokenism relies on the idea that the experiences of a person belonging to a historically excluded community make them an expert on that community. For example, a lesbian-identified student is called on to provide expertise on the experiences of all LGBTQ students taking introductory geography courses. Singling out the student and placing this burden of expert perspective on them can lead to negative personal and psychological effects, even without intention on the part of the faculty member (hooks 1994; Niemann 2011, 2016; Garibay 2014).

Finally, an inclusive classroom promotes critical thinking. In her book *Teaching Critical Thinking*, bell hooks put it simply: "Thinking is an action. For all aspiring intellectuals, thoughts are the laboratory where one goes to pose questions and find answers and the place where visions of theory and praxis come together. The heartbeat of critical thinking is the longing to know—to understand how life works" (p. 7). Encouraging critical thinking invites students to question assumptions, challenge biases, and think more expansively about the world around them (Hackman 2005).

In addition to the curriculum, pedagogical decisions are another key component of a socially just classroom. We suggest teachers spend time exploring differentiated instruction. By differentiated instruction, we mean the practice of reflecting on and tailoring the teaching practice to meet the varied needs of students in courses. By taking on differentiated instruction, teachers can better transform teaching practices in ways that meet overall goals while understanding the diversity of ways that students learn. In Box 11.3, we include various outcomes for each aspect of differentiated instruction. We also include ideas on how to implement each outcome in the classroom.

BOX 11.3 OUTCOMES OF AND SUGGESTIONS FOR DIFFERENTIATED EDUCATION

Outcome	*Description*	*Recommendations for Implementation*
Adaptive instruction	Meets the needs of all learners, adapting teaching methods and assessments to individual strengths and challenges.	Project-based learning with flexible options: Students choose a topic and decide in consultation with the instructor, projects that demonstrate mastery of concepts.
Active learning	Students are engaged in the learning process through dialogue, collaboration, and project-based activities.	Gallery walks: The instructor displays images/texts/artifacts around the classroom or in a slideshow for virtual classes. Students discuss each display in small groups and offer reflection.
Positive relationships	Teachers build strong, trusting relationships with students based on mutual respect and understanding.	Offering flexibility: Being flexible with varied assessments makes room for accommodating the diverse needs and circumstances that students experience.
Safe and inclusive	Free from discrimination, harassment, or bullying	Provide support: By curating and sharing a list of campus resources and support services (i.e., counseling services, student support groups, disability services, diversity and inclusion offices, emergency support services) in course materials and regularly through instruction, college teachers demonstrate and communicates the classroom is an inclusive space

(*Continued*)

(Continued)

Outcome	*Description*	*Recommendations for Implementation*
Collaborative and respectful	Students are encouraged to listen to each other, engage in respectful dialogue, and celebrate differences.	Encouraging interaction: Teachers can use the classroom to facilitate collaboration through group projects and peer learning activities, which increase interaction and shared learning among students.
Anti-bias practices	Proactively challenge stereotypes and biases, both within the classroom and the larger school community.	Challenge stereotypes: As stereotypes, biases, and misconceptions arise during discussion or in course content, instructors can use them as opportunities to critically evaluate the roots of beliefs and collaborate on reviewing evidence-based information that shifts ideas about people and places.

Socially just classroom as brave spaces

Our classes take place in a variety of sites: in traditional classrooms, in the virtual world, in the field, in labs, and in our offices. Understanding this variation means we can think holistically about the overall course climate. Reflecting on course climate means that we engage the physical environment of the classroom, whatever form that takes. But we also look to engage the intellectual and emotional factors that the course presents. Research on classroom climate informs us of the multifaceted nature that classroom climate can take:

> "Climate is determined by a constellation of interacting factors that include faculty-student interaction, the tone instructors set, instances of stereotyping or tokenism, the course demographics (for example, relative size of racial and other social groups enrolled in the course), student-student interaction, and the range of perspectives represented in the course content and materials."
>
> *(Ambrose et al. 2010, p. 170)*

As part of the process of creating a classroom climate that is inclusive of the wide variation in student and instructor positionality, we encourage the concept of brave space. Brave space differs from the common pedagogical idea of "safe space." Both "safe space" and "brave space" aim to create inclusive and supportive environments, but they differ in the level of risk and comfort they emphasize.

One of the primary challenges of employing "safe space" is in the term itself – the use of the word "safe" (Box 11.4). In any group setting, the promise of ensuring the safety of all requires a more critical lens: "commonly used ground rules include 'agree to disagree,' 'don't take things personally,' 'challenge by choice,' 'respect,' and 'no attacks.' We believe that, unexamined, these common ground rules may contribute to the conflation of safety and comfort and restrict participant engagement and learning" (Arao and Clemens 2013, p. 143).

BOX 11.4 SAFE SPACE VERSUS BRAVE SPACE

Safe Space Common Rules	*Brave Space Reframing*
Agree to disagree: Disagreements are welcomed in the college classroom. However, relying on this practice creates a culture of avoidance.	Controversy with civility: Conflict is inevitable and should be expected in heterogenous spaces. By not avoiding conflict, the group comes to a shared understanding on the importance of engaging through conflict.
Don't take things personally: As Arao and Clemons note, this is commonly used alongside "no judgements" and "it's okay to make mistakes." What is missing here is accountability.	Own your intentions and impact: We do our work as faculty with the best of intentions. That does not always mean we can avoid potentially negative impacts of our well-meaning behavior. By understanding that positive intentions =/= positive impact, we can be accountable to our intentions *and* our impacts.
Challenge by choice: Placing the decision to participate the education space is ultimately one that honors the boundaries of a student's comfort zone.	Question your choices: What is absent from challenge by choice is a more concrete externalization of student growth, in terms of their expression of the norms that influence their decisions and their mastery of course material. By encouraging students to question their choices, through activities or dialogue, they can be more participatory in knowledge exchange.

(*Continued*)

(Continued)

Safe Space Common Rules	*Brave Space Reframing*
Respect: Most of us can agree that respect is a basic ground rule. What most of us fail to do is spend time on operationalizing the term and what it means for the classroom environment.	Respect: The goal is not consensus but rather to create a classroom environment where increased attention is given to showing respect for each other and the classroom space. Ask the students, "What does respect look like to you?" and share your own perspective. Invest time in a classroom discussion, which serves to clarify the meaning of respect.
No attacks: Like respect, most of us can agree that attacks of any form are deeply disrespectful and inhibit the process and goal of a socially just classroom.	No attacks: As with respect, engaging in discussion about this ground rule is urgent and necessary. Having students differentiate between personal attacks and challenges to belief systems and preexisting ideas that cause discomfort is an important process for improving the classroom environment.

Arao and Clemens (2013).

Creating a classroom environment that functions as a brave space requires the development of guidelines, which should be completed collaboratively with students in the early days of the term. These guidelines can be created in person with index cards, poster paper, or shared documents, or in the virtual classroom through chat and whiteboard functions or discussion boards in the learning management system. Regardless of the method for creating guidelines, once they are decided upon by the classroom community, they should maintain a level of visibility throughout the term, whether as a pinned announcement in the learning management system or displayed in the classroom regularly.

In educational settings, a brave space is a classroom or discussion group that encourages open and honest dialogue, even on sensitive topics, while fostering mutual respect and understanding. It acknowledges the inherent challenges of discussing such topics and aims to create an environment where students feel valued in providing their perspectives and in the ability to express themselves fully. Arao and Clemens (2013) argue that "facilitators of social justice education have a responsibility to foster a learning environment that supports participants in the challenging work of authentic engagement with regard to issues of identity, oppression, power, and privilege" (p. 139).

Conclusion

It is important to remember that creating a socially just classroom is a process and goal. It means that you are committed to building a learning community that emphasizes setting ground rules, where you work with students to collaboratively establish guidelines for sharing the classroom space, in whatever form it takes. It means that your classroom resists avoidant behavior by developing a practice of facilitating difficult conversations. It means you make an active choice to engage in a regular practice of reflexivity and consistently examine the power dynamics at play in the practice of teaching.

Alderman et al. (2021) offer the possibilities that result from transformative educational transformation: "With a pedagogy of anti-racism comes opportunities to (1) ensure that our geography students become ethical and responsible citizens, (2) demonstrate the saliency of a geographic perspective to real-world issues and problems, (3) build solidarity with marginalized groups made even more peripheral by ongoing political changes, and (4) enhance the desirability of attracting a more diverse student body to the discipline" (p. 188). These same opportunities can be applied to us as teachers and educators. By adopting socially just pedagogical practices, we evolve into accountable and ethical educators who create respectful and supportive classroom environments. Our education practices serve as professional development opportunities, where challenges in the field, service, and research activities can mirror the challenges of society (Dill and Zambrana 2009), which are replicated in our classrooms. This interconnectedness opens opportunities for students to share their experiences. Importantly, it also means we encourage the use of the local community as an extension of our classrooms and a means to discuss the implications of the built and natural environment. Extending the previous statement, using the local community is key to building solidarities in exchanging knowledge and deciding who is included and how in the pedagogical process. Finally, by engaging in social justice practices, we develop a reputation for transformative learning environments that encourage a sense of place and belonging for students.

We applaud you for taking the first step in transforming your pedagogical practice and look forward to sharing this continuous learning journey (goal and process) in solidarity with you.

References

Adams, M. (2007). Pedagogical frameworks for social justice education. In M. Adams, L. A. Bell, & P. Griffin (Eds.), *Teaching for Diversity and Social Justice* (pp. 15–34), (2nd ed.). New York: Routledge.

Alderman, D., Narro Perez, R., Eaves, L. E., Klein, P., & Muñoz, S. (2021). Reflections on Operationalizing an Anti-racism Pedagogy: Teaching as Regional Storytelling. *Journal of Geography in Higher Education*, *45*(2), 186–200. https://doi.org/10.1080/03098265.2019.1661367

Ambrose, S. A., Bridges, M. W., DiPietro, Michele., Lovett, Marsha., Norman, M. K., & Mayer, R. E. (2010). In. Landreman, L. M. (Ed.). *How Learning Works: Seven Research-Based Principles for Smart Teaching*. (1st ed.). San Francisco: John Wiley & Sons, Incorporated.

Arao, B., & Clemens, K. (2013). From Safe Spaces to Brave Spaces: A New Way to Frame Dialogue Around Diversity and Social Justice. In *The Art of Effective Facilitation* (1st ed., pp. 135–150). New York: Routledge. https://doi.org/10.4324/9781003447580-11

Cochrane, S. V., Chhabra, M., Jones, M. A., & Spragg, D. (2017). Introduction: Cultural Literacy, Cultural Humility, and Reflection. In Cochrane, S. V., Chhabra, M., Jones, M. A., & Spragg, D. (Eds.). *Culturally Responsive Teaching and Reflection in Higher Education: Promising Practices from the Cultural Literacy Curriculum Institute* (1st ed.). New York: Routledge.

Collins, J. D., Staten, E., Long, S., Koeckes, A., Pham, G., Rochon, A., Dougherty, C., & Cho, A. R. (2021). *Anti-Racism Terminology Guide. SLP-CLDE KC, NASPA*. Retrieved from: https://www.naspa.org/files/dmfile/SLP-KC-AAP-2020-2021_-SLP-CLDE-KC-Collins_FINAL-SLP-CLDE-KC-Terminology-Guide.pdf

Dill, B. T. and Zambrana, R. E. 2009. Critical Thinking about Inequality: An Emerging Lens. In Dill, B. T. and Zambrana, R. E. (Eds.) *Emerging Intersections: Race, Class, and Gender in Theory, Policy, and Practice* (pp. 1–21). New Brunswick: Rutgers University Press.

Garibay, J. C. (2014). Diversity in the classroom. Retrieved from: https://equity.ucla.edu/wp-content/uploads/2016/06/DiversityintheClassroom2014Web.pdf

Hackman, H. W. (2005) Five Essential Components for Social Justice Education. *Equity & Excellence in Education*, *38*(2), 103–109, DOI: 10.1080/10665680590935034

Hernández, M. Á. 2018. More Than *Just* Words: Transforming Leadership Education Toward Liberty and Justice for All. In Guthrie, K. L. and Vivechkanand, C. S. (Eds.) *Changing the Narrative: Socially Just Leadership Education*. Charlotte, NC: Information Age Publishing, Inc.

hooks, b. (1994). *Teaching to Transgress: Education as the Practice of Freedom*. New York: Routledge.

hooks, b. (2010) *Teaching Critical Thinking: Practical Wisdom*. New York: Routledge.

Kinkaid, E., & Fritzsche, L. (2022). The Stories We Tell: Challenging Exclusionary Histories of Geography in U.S. Graduate Curriculum. *Annals of the American Association of Geographers*, *112*(8), 2469–2485. https://doi.org/10.1080/24694452.2022.2072805).

Kobayashi, A. 2020. Situated Knowledge, Reflexivity. In Kobayashi, A. (Ed.) *International Encyclopedia of Human Geography* (2nd ed., pp. 213–217). Amsterdam: Elsevier.

Ladson-Billings G. (2005). Culturally Relevant Teaching: The Key to Making Multicultural Education Work. In Grant C. A. (Ed.) *Research and Multicultural Education: From the Margins to the Mainstream* (pp. 102–118). Bristol, PA: Falmer Press.

McNair, T. B., Bensimon, E. M., & Malcom-Piqueux, L. E. (2020). *From Equity Talk to Equity Walk: Expanding Practitioner Knowledge for Racial Justice in Higher Education*. Hoboken, New Jersey: John Wiley & Sons, Incorporated.

NEA Center for Social Justice. 2021. Racial Justice in Education: Key Terms and Definitions. Retrieved from https://www.nea.org/professional-excellence/student-engagement/tools-tips/racial-justice-education-key-terms-and-definitions

Niemann, Y. F. (2011). Diffusing the Impact of Tokenism on Faculty of Color. *To Improve the Academy: A Journal of Educational Development*, *30*, 216–229. http://dx.doi.org/10.3998/tia.17063888.0030.020

Niemann, Y. F. (2016) The Social Ecology of Tokenism in Higher Education. *Peace Review*, *28*(4), 451–458.

Pappamihiel, N. E., & Moreno, M. (2011). Retaining Latino Students: Culturally Responsive Instruction in Colleges and Universities. *Journal of Hispanic Higher Education*, *10*(4), 331–344. https://doi-org.utk.idm.oclc.org/10.1177/1538192711410602

Sathy, V. and K. A. Hogan. n.d. "How to Make Your Teaching More Inclusive." The Chronicle of Higher Education. Retrieved from: https://www.chronicle.com/article/how-to-make-your-teaching-more-inclusive/?sra=true

Thayer, N. E. (2023). *Caring and Uncaring: White Supremacy, Diversity Work and Antiracist Pedagogies*. PhD Dissertation. University of Delaware.

University of Tennessee, Knoxville's Student Disability Services. n.d. "Instructor Information Guide." Retrieved from https://sds.utk.edu/wp-content/uploads/sites/88/2018/07/DisabilityServicesFolderOnlinev1.3-377322.pdf

12

TEACHING IN THE FIELD

Douglas W. Gamble, Peter J. Haproff and Chen Wu

One of the most rewarding and enjoyable aspects of teaching in geography and geoscience is field instruction. Designing and implementing field experiences can be both an exciting and overwhelming prospect. The reality is that field instruction requires diligent preparation, both in and outside the classroom, if it is to provide successful and meaningful learning experiences. Many geographers and geoscientists are committed to experiential, field-based learning and have written about both the practical and theoretical aspects of successfully teaching in the field. This chapter offers a summary of this helpful literature, while also discussing some of the authors' experiences of "lessons learned" from teaching in the field, and the impact of the recent COVID-19 pandemic on field instruction.

Field study in geography and geoscience

Field study has a long tradition as a core learning experience in the geography and geoscience curriculum*. At the most fundamental level, field instruction represents the first two phases of learning presented by Kolb (1981): do/feel and observe/record/reflect. If properly designed, field experiences offer students the opportunity to engage in the experiential learning cycle of concrete experience, reflection, formation of abstract concepts, and active experimentation. Generally, in geography and geoscience, this involves engaging with the field, making observations, generating ideas and hypotheses based on observation and academic theory, and then further engaging with the field to test ideas generated (Marvell et al. 2013; Fuller 2012; Chang and Taipei 2002; Compton 1985). For example, geoscience education involves the study of natural processes that have and continue to shape the landscape and internal structure of Earth and other planetary bodies. Thus, to fully understand the evolution of Earth and other planetary

DOI: 10.4324/9781003342953-14

bodies, students must directly observe and record geologic processes and their associated features in the field where they occur (Compton 1985). Opportunities to examine geologic/geomorphic features and/or processes take the form of field trips, which can be either short-term trips to local sites or long-term trips to potentially distant sites. Students on field trips can make their own observations, collect data, and apply the scientific method the same way researchers do. Field trips provide an unparalleled opportunity for hands-on, interactive learning, and can serve as the capstone experience wherein students apply knowledge, skills, and experiences from their previous coursework.

The advantages of field instruction as an active learning technique are that it increases student satisfaction and enjoyment, provides real-life examples of features and processes presented in a lecture, allows for application of material and collaborative problem-solving, increases knowledge retention, and creates a common frame of reference and observational base that can be built upon throughout a course or degree program (Compiani and Carneiro 1996; Frodeman 1996; Elkins and Elkins 2007; Fuller et al. 2000, 2006; Lonergan and Andresen 1988; Petcovic et al. 2009; Stokes and Boyle 2009; Thompson 1974, 1982). In short, more holistic learning and understanding of geography and geoscience can be achieved through field experiences. Specific student learning outcomes facilitated through fieldwork include the development of observational skills and respect for the environment, encouragement of students to take responsibility for their own learning, and exposure to "real" research techniques (Kozar and Marcketti 2008; Stokes et al. 2011; Wilson et al. 2017; Bednarz et al. 2013). In addition, students' presence in the field often leads to "unstructured" experiences, which can stimulate discovery and allow students to engage with relevant topics beyond the scope of the scheduled topic for the day, yet pertinent to the course objectives (Gibbes and Skop 2022; Tuan 2001; Salter 2001). Lastly, it should also not be overlooked that fieldwork often attracts students and is an effective recruiting tool for geography and geoscience programs (Morris et al. 2018).

Field instruction and experiences come in many forms, which can make it challenging to describe the different types of field experiences. But a fundamental difference in types of field experiences is duration, which limits the student learning outcomes and specific skills that can be taught or assessed. In addition, the complexity of pre-field trip preparation and logistical planning also aligns with the duration of the field experience. Short-term field trips typically occur over one class period or day, during which students are exposed to features, processes, and/or concepts discussed in a classroom setting. During short-term field trips, students typically make observations, collect data, and/or conduct descriptive analyses, which can be associated with laboratory assignments and/or small projects. For example, geoscience students can directly observe and measure an in-situ rock outcrop or fault zone that was previously introduced in a classroom setting via static images (see Tables 12.1 and 12.2 for more examples). Mid-length field experiences usually last several

TABLE 12.1 Examples of Short-to-Long Field Experiences in Physical Geography and Geoscience

Time Frame	*Description of Activities*	*Student Learning Outcome (SLO)*
Short length (single-day or weekend)	Groups of students observe and measure in-situ rock outcrops and geologic structures (e.g., fault zones, folds) at one or several local sites.	Learn to identify, describe, and measure Earth materials and 3-D geologic structures in the field, and understand their formation mechanism processes.
	Measure temperature in two thermally contrasting locations.	Observe and record air temperature using a thermometer; differentiate between different thermal environments based upon air temperature observations.
Medium length (several days to one week)	Individual or small groups of students perform daily geologic mapping and data collection across a large-scale study site.	Learn to locate oneself on a map; become familiar with common geologic data collection tools and techniques; construct a viable geologic map and cross section; use the scientific method to analyze and interpret large data sets and solve geologic questions regarding the study site.
	Map temperature at student homes and on campus.	Observe, record, and construct an air temperature data set for two different locations; identify patterns in time series of air temperature data; create a map that displays air temperature at multiple locations.
Long-term capstone course or field program (several weeks)	Field camp/course wherein individual or small groups of students perform daily geologic mapping and data collection across one or more large-scale study sites.	Master skills related to locating oneself on a map, identifying and describing Earth materials and geologic structures, and using common data collection tools and techniques; construct one or more professional-quality geologic maps and cross sections; read and critically review relevant scientific literature; use the scientific method to analyze and interpret large, new, and existing data sets and solve geologic questions regarding one or more study sites.
	Semester-long assessment of urban heat island.	Middle-length SLOs + create a map that displays air temperature at multiple locations; interpret and explain differences in air temperature patterns based upon location in a rural or urban environment.

TABLE 12.2 Examples of Short-to-Long Field Experiences in Human Geography and Related Fields

Time Frame	*Description of Activities*	*Student Learning Outcome (SLO)*
Short length (single-day or weekend)	Urban Land Use Survey. Students identify and inventory land use in the neighborhood or town where they live.	Students gain experience identifying, describing, and categorizing land use into residential, commercial, industrial, recreational; estimating the intensity of use; and mapping the spatial distribution and density of each type.
	Food Access Study: Ask students to study the local food system: where do people get their food; how accessible are these sources; and what types of foods are available.	Students can map the grocery and convenience stores, restaurants, farmers markets, and other food sources where they live. They can focus on the types of foods available (fresh, processed, etc.), price, quality, and accessibility.
Medium length (several days to one week)	Cultural Heritage Conservation Project: Students identify sites of cultural, historical, and social value that are threatened by neglect or redevelopment.	Students can spend several days working together to map cultural heritage sites, document their histories and architectural features, and estimate the threats to their preservation. Local stakeholders might be involved to outline the conservation plans and strategies they might be developing for the management of this heritage.
	Natural Hazards Vulnerability Assessment: Assign students to study the vulnerability of their community to a natural hazard that occurs in their region such as floods, earthquakes, wildfires, tornadoes, or hurricanes.	Students can spend several days surveying the physical landscape, infrastructure, and population distribution, and interviewing residents and local officials about their experiences with past disasters. Depending on the scope of the project, students can analyze the factors contributing to vulnerability, assess preparedness and mitigation efforts, and recommend strategies for enhancing resilience.

(Continued)

TABLE 12.2 (Continued)

Time Frame	*Description of Activities*	*Student Learning Outcome (SLO)*
Long-term capstone course or field program (several weeks)	Social Justice and Inequality Assessment: Task students with investigating social justice issues and inequalities within a specific urban or rural context.	This could involve studying housing disparities, access to education, healthcare, and employment opportunities. Students can analyze how these factors contribute to spatial and social inequities. In preparation, students can read, discuss, and critically review relevant literature.
	Climate Change Impact Assessment Using Participatory GIS (PGIS): Assign students to study a region's vulnerability to climate change. They can assess factors such as sea-level rise, extreme weather events, decreasing water supplied, and changing temperatures, and analyze how communities are adapting or mitigating the impacts.	Students can learn to use PGIS to involve community members in mapping spaces and places that are vulnerable to the effects of climate change. Students work in collaboration with local residents to identify and estimate threats. Students can analyze the results in terms of community perspectives and spatial patterns.

days to two weeks and are designed to teach multiple field skills and fulfill several student learning outcomes. During these field experiences, students learn via an immersive experience where they directly observe and touch the landscape. A common example of such a field experience is a "destination" experience wherein a main goal is for the students to experience and immerse themselves in a new and unique landscape. Examples include "The Geography of the Outer Banks, NC," or "The Geology of Death Valley, CA."

Long-term, capstone field camps and courses typically involve several weeks of fieldwork in one or multiple locations and often serve as the capstone course for an undergraduate program (Mayer 1970). Mastering of fields skills is the main student learning outcome, and the final product is a large-scale project that requires students to complete independent field observations, data collection, and analysis and interpretation. The key distinction from short-term field trips is the expectation of students to use acquired skills, experiences, and knowledge from all previous courses to complete projects and ensure success. Students use skills including reading comprehension, verbal communication, critical thinking, and a developed method. For example, students will follow the scientific method by making hypotheses and testing predictions during each day and throughout the entire trip. In addition, a field camp or course can investigate previous findings in academic literature and test published theories or models in the field, thereby learning the research process through firsthand experience. A key component in the design of such a field experience is determining the amount of pre-trip knowledge and preparation provided to students. Generally, students with little field experience require more pre-field knowledge and preparation to achieve a learning outcome and have a meaningful educational experience. Students who are further along in their academic career typically require less pre-field trip knowledge and preparation. For example, more experienced students may be assigned a "site-unseen" field exercise, wherein students work in a field area without prior knowledge of the location and use all fundamental skills acquired from previous coursework and experiences. In addition, student success is hinged on the constant development and testing of hypotheses through field observations. Such a field experience allows an instructor to assess students' knowledge, skill, and analytical abilities, and determine how successful an academic program is in imparting knowledge and skills to students.

The recent expansion of the use of digital instruction, due to technological advances and adaptations during the COVID-19 pandemic, has started a discussion in geography and geoscience of how to integrate digital instruction into field experiences. This discussion has led to the creation of new labels for field experiences such as "hybrid," "online," "remote," or "virtual" field teaching (Walshe and Healy 2017). We will discuss this case in more depth later in this chapter, but for now, we would like to note that digital instruction should be considered as a new, more inclusive avenue for information delivery and collaboration in field experiences, which can result in better experiences and involvement of a wider student body. However, we stress that field experiences must include a significant, traditional field component and cannot be fully replaced with online teaching.

Beyond the structure of a field experience, the structure of an academic program and how field experiences fit into the academic curriculum are important considerations for instructors as they design field experiences. If the learning outcomes of a given academic program require students to be proficient in field inquiry and research, students must know not only how to observe and collect field data but also how to design and manage field research. Such skills and proficiencies must be taught across a series of courses, typically beginning with an introduction to basic observations and interpretations in lower-level courses, followed with discussions of more specific techniques of collecting observations and data and technical communication in upper-level courses. As a final step in the academic program, students complete a capstone course/experience during which students implement and are assessed on the design and completion of an independent field research project.

The student learning outcomes, structures, and challenges of field experiences are comparable for geography and geoscience programs across the world (Yan et al. 2009; Wu et al. 2017; Wilson et al. 2017; Scott et al. 2012; Elkins and Elkins 2007). For example, many geoscience programs in the People's Republic of China require undergraduate and graduate students to participate in both short- and mid-length field trips as part of their coursework (Yan et al. 2009; Wu et al. 2017). The purpose and activities of field experiences in China include learning to identify, describe, and measure Earth materials and structures, often through geologic mapping, and developing a better understanding of rock formation and deformation processes. A key universal goal of field experiences in China is students' use of the scientific method to analyze and interpret their own observations and solve geologic questions. However, the continued use of field studies in China faces similar challenges as other geoscience programs, including the decreased use and focus of field trips in curricula over the last few decades. This issue in China is particularly evident regarding the long-term field camp/course model, as only few geoscience programs still offer capstone field experiences during which students use the skills, experiences, and knowledge acquired from previous courses. For geoscience programs that continue to offer the capstone field experience, the most effective field camp or course model involves several weeks of dedicated study within a large-scale area, such as geologic mapping within the well-known Zhoukoudian area of Beijing and Beidaihe area of Heibei Province (Wei et al. 2019; Cao et al. 2015). The success of the capstone field camp/course within the Zhoukoudian and Beidaihe areas can partially be attributed to the joint management and student involvement of the Beijing and Wuhan campuses of the China University of Geosciences over the last three decades (Wei et al. 2019; Cao et al. 2015). Whereas the current biggest obstacle to offering student field experiences in China is the reduced focus of field trips in the curriculum, the recent advancements in technology have created new opportunities to complement traditional field studies, including the use of "remote" or "virtual" field trips, mapping via remotely sensed imagery, and training in mapping software.

Advice for field experience design and instruction

Once the duration, the student learning outcomes, and appropriate placement within an academic program have been determined for a field experience, several important factors cannot be overlooked. Based upon our combined 35+ years of field instruction experience, we offer the following advice for designing field experiences.

1 *Pre- and Post-Trip Activities*: Do not forget the time before and after venturing into the field. As a rule of thumb, instructors must schedule, at the least, an equal amount of time for both pre-field preparation/post-field reflection

and actual field instruction. The quality of the student learning experience is just as dependent upon this preparation and reflection as it is on the time spent in the field. Proper preparation and meaningful analysis/reflection are often critical to meeting the student learning outcomes of a given field experience. As a result of the COVID-19 pandemic, many instructors have determined that online preparation is effective for pre-field preparation and increases access for students who are employed and may have limited time to participate in a field experience (Day et al. 2023). In addition, while on an extended field trip, unstructured "free" time is necessary to maintain group morale. This free time can occur as breaks during fieldwork, or in evenings following fieldwork. Such time allows the students to recharge, put the day in perspective, socialize, and monitor themselves mentally and physically. In addition, the free time allows the instructor some time to rest and disengage from instructor duties.

2 *Beware of the "exotic landscape syndrome"*: Textbook and their supplementary teaching resources often offer idealized examples of landscapes that can be quite useful in illustrating concepts (i.e., the Himalaya as an orogen developed via continental collision, New York City as an urban environment, or the Outer Banks of North Carolina as barrier islands). However, the use of these exemplar or "exotic" landscapes can create an under-appreciation of the local landscape. There are plenty of examples within the local landscape or community that represent sites described in course material just as effectively as publisher provided graphics. An added benefit of such local field studies is that they can provide data that can be used by local governments or university officials to develop plans and policies, thereby creating an opportunity for students to apply classroom material to real-world problems.
3 *Equipment and Materials, Safety, Health, and Inclusion*: The equipment and materials required for fieldwork can be quite extensive. Ensuring that all equipment and materials are available and accessible is key to fostering a positive and safe learning experience for all participants. For example, a multi-week capstone field camp/course requires students to have appropriate outdoor clothing, hiking boots, sufficient water, a Brunton compass, mapping and drafting supplies, field notebook(s), and a first aid kit. Making sure that each participant has these items is imperative for not only their learning but also their safety during the experience. Other equipment and materials that may be shared by the class include global positioning system units, satellite/aerial imagery, and a personal computer for report writing. Furthermore, camping equipment as well as four-wheel-drive vehicles, vehicle recovery gear, and extra spare tires to access remote locations may be required. This example clearly illustrates the need to be thorough and meticulous in planning for a field camp/course, as proper equipment and materials are key to the success of a field experience.

4 Safety for students and instructors is paramount in the field. Consequently, personnel and students must be aware of the environmental conditions they may be exposed to in the field so that they can prepare adequately. Necessary components include access to ample clean water, first aid supplies, and a means to getting to first responders and/or hospital. Walkie talkies and satellite phones are helpful for remote and large areas. Beyond traditional physical protection concerns, safety issues related to race, gender, religion, and sexuality must also be considered. Without consideration of these issues, students may feel unsafe, which diminishes their ability to engage, participate, and learn. A proactive field strategy should be developed and may include mandatory risk assessment of field experiences, anti-discrimination training, encouragement of bystander interventions and allyship training, reaching out to local authorities before the field experience, identifying and sharing cultural norms of field areas, and documenting hostile encounters (Ackerman et al. 2023; Anadu et al. 2020).
5 Beyond safety, key forms of inclusivity during field trips include accommodating the time, physical, and dietary needs of all students and personnel (Emery et al. 2020). We must also provide equitable field experiences to differently abled persons, which may entail travel to accessible locations with infrastructure in compliance with the Americans with Disabilities Act. Documents such as digital maps can be provided to students with eyesight disabilities (Carabajal et al. 2017). Access to remote field sites has traditionally been an issue for inclusiveness, so providing a means for field trip transportation and equipment for students who otherwise may not have the ability to access these resources is necessary. Furthermore, inclusivity should not be associated with access for differently abled students only. Reduced costs and financial aid are important considerations for providing field access to a broad range of students. An instructor must ask themself, does the time length of a field experience exclude some students due to their work or family commitments? Are alternative experiences provided? In this case, instructors may consider using a series of shorter field trips that cumulatively equal the time spent on longer field experiences. Such student needs must be accommodated for the field experience to be effective to everyone in an academic program.

Impact of COVID-19 on field studies

The COVID-19 pandemic caused many academic programs and instructors to modify existing, time-tested types and modes of field instruction, including developing alternatives to face-to-face instruction and/or travel (Bursztyn et al. 2022; Gregory et al. 2022; Rader et al. 2021). Most instructors offered some semblance of a traditional field experience, typically through fully online options, either synchronous or asynchronous, although student learning varied. Day et al. (2021, 1) characterized the instructional pivot caused by the COVID-19 pandemic as "the largest unplanned educational experiment ever undertaken." Many researchers and instructors have documented how the pandemic and resulting lockdown impacted physical science education, including field education, and how we have learned key lessons from this curricular pivot (Day et al. 2021, 2023; Gibbes and Skop 2022). Positive lessons learned include students' value of continued course offerings during times of disruption, such as asynchronous course formats that allow for greater student participation (Day et al. 2023; Gibbes and Skop 2022). However, the pivot to online instruction showcased students' weaker abilities to critically reflect on learned topics and synthesize broader ideas, despite being competent in making observations (Gibbes and Skop 2022). Such results have been documented for experiences during non-pandemic times when online experiences substituted traditional field experiences (Kolivras et al. 2012). Thus, the integration of both online learning and field experience is an important lesson in recasting field education from the COVID-19 pandemic (Larsen et al. 2021).

The implementation of face-to-face, hybrid, or fully online teaching formats to supplement traditional field experiences has been a product of not only the COVID-19 pandemic, but also the confluence of other factors. Many school administrators have begun to question the educational benefits of field experiences, arguing that when fieldwork occurs, the experiences tend to be restricted to long-term field camps and courses rather than several short-term trips integrated throughout the program of study (Wilson et al. 2017). Tight university budgets, institutional focus on research, and concerns over litigation, safety, and insurance have also hampered efforts to offer field experiences (Scott et al. 2012; Cook et al. 2006). In addition, untenured faculty who are facing escalating pressure to publish research papers and acquire grant funding have questioned whether they should invest time and resources in a course that may not be rewarded in terms of salary or promotion (Mullens et al. 2012). Furthermore, many academic programs are facing difficulties in offering field experiences due to the burdens these courses put on departmental resources including costs, faculty time, and a need for larger class sizes (Mullens et al. 2012). As a result, the number of field camps/courses offered is declining as degree programs in the United States, Canada, the United Kingdom, Australia, New Zealand, and China remove such courses from their required curriculum (Wilson et al. 2017 and references therein; Scott et al. 2012).

Although it remains easy for an instructor to get caught up in these external factors, critical reflection and decisions on how to offer future field experiences provide an opportunity for instructors to assess which teaching method best serves the student learning outcomes of their courses. A common outcome of such reflection has been shifting to online instruction for pre- and post-field activities, which makes field experiences accessible to a wider student population (Day et al. 2023). This outcome capitalizes on the primary advantage of online learning, specifically the increased access and student-paced engagement with the course content (Appana 2008). Such a hybrid approach to field experiences offers a positive outcome, which is an excellent example of how recent challenges of offering traditional field experiences can lead to the improvement of student learning and success.

Conclusion

Teaching in the field is a rich and rewarding experience for both students and instructors. There are common themes that are salient to all field instruction. These issues include thorough pre-field trip preparation, development of clear learning objectives and themes in field instruction, clear demonstration of tools and techniques, implementation of local field trips, and use of a post-field "debriefings" that links student activities to course objectives, assessment, and expected outcomes. In short, an instructor should view themselves as an enabler of field instruction. Their responsibility is to take care of the logistics, mitigate safety concerns, and clearly define objectives so that students can focus solely on their field experiences. Even with great preparation, however, field experiences will sometimes not go as planned. Successful fieldwork is a "trial-and-error" process that can be impacted by forces beyond our control (e.g., extreme heat, a flat tire, a closed museum or park, or unideal group dynamics). In addition, resistance to continuing traditional field trips and the online pivot resulting from the COVID-19 pandemic have provided an opportunity to rethink how field studies can involve online components to offer a more successful inclusive experience. Beyond development of such "hybrid" experiences, the challenge is to maintain an enthusiasm for and commitment to teaching in the field despite heavy time and effort requirements. When field experiences work, the rewards are unparalleled for both students and instructors.

References

Ackerman, A., K. Yarincik, S. Murphy, I. Cetinić, A. Fundis, A. Miller, E. Shroyer, A. Busse, Q. Covington, A. DeSilva, A. Haupt, L. Johnson, C. Lee, L. Lorenzoni, B. Murphy, J. Ramarui, B. Rosenheim, and D. Steinberg. 2023. Know before you go: A community-derived approach to planning for and preventing sexual harassment at oceanographic field sites. *Oceanography* 36 (1): 38–43. https://doi.org/10.5670/oceanog.2023.112

Anadu, J., H. Ali, and C. Jackson. 2020. Ten steps to protect BIPOC scholars in the field. *EOS* 101. https://doi.org/10.1029/20209EOS150525

Appana, S. 2008. A review of the benefits and limitations of online learning in the context of the student, the instructor, and the tenured faculty. *International Journal of E-learning* 7 (1): 5–22.

Bednarz, S., S. Heffron, and N. Huynh. 2013. *A road map for 21st century geography education: Geography education research*. Washington, DC: American Association of Geographers.

Bursztyn, N., Sajjadi, P., Riegel, H., Huang, J., Wallgrün, J. O., Zhao, J., Masters, B., and Klippel, A. 2022. Virtual strike and dip–advancing inclusive and accessible field geology. Geoscience Communication 5 (1): 29–53.

Cao, X. H., X. W. Wang, and J. P. Lin. 2015. Ponder on field teaching for fundamental geology practice in Beidaihe area. *Chinese Geological Education* 24 (4): 42–45.

Carabajal, I. G., A. M. Marshall, and C. L. Atchison. 2017. A synthesis of instructional strategies in geoscience education literature that address barriers to inclusion for students with disabilities. Journal of Geoscience Education 65 (4): 531–541.

Chang, C. Y., and Y. H. W. Taipei. 2002. An exploratory study on students' problem-solving ability in earth science. *International Journal of Science Education* 24 (5): 441–451.

Compiani, M., and C. D. R. Carneiro. 1996. The didactic roles played by geological excursions. In *Geoscience education and training: In schools, universities, for industry and public awareness*, ed. D. A. V. Stow and G. H. J. McCall, 233–242. Rotterdam: A.A. Balkema.

Compton, R. R. 1985. Geology *in the field*. New York: Wiley.

Cook, V., D. Phillips, and J. Holden. 2006. Fieldwork in a 'risk society'. *Area* 38: 413–420.

Day, T., I. C. C. Chang, C. K. L. Chung, W. Doolittle, J. Housel, and P. McDaniel. 2021. The immediate impact of COVID-19 on postsecondary teaching and learning. *The Professional Geographer* 73 (1): 1–13. https://doi.org/10.1080/00330124.2020.1823864

Day, T., C. Chung, W. Doolittle, J. Housel, and P. McDaniel. 2023. Beyond COVID chaos: What postsecondary educators learned from the online pivot. *The Professional Geographer* 75 (1): 14–30.

Elkins, J. T., and N. M. L. Elkins. 2007. Teaching geology in the field: significant geoscience concept gains in entirely field-based introductory geoscience courses. *Journal of Geoscience Education* 55: 126–132.

Emery, N. C., E. K. Bledsoe, A. O. Hasley, and C. D. Eaton. 2020. Cultivating inclusive instructional and research environments in ecology and evolutionary science. *Ecology and Evolution* 11: 1480–1491. https://doi.org/10.1002/ece3.7062

Frodeman, R. L. 1996. Envisioning the outcrop. *Journal of Geoscience Education* 44 (4): 417–427.

Fuller, I. C. 2012. Taking students outdoors to learn in high places. *Area* 44 (1): 7–13.

Fuller, I., S. Rawlinson, and R. Bevan. 2000. Evaluation of student learning experiences in physical geography fieldwork: paddling or pedagogy? *Journal of Geography in Higher Education* 24: 199–215.

Fuller, I., S. Edmondson, D. France, D. Higgitt, and I. Ratinen. 2006. International perspectives on the effectiveness of geography fieldwork for learning. *Journal of Geography in Higher Education* 30: 89–101.

Gibbes, C., and E. Skop. 2022. Disruption, discovery, and field courses: A case study of student engagement during a global pandemic. *The Professional Geographer* 74 (1): 31–40.

Gregory, D. D., H. E. Tomes, S. L. Panasiuk, and A. J. Andersen. 2022. Building an online field course using digital and physical tools including VR field sites and virtual core logging. *Journal of Geoscience Education* 70 (1): 85–100.

Kolb, D. A. 1981. Learning styles and disciplinary differences. In *The American College*, ed. A. W. Chickering, 232–255. San Francisco, CA: Jossey-Bass.

Kolivras, K. N., C. R. Luebbering, L. M. Resler. 2012. Evaluating differences in landscape interpretation between webcam and filed-based experiences. *Journal of Geography in Higher Education* 36 (2): 277–291.

Kozar, J. M., and S. B. Marcketti. 2008. Utilizing field-based instruction as an effective teaching strategy. *College Student Journal* 4 (2): 305–311.

Larsen, T., L. Tabor, and P. Smith. 2021. End of the field? Hacking online and hybrid environments for field-based learning in geography education. *Journal of Geography* 120 (1): 3–20.

Lonergan, N. and L. W. Andresen. 1988. Field-based education: Some theoretical considerations. *Higher Education Research and Development* 7: 63–77.

Marvell, A., D. Simm, R. Schaaf, and R. Harper. 2013. Students as scholars: Evaluating student-led learning and teaching during fieldwork. *Journal of Geography in Higher Education* 37 (4): 547–566.

Mayer, V. J. 1970. A field course for undergraduate earth science education majors. *Journal of Geological Education* 18 (3): 129–131.

Morris, P., C. Gibbes, and S. Jennings. 2018. An examination of student veteran education pathways at an America university. *Journal of Geography in Higher Education* 42 (2): 276–292.

Mullens, J. B., R. Bristow, and P. Cuper. 2012. Examining trends in international study: A survey of faculty-led field courses in American departments of geography. *Journal of Geography in Higher Education* 36: 223–237.

Petcovic, H. L., J. C. Libarkin, and K. M. Baker. 2009. An empirical methodology for investigating geocognition in the field. *Journal of Geoscience Education* 57 (4): 316–328. https://doi.org/10.5408/1.3544284

Rader, E., R. Love, D. Reano, T. A. Dousay, and N. Wingerter. 2021. Pandemic minecrafting: An analysis of the perceptions of and lessons learned from a gamified virtual geology field camp. Geoscience Communication 4 (4): 475–492.

Stokes, A., and A. P. Boyle. 2009. The undergraduate geoscience fieldwork experience: Influencing factors and implications for learning. In *Field geology education: Historical perspectives and modern approaches*, ed. S. J. Whitmeyer, D. W. Mogk, and E. J. Pyle, 291–311. Boulder, CO: Geological Society of America. https://doi.org/10.1130/2009.2461(23)

Stokes, A., K. Magnier, and R. Weaver. 2011. What is the use of fieldwork? Conceptions of students and staff in geography and geology. *Journal of Geography in Higher Education* 35: 121–141.

Salter, C. L. 2001. No bad landscape. *Geographical Review* 91 (1–2): 105–112.

Scott, G. W., R. Goulder, P. Wheeler, L. J. Scott, M. L. Tobin, and S. Marsham. 2012. The value of fieldwork in life and environmental sciences in context of higher education: A case study in learning about biodiversity. *Journal of Science Education and Technology* 21 (1): 11–21. https://doi.org/10.1007/s10956-010-9276-x

Thompson, D. B. 1974. Types of geological fieldwork in relation to the objectives of teaching science. *Geology* 6: 52–61.

Thompson, D. B. 1982. On discerning the purposes of geological fieldwork. *Geology Teaching* 7: 59–65.

Tuan, Y. 2001. Life is a field trip. *Geographical Review* 91 (1–2): 41–45. https://doi.org/10.1111/j.1931-0846.2001.tb00456.x

Walshe, N., and G. Healy. 2017. *Geography education in the digital world: Linking theory and practice*. Milton Park: Taylor and Francis.

Wei, Y. S., J. P. Chen, G. H. Wang, D. B. Xu, and S. Li. 2019. Simulation of Geological survey field trip in Zhoukoudian of Beijing. *Chinese Geological Education* 28 (2): 81–84.

Wilson, H., J. Leydon, and J. Wincentak. 2017. Fieldwork in geography education: Defining or declining? The state of fieldwork in Canadian undergraduate geography programs. *Journal of Geography in Higher Education* 41 (1): 594–105.

Wu, C., G. S. Wang, Z. G. Zhou, C. F. Liu, W. C. Liu, and T. Jiang. 2017. Contrasts and enlightenments from field capacity trained of geological students by regional geological mapping between China and America. *Chinese Geological Education* 26 (4): 94–97.

Yan, D. P., W. C. Zhang, G. H. Wang, G. C, Zhao, D. B. Xu, W. J. Zhang, and P. Robinson. 2009. Construction of geological practice teaching system at variable level of the undergraduate stage. *Chinese Geological Education* (1): 65–68.

13

INTEGRATING WEB MAPPING AND GEOSPATIAL TECHNOLOGIES

Joseph J. Kerski

Introduction

Education, society, and technology are simultaneously and rapidly evolving. Which tools, methods, and approaches should instructors choose to foster student engagement and equip students to be lifelong learners? In 2004, the US Department of Labor described three top high-growth technologies on which to build careers: nanotechnologies, biotechnologies, and geotechnologies, or "geospatial technologies." Geospatial technologies are the focus of this chapter: web mapping, geographic information systems (GIS), remote sensing (examining the world from space, aircraft, or drones (UAVs, or unmanned/uncrewed aerial vehicles)), and GPS (global positioning systems or GNSS (global navigation satellite systems)). Teaching with geospatial technologies enables students to think spatially and critically, examining the "whys of where" as they examine Earth changes across space and time. Geospatial tools, data, and methods are applicable far beyond their home disciplines of environmental studies, geography, and geographic information science: sociology, civil engineering, economics, data science, language arts, mathematics, history, health, criminal justice, others. Why? Because all disciplines are concerned with place, space, and change.

Incorporating geospatial technologies in classrooms, courses, and programs provides unique and engaging opportunities for students to grapple with key 21st-century issues, from local to global scales. These issues, including all UN Sustainable Development Goals, include equity, water, natural hazards resiliency, smart city planning, sustainable tourism and agriculture, energy, and education. This chapter evaluates specific geospatial tools, data sets, and web mapping applications. It demonstrates how to implement these

DOI: 10.4324/9781003342953-15

powerful and accessible tools and their associated rich data sets for meaningful problem-solving. It shows how these tools can empower students to be positive change agents of tomorrow.

Common goals

The use of geospatial technologies connects to instructional, institutional, and personal goals that all faculty hold dear. Instructors seek meaningful engagement with students. They want students to care not only about course content but even about their own *learning*. When using geospatial technologies, students use tools that climatologists, wildlife biologists, or city planners use in their day-to-day work. Students aren't just learning *about* urban greenways, natural hazards, soil erosion, or habitat loss. They emulate what people in the workplace are doing with these same tools. They ask questions, examine real issues affecting real places, deal with uncertainty, propose solutions, and communicate results.

Every instructor also wants to contribute to their educational institution's goals. These institutions seek graduates who will make positive changes happen in nonprofit organizations, government agencies, academia, and private industry. Geospatial technology is used daily for positive purposes by thousands of organizations at all levels: global (World Health Organization, Nature Conservancy), national (NOAA, US Health and Human Services), and local (parks and recreation and water resources departments). Using geospatial technologies is also key to every educational institution's resiliency, equity, and sustainability plan. This same technology is used by campus facility managers, planners, and recruitment officers to manage electrical and water systems, and to make the campus safer, accessible, equitable, and energy efficient.

Every faculty member wants to grow personally and professionally. Geospatial technologies have evolved into a web-based SaaS (Software as a Service) model, opening it to students, faculty, researchers, and practitioners who can use it and communicate with it via maps, dashboards, story maps, infographics, and other multimedia. As geospatial technologies evolve, they represent "lifelong learning" for faculty – always offering something new to discover. Faculty use it as an instructional tool *and* as a tool for enhancing research that is place-based and that benefits from spatial analysis. Moreover, faculty shape the future of geospatial technologies as they apply it in new ways. Every faculty member also wants to feel that they are making a positive contribution to society. Geospatial technologies are rapidly becoming the "language of the planet" – bringing people together around common goals.

What is modern geospatial technology?

Geospatial technologies gather, represent, visualize, analyze, and communicate place-based information tied to specific locations over, under, and on lands and

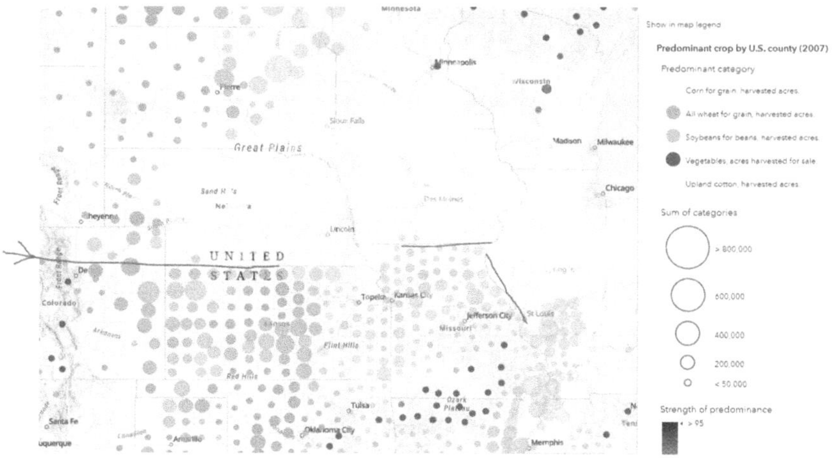

FIGURE 13.1 Mapping crop types and amount in the USA using geospatial technologies.

Source: US Department of Agriculture data mapped by Joseph Kerski using Esri ArcGIS Online software.

oceans. Spatial data layers are represented as vectors (points, lines, and polygons), rasters (grids and images), and themes (zoning, land use, vegetation, water, demographics, others). Layers are joined in geographic space in a relational, topologically connected set of geodatabases. The power of geospatial technology derives from the ability of the analyst, a human decision-maker, to "see across and through" these layers to holistically understand a place (Figure 13.1) as it exists, as it was, and how it could be in the future. Geospatial technologies include a powerful set of analysis tools – statistics, overlay, proximity, routing, and others. Data are represented as understandable 2D and 3D maps. Web mapping applications communicate visualization and analysis as actionable results, via dashboards, story maps, and infographics. Because geospatial technologies are web-enabled, their maps can be customized, shared, and embedded in other applications, greatly expanding their influence to help people make wiser decisions.

Over its 60-year history, geospatial technologies have endured numerous evolutionary changes in information technology (IT), from mainframes, microcomputers, personal computers, to web and mobile technologies (Sinton and Kerski 2020). Geotechnologies have not only *endured* as a persistent set of problem-solving methodologies through these changes, but they have *thrived* through these changes. Each change brought new capabilities and applications. As it migrated to the web, geospatial is connected with mainstream IT (Information Technology) trends: cloud computing, machine learning, distributed architecture. The web hastened the adoption of geospatial technologies through an increasing diversity of disciplines and users, inside and outside of academia.

The advent of geospatial technologies in education

As geospatial technologies evolved, societal expectations placed on the value and outcomes of education also changed. In an environment where employable graduates are in demand, community, tribal, and technical colleges, with their focus on workforce development, have embraced geospatial technologies as providing students with job skills. Web GIS led to an increased demand for graduates with coding and computer systems skills. Geospatial technologies are seen as a key component of university innovation in data science, geodesign, sustainability, and other new fields. The COVID crisis made clear that interactive maps, dashboards, and spatial analysis are extensively relied upon in health and other fields. As geospatial technologies have become key for ethical and socially responsible operations of modern businesses and to resiliency initiatives in tourism, transportation, energy, and agriculture, it is used as an instructional tool and method. It is also used as a research data source and as a tool for teaching racial and social equity themes.

Simultaneously, in primary and secondary institutions, geospatial technologies are expanding. They can be taught as a stand-alone course, substituted for advanced geography or coupled with a computer and technology education (CTE) focus. More frequently, it is viewed as an instructional tool for social studies, biological and earth sciences, geography, history, and language arts. The analytical, problem-solving, critical-thinking aspects of GIS are considered valuable for students of all ages (Kerski 2003; Jo, Hong and Verma 2016). Primary and secondary educators typically must teach according to national educational content standards. Educators see that the tools with which they can teach core *concepts* – such as change, scale, or ratios – or core *topics* – such as historical events or biomes – are tied to mandated content standards. *Learning to Think Spatially* from the National Academies of Sciences (2006) supports the use of geotechnologies in schools, as does research showing its effectiveness in learning (Schulze 2020). Students using geotechnologies in schools bring new skills and perspectives to the university upon their arrival.

Why teach with geospatial technologies?

As SaaS architecture took hold in the 2010s, cloud-based geotechnologies enabled key issues of the 21st-century world – climate, hazards, social, economic and racial equity, sustainable agriculture, health, water quality and quantity, energy, public safety, supply chain management – to be taught *alongside* their spatial characteristics. These complex, serious problems from global to local scales that affect people's everyday lives exhibit spatial patterns and trends. These problems transcend physical boundaries (watersheds, ecoregions), political boundaries (states, countries), and disciplinary boundaries. Educators can thus use geospatial technologies to teach geography, environmental studies, planning, history, criminal justice, geology, civil engineering, data science, business, health, sociology, and other subjects in an *integrated holistic manner* just as they are applied in the workplace. Solving 21st-century problems requires collaboration enabled by geotechnologies and a wide diversity of thought leaders and practitioners from every discipline (Kerski 2019).

Learning best takes place in community settings, collaborating with others even outside the classroom. Through dashboards, story maps, audios, and videos, coupled with the attraction that maps have always had, the number of and diversity of the geotechnology instructional community is increasing. Web GIS is part of the open science movement, in which scientific advancements are shared (National Academies of Sciences 2018). People can also *contribute their own data* – on community gardens, dangerous intersections, litter, public art, and more. A community of practice is emerging – a domain of interest and a commitment to it, wherein participants learn with each other, producing a shared repertoire of resources (DeMers, Kerski and Sroka 2020).

Web-based geospatial technologies enable data collection, mapping, and spatial analysis to be done in the cloud, on any device – tablet, laptop, smartphone – with nothing to install, requiring only modest bandwidth. Data-as-services through portals, ArcGIS Hubs, and other data provisioning technologies provide instant access to spatial data, scalable from local to global, on themes from agriculture to zoology. Data can be collected in the field, and fed instantly to maps for analysis and understanding. ArcGIS Survey123 and iNaturalist can be crowdsourced by anyone, even in disconnected mode. Students and instructors can move from collecting to analyzing invasive species, water quality, litter, historical homes, birds, or other phenomena. Multimedia web mapping applications allow for stories to be created and told. Story maps, dashboards, and infographics can be configured with drag-and-drop tools and fed from real-time data streams, and shared. As lines between maps and visualizations become blurred, the array of people engaged with them increases. These tools fit perfectly into peer-to-peer and student-to-instructor presentations, portfolios, assessment instruments gauging student work, and tools to communicate beyond the classroom.

Learning any system to understand our complex world is complex and requires effort. Educators using geotechnology software of the past were faced with challenging user interfaces, limited data, limited computer power, and clumsy storage media. Modern geotechnologies enable instructors to use specific *components* pertinent to their needs rather than having to master the entire platform. While certain fundamentals (symbology, classification, metadata) are important, a business instructor can teach vehicle routing, map consumer behavior, or determine drive times to competitors, specific components can be used to teach marketing, risk assessment, or supply chain management. A sociology professor can use specific tools to teach patterns of behavior. A history professor can map specific data to teach about changes in demographic characteristics across a city over time. And the list of examples goes on.

Challenges exist, including uncertainties about what to teach and how to teach (Sinton and Kerski 2020). Teaching with geospatial technologies requires instructors to be comfortable posing "what if" questions that may have multiple answers. Teaching with geotechnologies engages a professional toolkit for instruction, rather than using "educational software." Instructors must also have "Plan B" when a map layer is slow to draw or is offline. Instructors must also be comfortable "learning alongside" the students when working with new, unfamiliar tools. Another challenge is frequent updates and developments, for example, drones (UAVs), field collection methods, 3D voxels, LocateXT for unstructured data. To meet challenges, the geospatial education community is generous about sharing best practices for effective instruction.

Geospatial technologies change the way hands-on exercises can and should be structured and delivered. The technology long depended upon scripted, step-by-step exercises laden with screenshots. Intuitive tools and data-as-services allow students to learn more quickly, using videos, graphics-enriched documentation, short hands-on activities, and the geospatial user community. Modern intuitive geospatial tools reduce the need for overly scripted activities. Instructors can give students place-based problems, direct them to relevant data, and make assignments brief and focused.

Teaching with geospatial technologies is never *only* about the technologies: higher goals always exist when teaching about water, equity, or climate. Geotechnologies have always involved using real-world data about the Earth and its people – solving real problems in real places. When instructors teach a proximity tool to create riparian zones, they teach about urban greenways or the value of wetlands. When coordinates are geocoded, students can visualize results as sales volumes in business courses, traffic accidents in public safety courses, or earthquakes, in earth science courses. Geotechnology has always been a ***thinker's tool*** – the technology doesn't solve the problem – it is the ***user*** of the tools, thinking spatially, holistically, and critically, who solves them. These make it appealing to today's faculty – they want their students to ***think***.

Geospatial technology instruction encourages geoliteracy: the "geoliteracy stool" is held steady by content knowledge, skills, and the geographic perspective (Figure 13.2).

Content knowledge encompasses disciplinary theory, tenets, and themes. Any discipline grappling with space, place, patterns, and the dynamic Earth can make use of geospatial technologies. Content knowledge includes scale; cycles (carbon cycle, hydrologic cycle); the interconnectedness of the lithosphere, biosphere, atmosphere, hydrosphere, and anthroposphere; climate and weather; population dynamics; energy; water; hazards; and human–environment interaction. Skills include geotechnologies (projecting, georeferencing), applying

FIGURE 13.2 The stool of geoliteracy (Kerski 2015).

spatial statistics, and communicating using visualizations. The geographic perspective ties the other two together – asking questions, framing research problems, synthesizing information about spatiotemporal change, and articulating recommended actions. Web geospatial technology enables instructors to devote more course content to analysis, rather than preparing data or teaching software. Instructors use interactive map layers in libraries like the ArcGIS Living Atlas of the World to tie potential sea-level rise to population density in climate and sustainability lessons. Students use the trace downstream tool to understand hydrologic systems. Teaching with geotechnologies is conducive to problem-solving learning environments using real data (Milson and Kerski 2012). Data are often messy, laden with uncertainty: using geotechnology fosters data fluency. Issues are examined using geotechnology not *because* the technology exists but because the technology is *useful* to solve problems. And because of the media and data fluency these tools foster, spatial skills instruction can improve science, technology, engineering, and mathematics (STEM) outcomes (Sorby, Veurink and Streiner 2018).

Educators want their students to develop fluency in using relevant, meaningful data and to develop critical thinking about it. Geospatial technology instruction has always relied on a wide variety and volume of data – satellite imagery, historical maps, spreadsheets, videos, audios, ground photographs, current maps and plans, and more. Open data policies have steadily advanced globally, bringing data to the fingertips of instructors and students. From international to local organizations (the UN Food and Agriculture Organization, to city parks and transportation), nonprofit organizations, and private industry, geospatial data are available and accessible. The ArcGIS Living Atlas of the World offers over 8,500 data layers on themes central to instruction, including population change, natural hazards, consumer behavior, economic data, watersheds, river systems, energy, health, ocean currents, ecoregions, landforms, weather, and climate. The results are transformative: a greater amount of time in any lesson or research project can now be on analysis, rather than on manipulating data. In addition, students and instructors can collect their own data, even on their own campus.

How to teach with geospatial technologies

Because its data, tools, and maps reside in the cloud, geospatial technologies can be used in face-to-face courses, hybrid courses, and online courses. These tools are also conducive to new types of certificates, microcredentials, and certificate programs. Deans, instructors, and curriculum developers can choose tools as needed as they build courses and programs, rather than having to choose **all** of geospatial technologies as they had to in the past. For example, field surveys, interactive maps, dashboards, and infographics can be combined for studies on public art, water quality, noise, walkability, or other topics (Kerski 2021). These tools and their lessons can be nimbly adjusted as the technology and the institution evolve and to mesh within face-to-face, hybrid, and online teaching. Learning management system (LMS) can embed multimedia web maps (Lukinbeal 2022). LMS and geotechnologies are both cloud-based and thus are "the perfect union." Instructors can use VMware or Citrix so that students can run lab-installed geospatial software on their own devices if needed (Geraghty and Kerski 2020). Tools to visualize, classify, symbolize, and analyze data can be used on any device anytime, using a standard web browser. Cloud infrastructure eliminates the need to transfer map files via physical media or email. Maps and applications can be saved and shared, enabling collaborative learning.

Which tools should be included in the curriculum, where should they be included, and how should they be taught? These questions can be answered if the instructor first examines the points in their curriculum in which they ask

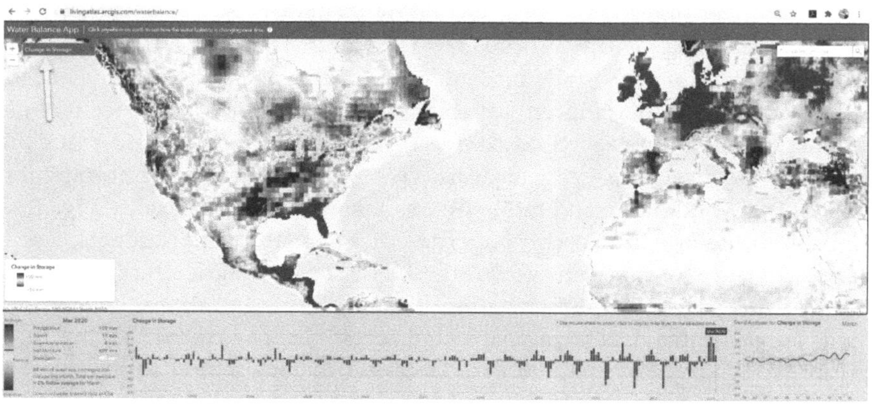

FIGURE 13.3 Exploring the ArcGIS Water Balance web mapping application. Care should be taken in interpreting this black-and-white version of the color version of this Water Balance application. Please visit the website to explore it, and interpret the data correctly.

the "where" question, and what tools they use to answer those questions. If the tools are an outdated, small-scale, or static map, it is at that point where geospatial technology could play a role. By identifying other such points in the curriculum, geographic thinking can be introduced in a sequenced manner, adhering to methods advocated by Bednarz, Heffron, and Huynh (2013). The approach should be to use geospatial technologies when and where they are *appropriate* and *meaningful*, and not to "turn every activity into geospatial" and not "just because interesting tools exist."

The tools could be introduced in a scaffolded manner. Start with web mapping applications, which provide rich content and approachable immersion. These include the ArcGIS Living Atlas of the World applications, such as the water balance (Figure 13.3), the Wayback imagery, the Landsat Explorer, and the Drought Aware. The Living Atlas Live Feeds provide a "heartbeat" of dozens of variables in real time, including wildfire, air quality, and cyclones and can provide a check on the "state of Planet Earth" throughout the semester. Each week, examine a variable, allowing "what's where, why is it there, and why should we care?" to be appreciated and examined. Through the Ecological Marine Units Explorer, students investigate temperature, salinity, dissolved oxygen, and other variables from surface to floor at ten-kilometer resolution, across all oceans. Through the 3D migration app, students analyze migration from *and* to all the countries of the world over time, considering the push and pull factors (economies, immigration laws, famine, natural hazards, political instability) and how those factors change.

Through the water balance app, examine the relationship among precipitation, evapotranspiration, and other variables, across multiple scales. Investigate

Texas tropical storms, determine the number of months without rain in the Sahara, and examine the India and Amazon wet-and-dry cycles. Students learn about the dynamic nature of the Earth and about geospatial technology by continually updating data.

Next, work with the ArcGIS Online 2D map viewer and 3D scene viewer to access, symbolize, classify, and filter mapped data. This web-based GIS allows students to add data, create maps, share maps, and create web mapping applications. By becoming familiar with a short list of ten skills, students gain superpowers in their ability to use any map in the platform (see skills below). Filtering and sorting data is key to focusing on what is important and gaining a deeper understanding. Working with the table (the "I" part of GIS) allows students to investigate the most intense earthquakes over the past month, which nearby lakes contain the lowest pH, and which neighborhoods have the lowest median age or income. Students create scatterplots to determine which variables may be related (such as median home value or household income) or unrelated (such as height of mountain peaks vs. difficulty of climb). Students thus learn content, develop web mapping and analysis skills, and apply the geographic perspective to understand change over space and time.

Next, work with spatial analysis tools in ArcGIS Online, applying geographic thinking in different settings. These tools, such as routing, overlay, and aggregating points, change students' views of maps from what is *where* to *why* things are where they are. They change student perceptions of maps *from* static reference documents ("Where are forests? Where is Knoxville?") to *analytical tools* with a focus on relationships, patterns, and trends. Analysis functions are accessed through menus as wizard-driven tools with helpful prompts. Students use these tools to determine ideal locations for studying plant species based on slope, elevation, direction of slope, and land cover. They integrate mathematics and geography to determine the direction, distance, and time required for water to travel from their campus to the ocean. They determine regions susceptible to more than one kind of hazard. Students then ask deeper questions: "What distance should I use to buffer the stream to create a riparian zone?" "What data layer do I need to analyze to make my investigation more complete?"

Each geospatial component should be taught in a scaffolded manner. Start by adding data to an existing field survey, such as the author's walkable neighborhoods. Next, create a survey of something else, such as trees (species, height, health) for the campus. Ask each student to collect data into this survey over the course of the next week. The field-collected survey points flow in real time to the instructor-created dashboard and web map. Next, ask students to create their *own* survey on a topic they care about, using the same tools. Ask students to add data to their survey *and at least two of their classmates' surveys* over the following week. Next, students map their results in a dashboard, create a story map, and explain their findings in an oral presentation in a face-to-face-course or via an LMS in an online course. In this scaffolded manner, students learn the tools, learn why they are relevant, participate in someone else's survey, and create their *own* map and survey. Creating one's own content is powerful. Formerly, only scientific and mapping agencies could create content; now, *everyone* can be a mapmaker and an analyst, grappling with problems of concern to *them and to their communities.*

Many educators start with story maps. First, use the existing story maps library to interactively teach content, the removal of dams, Indigenous land rights, historical redlining, or other issues. Next, ask students to conduct their own research and build a story map on that issue. Asking students to build their own story map as their very first activity exposes too many uncertainties: "What should I study? Do I have permission to use that image? How do I create a map?" The scaffolded approach ensures that students understand how the tools and methods mesh.

Readings and discussions can be effectively tied to interactive hands-on activities. For example, let us say three course objectives are to identify ways in which maps and geo-visualizations provide a common language for communication and solving problems, analyze data spatially using a variety of techniques, and identify how society influences mapping and how mapping influences society. To meet these objectives, ask students to read Doug Laney's big data definition – volume, velocity, veracity, and variety. Then, ask students to read how big data encourages software vendors to develop tools to handle big data, changes people's expectations of what and how they can analyze, and alters how organizations create resiliency plans. Students read how geospatial technology handles big data from satellites and the Internet of Things (IoT) network of traffic, streamflow, weather, and the 8-billion strong human sensor network. Ask students, "What is one set of big data that you are interested in examining?" Then ask students to examine the Microsoft geodatabase of every USA building in ArcGIS Online, including their own neighborhoods. How are buildings extracted from imagery and segmented using deep neural networks via machine learning and artificial intelligence (Figure 13.4)?

Compare building outlines to a satellite image – Which were captured? Which buildings were missing? Ask students how they will apply what they have learned about big data, geospatial data, maps, and data quality.

As decisions hinge on mapped information, instructors have many ways to teach critical thinking about data. As an example, ask students to read sections of a public domain data textbook and blog (https://spatialreserves.wordpress.com), asking, "Where can I find geospatial data?" and "How can I know whether I can *trust* the geospatial data that I find?" Use the blog's examples of maps that have data collection errors or faked data. Ask, "What are some key societal issues surrounding spatial data?" Using the spatial reserves content on location privacy, ask: "How many apps on your phone are tracking your movements right now? Do you know how to turn them off?" Discuss the ethics of

FIGURE 13.4 Examining Microsoft's big data layer of 100,000,000 buildings using ArcGIS Online.

choosing appropriate symbology, map projection, and classification to avoid misleading an audience. Use work scenarios of various ethics-challenging situations. Instead of giving students canned "perfect" data sets, require students to work with a wide variety of data portals, including some that may be challenging to use. Today, when everyone is a map user *and* a map *creator*, fostering critical data thinking through these scenarios provides students with benefits in other courses and in the workplace. Include personal competencies from the Geospatial Technology Competency Model: "Are you organized? Are you ethical? Can you work with data?"

What are ten skills to nurture when teaching with geospatial technologies?

When faculty and students become familiar with ten skills, they can do virtually anything with geospatial technologies. It is analogous to teaching students "how to fish" – they develop the perspectives and problem-solving abilities that students can *transfer* to different issues and regions.

(1) Working with maps and layers: Searching for, opening, and saving maps, layers, and 3D scenes; managing content; creating metadata. (2) creating and sharing map content: feature services, spreadsheets, multimedia, making maps, layers, models, and applications. (3) Map navigation: changing scale and projection, finding places, measuring, bookmarking, selecting. (4) Changing symbology, classifying, clustering, filtering, rendering imagery. (5) Attributes: creating fields, sorting, summarizing, creating charts. (6) Collecting field data from many different sources. (7) Drawing on maps and configuring popups. (8) Creating and using expressions in filtering, labeling, symbolizing. (9)

Creating web mapping applications: story maps, infographics, dashboards. (10) Analysis: proximity, overlay, map algebra, routing, spatial statistics.

These skills could be assembled into three groups: *create* geographic data, *analyze* data, and *visualize* data. Instructors can use these skills to create assessment tools and pre-test/post-test course evaluations.

What are ten key strategies for teaching GIS?

What *strategies* or *approaches* are suitable for instruction with geospatial technologies? The following have been tested in many courses and in secondary schools and universities.

1 Make the strategy **anchored** to program goals. Fostering critical thinking, data fluency, spatial thinking, communication skills, community connections, field methods. While geospatial skills are important, maps are a *means to an end*. "Putting your data on a map" is not the end point, but rather the *beginning*. The higher goal is not to make maps, but to **understand** something better, in a deeper, richer way, and then to act on the issue.

Examine course goals: How can teaching with geotechnologies help achieve those goals? How can goals be articulated into course learning objectives? How can readings, discussion, activities, and assessments be created so that students will learn content, skills, and perspectives?

2 Make it **holistic.** A geographic perspective through geospatial technology fosters the consideration of the Earth as an interacting "system of systems." One cannot model or understand the complex world in any single course. However, instructors **can** encourage students to consider the "bigger picture" – decisions in one sphere create ripple effects in other spheres. With themes-as-layers, multivariate maps, and animations, instructors can foster the holistic view in topics from water to equity. By varying the themes (see Strategy 6), students will emerge with a sense that everything is spatial and is interconnected.

3 Make it **focused**. Focus each week on one problem in society such as historic structures, biodiversity, and supply chain management. Tying course sections to readings, videos, discussions, and activities builds content knowledge around a specific subject or domain. Doing so also enables students to realize that the spatial perspective is relevant to *every* 21st-century problem we face.

This advice meshes with strategy about holistic consideration as instructors pose thought-provoking questions to students in online discussion boards, face-to-face conversations, and quizzes that "go beyond" the lesson. Ask, "If you were *really* doing site selection as a convenience store chain consultant, what other data and themes would you consider?" Students use traffic, demographics, consumer preferences (fuel, coffee), and drive times to competitors to determine the ideal location for a store. Students thus consider variables that class time did not permit: zoning, left versus right turns, employment, and commuting patterns.

Similar questions should be asked after other activities: "If you were *really doing* this study as a consultant, what variables, themes, and other considerations would you pursue?" Students consider aspects of the physical or cultural environment – the world is complex; variables change over space and time and are connected. They consider the constraints of time, budget, and staffing that they will face on the job.

4 Make it **multiscale**. Many issues are global in nature and foster planet-scale investigation. But faculty should focus on some activities in their own community. Students will have more knowledge of and interest in what occurs in their **own** area. Oftentimes, students have never been asked what they care about and may feel as if they do not have a voice in their local community. These tools provide opportunities for students to investigate, and care: a skateboard park, urban greenway, farmers market, historical homes, dangerous intersections, equity, access. Many geospatial lessons online can be modified for instructors' own area. If students are studying far from the institution, they can transfer this local knowledge to, "How is my *own* community dealing with <floods> <crime> <clogged transportation arterials>?"

5 Make it **varied** and **interesting**! Despite the many ways people learn, instructors still have a large influence on student engagement. If instructors demonstrate interest in exploring the world through geotechnologies, most students will engage. By *varying* instructional methods, discussion, activities, quizzes, video, group versus individual projects, courses remain lively. Asking students to investigate a fun map (a "hats around the world"), or visualization (the NFL Superbowl dashboard), or tool (the Mars 3D viewer) will ensure that these courses will be anything but boring!

6 Make it **relevant**. With local to global issues abounding, this strategy may be the easiest strategy to implement. Creating a weekly lesson out of a current event in class is a powerful way of stoking interest and making courses timely. Ask a team of two students to teach an in-the-news topic with maps in a "geonews-"style presentation. The Living Atlas layers and maps can be accessed with a click, providing springboards for content and discussion.

7 Make it **field-based.** Geospatial is inherently tied to space and place. Working with imagery and visualizations fosters the love of place (Tuan 1990). To encourage students to use all senses, include activities to collect, map, and analyze something in the physical or cultural field environment. Collect changing phenomena (noise, pedestrian counts, weather) so that comparisons can be made between different places and over different days or seasons. Collecting similar themes each semester enables building

long-term databases. Field work on one's own campus is valuable. For virtual courses, require field work in the students' own neighborhoods. International students' local data on housing types, plants, and weather will be fascinating to compare and discuss.

8 Make it **multilevel**. Teaching with modern geotechnologies offers multiple levels of engagement. For example, using the Wayback imagery in swipe mode to compare land cover change fosters analysis of change-over-space-and-time. No sign-in is required as students compare urbanization, glacial and urban extent, coastal erosion, and other themes in this powerful level one mode. In level two, students use the data in ArcGIS Online with spatial analysis tools. In level three, students use image classification tools in ArcGIS Online or ArcGIS Pro to assess changes on the landscape in 2D or 3D.

9 Make the course a place where students can **shine, explore, and grow**. Ask students to reflect upon their learning with questions, even in quizzes, such as, "What is the most valuable thing you learned this week? What was the most frustrating thing about this week in class? What is 1 thing that you read about this week that you would like to learn more about?" (Brookfield 1995). Students share content with their instructor and peers. Students show it online or present face-to-face presentations to their classmates or even with external stakeholders. Story maps become a part of the students' professional portfolio.

10 Make it **visionary.** Touch space-time cubes, artificial intelligence, mapping and visualizations, 3D analytics, combining inside-building with outside mapping, Jupyter Notebook coding, virtual reality, and other cutting-edge trends in courses. Again, not just because these tools exist, but to affirm that the tools rapidly evolve, and to encourage lifelong learning. Include interviews with visionaries (such as the *GeoInspirations* series in *Directions Magazine*). Help students consider how they can contribute to society through geospatial technologies. Keep higher, noble goals at the forefront: Using geotechnologies is about building a better world. You and your students have a key role to play in that world!

Teaching with geotechnologies presents educators with a stunning array of resources with which to create engaging content and hands-on activities to engage students, enriching teaching and learning.

References

Bednarz, S. Heffron, S. Huynh, N. 2013. A road map for 21st Century geography education: Geography education research. A report from the Geography Education Research Committee of the Road Map for 21st Century Geography Education Project. Washington, DC: American Association of Geographers.

Brookfield, S. 1995. *Becoming a Critically Reflective Teacher.* San Francisco, CA: Jossey-Bass, 320 p.

DeMers, M. Kerski, J., and Sroka, C. 2020. The teachers teaching teachers GIS institute: Assessing the effectiveness of a GIS professional development institute. *Annals of the American Association of Geographers* 111 (4): 1160–1182.

Geraghty, E., and Kerski, J. 2020. The impact of COVID-19 on geography, GIS, and education. *J-Reading Journal of Research and Didactics in Geography* 2: 53–66.

Jo, I. Hong, J., and Verma, K. 2016. Facilitating spatial thinking in world geography using web-based GIS. *Journal of Geography in Higher Education* 40 (3): 442–59.

Kerski, J. 2003. The implementation and effectiveness of GIS in secondary education. *Journal of Geography* 102 (3): 128–37.

Kerski, J. 2015. Geo-awareness, geo-enablement, geotechnologies, citizen science, and storytelling: Geography on the world stage. *Geography Compass* 9 (1): 14–26.

Kerski, J. 2019. GIS in higher education. *xyHT Magazine.* https://www.xyht.com/spatial-itgis/gis-in-higher-education/

Kerski, Joseph J. 2021. Online, engaged Instruction in Geography and GIS Using IoT Feeds, Web Mapping Services, and Field Tools within a Spatial Thinking Framework. *The Geography Teacher* 19 (3): 93–101.

Lukinbeal, C. 2022. From emergency remote teaching to academic strategic digitalization planning: How a college responded to COVID-19 and what departments need to consider implementing. In *COVID-19 and a World of Ad Hoc Geographies.* Cham, Switzerland: Springer, pp. 2119–2132.

Milson, A. Demirci, A. Kerski, J. 2012. *International Perspectives on Teaching and Learning with GIS in Secondary Schools.* New York, NY: Springer. 392 p.

National Academies of Sciences. 2006. *Learning to Think Spatially.* Washington, DC: National Academies Press.

National Academies of Sciences. 2018. *Open Science by Design: Realizing a Vision for 21st Century Research.* Washington, DC: National Academies Press.

Schulze, U. 2020. GIS works—But why, how and for whom? *Transactions in GIS* 25 (2): 768–804.

Sinton, D., and Kerski, J. 2020. GIS&T education and training. In *The Geographic Information Science & Technology Body of Knowledge*, ed. J. P. Wilson. Ithaca, NY: University Consortium for GIS&T.

Sorby, S. Veurink, N., and Streiner, S. 2018. Does spatial skills instruction improve STEM outcomes? The answer is 'yes'. *Learning and Individual Differences* 67: 209–222.

14

USING TECHNOLOGY TO PROMOTE CRITICAL THINKING

Kenneth Foote and Chris McMorran

Promoting critical thinking among students is often at the top of our teaching goals. Indeed, some argue it should be at the core of a university education (Brookfield 2012; Cáceres, Nussbaum, and Ortiz 2020; Halpern 2014, Paul and Elder 2020). We want students to pose and answer questions by collecting and analyzing evidence of many kinds, particularly spatial or geographical data. We encourage close readings of key texts, engage students with primary and secondary data sources, challenge them to collect data on their own, and ask them to qualify their findings. Indeed, we want students to be *critical* about the sources they use and the answers they reach. This means helping students understand how implicit and explicit biases can shape data, the ways they analyze the data, and the answers they find. Fundamentally, a focus on critical thinking encourages students to develop a careful, questioning approach to their work, leading them to ask, "What confidence do I have in this information" and "What are the limitations of my solution"?

Information and network technologies, along with the expansion of information (and misinformation) online, open possibilities and create challenges for students as they pursue research. Effective use of online technologies has attracted considerable attention in terms of both research and practice (Hobbs 2011, Major 2015, Murphy, Walker, and Webb 2013, Nilson and Goodson 2021). A major focus of this work has been developing strategies targeted toward critical thinking and reasoning skills (Abdul Razzak 2016, Allen 2008, Ertmer, Sadaf, and Ertmer 2011, Hobbs 2011, Major 2015, Robertson and Scheidler-Benns 2016, Schmidt 2012, Smith, Rama, and Helms 2018, Wardani et al. 2019). But these strategies seem to be challenged at every turn. Even as we write this chapter, educators are wrestling with the emergence of artificial intelligence (AI) systems like ChatGPT, with some excited by the possibilities

DOI: 10.4324/9781003342953-16

and others feeling threatened (McKnight 2022, Mollick and Mollick 2022, OpenAI 2023). It is worth recalling that 20 years ago, Wikipedia spurred similar conversations among educators. Clearly, technology is not the problem. The challenge lies in supporting students as they explore the potential and limitations of new technologies and develop core critical thinking skills that will be valuable to them regardless of what the future holds.

Troubling tech

What we face is a tension between the enabling and disruptive potential of technologies that can be used in research. This stretches from the seemingly innocuous use of search engines to start a project, to the use of government databases or online mapping tools. Thanks to so much information being shared online, we can instantly access documents and data sets that were previously unavailable, inaccessible to the public, or housed in distant archives. If combined and carefully analyzed, this information has tremendous potential to address important global concerns. But the sheer volume is overwhelming. Where to begin? And what can be trusted? How does easy access to materials change how we judge their quality and provenance? In other words, students may believe they can find everything online, but do they understand the limitations of what they can find, and can they distinguish quality scholarship from materials that have not been peer reviewed or published in reputable sources? Moreover, do students understand how these tools work or the underlying algorithms that impact what we can access? One must pay attention to the completeness and quality of what one finds, or risk misinterpreting it or assuming it is unbiased and error-free.

Take the example of online mapping. In our lifetimes, spatial data have moved from an expensive commodity requiring expertise to create, access, and analyze, to something free and seemingly easy to use. Systems like Google Maps, Google Earth, Bing Maps, and OpenStreetMap provide maps, satellite images, and street-level perspectives that can be viewed on phones, tablets, and laptops. They give us a bird's eye view of remote landforms, complex highway networks, agricultural patterns, and suburban housing, or drop us to ground level. Yet, users may assume too much from what they see, unaware that these tools use a mashup of thousands of images and data sets that differ in age, quality, and data type. Features like buildings and roads may appear and disappear as users zoom and pan. Moreover, the Sun never seems to set, thus creating an impression that every point on the planet is easily knowable. In other words, these systems mix fact with fiction and create half-truths that seem harmless, until they lead to misperceptions and mistakes. When we use these materials in class, we often rely on their convenience without discussing how they are created, their flaws, or how they can both enable and disrupt learning.

Geographers also face unique challenges related to helping students understand spatial, geographical, and environmental patterns and processes. Yet, current technologies don't always allow users to explore the spatial dynamics of demographic, economic, political, social, and cultural data, or may even distort these patterns. For instance, many online mapping systems use the Mercator projection, even though it can present a very misleading view of global patterns, which might lead to misinterpretations. Furthermore, the issues Monmonier presents in *How to Lie with Maps* (2018) are as important as ever (see below). Getting students to recognize how spatial patterns and processes can be manipulated can be just as important as helping them think spatially and geographically in these new digital worlds.

Equally worrying is the black box of search engines. While we seem to be able to access anything on the internet, we often fail to appreciate how little we know about how the underlying algorithms rank and display information or narrow a search in response to user choices. For example, a search for maps of the Caribbean may result in maps of several regions, islands, and nations. However, once a person clicks on Cuba, the algorithm begins to narrow down the results, thus leaving off other areas of the Caribbean, even if the user intended to look more widely. In some engines, a user's choices are stored on their computer as cookies that potentially skew all future searches. This can be a useful feature if the user is researching only Cuba, but not helpful otherwise. In addition, some search engines mix "clickbait" articles into results, hoping to entice users to follow links with misleading, deceptive, and provocative graphics and captions. Finally, users may be led into "search bubbles" or "echo chambers." Confirmation bias, or people's tendency to look for information consistent with their own beliefs, can hook users into staying in these bubbles rather than looking for contrary information. In other words, despite the vastness of information available online to students, the algorithms underlying search engines can actively discourage the kind of critical thinking educators aim to support.

Technologies can do more than create echo chambers. They can also spread disinformation. Social media technologies were once praised for creating online communities and democratizing information. For instance, Twitter's role in helping upend governments during the Arab Spring movements of the early 2010s demonstrated how social media could turn a critical eye on powerful institutions, counter the narratives of corporate and state-run media, and help build a different future. Consequently, social media platforms like Twitter, Facebook, and YouTube became essential research tools for understanding political and social movements. However, within a decade, those same platforms were being used to spread dangerous lies that fostered conspiracy theories, anti-vaccination movements, and religious and ethnic violence. In such cases, technology enables the instantaneous, global spread of (dis)information, without any of the previous guardrails of editorial review, peer review, domain expertise, or simple fact-checking.

All of this is to say that the technologies of the 21st-century university have vastly reshaped how and where students think and learn. As geographers, we are acutely aware of how the spatial ordering of learning and research materials has changed, thus requiring better critical thinking skills than ever. In the past, students bought textbooks in the bookstore and visited libraries to access books, journals, newspapers, government data, sound recordings, and more, all of which were carefully acquired, cataloged, and shelved by professionals. Today, when every search begins online, students face a tsunami of links to journals, books, Twitter posts, Wikipedia pages, broadcast news, and Instagram posts, all of which can seem equally legitimate. In this mix are materials one would have never found 20 years ago, but the jumbled and flattened results lack the guardrails and expertise that once helped us identify trustworthy materials. Today, the university library is just another window on a screen, with email and AI-powered chatbots often replacing an in-person interaction with a librarian or a slow-walk among the stacks. Most troublingly, the range and depth of materials available online is so great that it can give students the impression that there are no longer gaps in human knowledge and that a Google search can answer any question. After all, if students can find reasonable-sounding answers online, it may be difficult to convince them to question what they find or to believe that they may have the capacity to conduct their own unique research.

Underlying all of these issues is the need to address ethical issues that arise in these new digital worlds of learning. These include data privacy; data ownership; hacking and misuse of data; inequities related to the digital divide at the local, regional, national, and international levels; the impact of information technologies on political life; and their environmental impacts. Weaving these issues into the curriculum can help students understand and make informed choices about their use of information technologies.

So, what do we do?

Given these challenges, how can we encourage students to develop a careful, questioning approach to their work and use technologies thoughtfully? The questions we raised at the start of this chapter remain: What confidence do we have in information we find on the web and what are its limitations? Our strategies for answering these questions must evolve with the technologies. Some of our colleagues would rather ignore the technology than face the changes that are underway: No laptops in class! Lockdown browsers for exams! Turnitin for papers! No chatbots! We understand these reservations, but we suggest a middle ground: one that accepts and engages creatively with the changes we are experiencing now and that are likely to continue into the future, while also sustaining a healthy degree of skepticism.

This position doesn't involve reinventing the wheel. Many activities we already use can be adapted or extended to address the points we've raised above. Moreover, other academics, universities, non-profits, and educational associations have developed excellent materials that can be readily customized (Table 14.1). We suggest activities we have used in our classes or adapted from other sources like those in Table 14.1 and the reference list, particularly Major (2015), Murphy, Walker, and Webb (2013), and Nilson and Goodson (2021). If you want to learn more, we suggest Mackey and Jacobson's *Metaliteracy* (2014), which provides a particularly effective framework for interlinking the many skills and abilities needed to master contemporary information environments.

TABLE 14.1 Sources of Learning and Teaching Ideas and Activities from Universities and Non-profit Organizations

1	Association of College and Research Libraries and American Library Association, Information literacy in the disciplines guide, https://acrl.ala.org/IS/is-committees-2/committees-task-forces/il-in-the-disciplines/information-literacy-in-the-disciplines/
2	Common Sense Education, https://www.commonsense.org/education
3	Harvard University, Berkman Klein Center for Internet and Society, Digital Citizenship+ (Plus) Resource Platform, https://cyber.harvard.edu/projects/digital-citizenship-resource-platform
4	Indiana University Bloomington, Libraries, Media Services, Integrating media literacy into your curriculum, https://libraries.indiana.edu/integrating-media-literacy-your-curriculum and https://dcrp.berkman.harvard.edu/
5	Media Smarts, Canada's Centre for Digital and Media Literacy, https://mediasmarts.ca/
6	News Literacy Project, https://newslit.org/
7	NewseumED, https://newseumed.org/
8	Northeastern University, Graduate Programs, Five media literacy activities for college students, Kristin Burham, https://graduate.northeastern.edu/resources/media-literacy-activities-for-college-students/
9	PBS Newshour Classroom, Lesson Plans, https://www.pbs.org/newshour/classroom/lesson-plan/
10	Stanford University, Stanford History Education Group, Civic Online Reasoning, https://cor.stanford.edu/
11	University of California Berkeley, School of Information, Strategies for Teaching Data Literacy, https://ischoolonline.berkeley.edu/blog/strategies-teaching-data-literacy/
12	University of Minnesota, Libraries, Integrating media literacy concepts and skills into teaching, https://libguides.umn.edu/c.php?g=1137999
13	University of Rhode Island, Harrington School of Communication and Media, Media Education Lab, Teaching resources, https://mediaeducationlab.com/curriculum/materials
14	University of Washington, Center for an Informed Public, Resources, https://www.cip.uw.edu/resources/

In framing activities relevant to geography and nearby disciplines, we suggest possibilities in six areas, starting with a focus on search engines and content review standards as a foundation for activities addressing misinformation and bias in online sources, and issues relating to the use of online data. The final two areas present activities that draw attention to the critical role maps play in online media and, finally, some of the ethical issues underlying online materials. All activities can be adapted for courses across physical and human geography and GIScience. Most can be adapted for small and large classes, used in a single class period, or extended over multiple sessions. All instructions mention individual students but can be done in groups.

Search engines, filter bubbles, and echo chambers

Search engines are fundamental to navigating the web. These activities help students explore differences among search engines as well as the ways search algorithms can be both beneficial and detrimental: beneficial by finding sources that might otherwise be missed; detrimental by leading users into filter bubbles and echo chambers that can hinder searches.

Test Drive a Search Engine. Assign students different search engines to research the same topic. Have them try a search engine they haven't used before, like Bing, DuckDuckGo, Yahoo!, Ask, or Yandex. Begin with the same keyword, but then let students get creative in their search. After a few minutes, discuss the results – by sharing screens if possible. Share any follow-up terms used, and discuss why results might differ on different sites.

Filter Bubbles and Echo Chambers. Explain how search engines can create filter bubbles and echo chambers by using algorithms that tailor results to suit user choices. Ask students to spend 10–15 minutes searching for information on the same controversial topic. Compare search results and discuss variations. Raise the issue of how search results can reinforce existing beliefs, create echo chambers, and limit exposure to opposing viewpoints.

Content standards on the web: fact, fiction, and everything in between

Perhaps the most important issue students need to understand is that there are no standards of accuracy, truthfulness, fairness, and balance on the web. Standards may be employed by information providers, but few legal and professional guardrails prevent people and organizations from posting just about anything they like online. Opening the online world to anyone who can connect gives voice to billions of people and organizations around the world, but users must constantly make judgments about what sources can be trusted.

It is important for students to understand the process of fact-checking as it is carried out in different publishing, professional, and institutional settings, as well as how peer review is used to vet books, articles, and other publications. Instructors can introduce other concepts, such as the principle of free speech as interpreted within particular national contexts; definitions of defamation, libel, and slander; as well as codes of professional ethics. These concepts and processes are not always easy to understand, so providing practice with activities like the ones mentioned help students grasp the differences.

Comparative Analysis. Have students select a current event and find content from social media, journalistic sources, and academic sources. Ask students to critically evaluate each source based on criteria such as tone, language, author credibility, source reliability, use of evidence, citation practices, objectivity, intended audience, and bias. Discuss the results, highlighting the key differences students observe across the different mediums.

Preaching to the Choir. Have students write a short one- or two-paragraph news release or article on a current controversial geographical issue. Have them pick (and study) the website where they would like to submit their article. The posts need to be shaped to fit the perspective, tone, and audience of the website. Have students explain how they would rewrite the article to suit a different venue.

Peer Review. Explain the role of peer review in academic publishing, and then take students through the process. Share part or all of a paper you published, and then have students read and discuss the original reviewer reports you received. Explain how you addressed the reviewer reports, including any changes you made. Finally, have students do their own peer review, reviewing the first drafts of several classmates, before submitting a final draft. Additional topics worth discussing include reviewer etiquette, the value of rewriting, the potential challenges of peer review (e.g., how to address conflicting reviewer reports), and the collaborative nature of academic publication (Kitchin and Fuller 2005).

Dealing with misinformation and bias

Identifying misinformation and bias is the next step in working with online materials. Here, we emphasize the value of "lateral reading," comparing information collected from a wide variety of sources and search strings. The point is to see how much the information varies from site to site and in what ways. In a sense, this is a type of fact-checking that seeks to verify and contextualize information found online by looking for bias and weaknesses.

Lateral Reading. Have students watch a video about lateral reading such as this one from the University of Louisville: https://www.youtube.com/watch?v=GZvsGKvqzDs. Then select an article on a controversial environmental or geographic issue and have students contextualize it through lateral reading: research the article's hosting site, the author, the origins of the story, and more. Then have students find other articles by the same authors and from a variety of web and social media sites. Discuss the coverage, language used, and potential biases evident in the materials they discovered. This activity promotes awareness of media bias and helps students understand the importance of diverse sources for obtaining a balanced perspective.

Using AI to Mis/Inform. Have students use ChatGPT or a comparable chatbot to create contrasting statements for and against an environmental or geographical issue. Have students present and discuss the structure of each argument and the evidence. The statements can also be used to frame an in-class debate, with each side using the chatbot text as its source. Discuss the strengths and limitations of using AI tools to debate such issues.

Confronting Stereotypes. Have students search for stereotypes used online to depict particular countries and peoples. For instance, project https://blog.nus.edu.sg/wjproject/inspiration/ focuses on how Japan is depicted as weird or "wacky" in English-language websites, YouTube channels, and more. Choose different countries and discuss the stereotypes they discover, how often they appear in searches, and how they are related to cultural, economic, and historical forces. This exercise can also be used to explore how other stereotypes and biases appear online, such as those relating to race, gender, sexuality, ethnicity, nationality, and (dis)ability.

What's In the News? Pick a high-profile global issue from this week's news and have students examine how it is reported in major newspapers around the world. Have students report what they find, including what other stories are featured and the balance of local, national, and international news in these papers. Lists of major newspapers can be found at https://en.wikipedia.org/wiki/List_of_newspapers_by_circulation. For newspapers in other languages, have students use http://translate.google.com or other online translation systems to read the headlines and major stories. This activity can also be done with podcasts or videos. This media bias chart can also be used to create interesting activities: https://adfontesmedia.com/interactive-media-bias-chart/.

Ins and outs of online data

The internet is now the principal access point for data from agency providers all around the world, yet the data can vary considerably in quality, age, coverage, and completeness. The issues mentioned above about differing content standards and the problems of misinformation and bias are as relevant to data searches as they are for other online materials. The following examples encourage students to develop a healthy skepticism toward online data. These activities can also guide students to major in international, national, state, and local information providers that may prove useful to them in the future. See also Tyrell (2023) for excellent lists of government statistics sites for dozens of nations, along with other organizations with quality data.

What Data Do You Trust? Have students compile a data set on a specific geographic phenomenon. This might be a demographic (what are the largest cities in the world?), economic (what counties have the highest/lowest GDP?), or environmental (what countries are the lowest emitters of CO_2?) issues for which multiple sources are available. Have students present the data and discuss what sources were available and how they varied, as well as which source they chose to use and why. How do they account for the differences they found among these data sources?

Data Cleaning. Provide data sets related to geographical research and ask students to analyze the data. Include data sets that intentionally contain errors or are flawed or biased in other ways. This activity can be used as a competition, with students given points for identifying errors in the data sets. The aim is to demonstrate how flaws in a data set can bias conclusions and to emphasize the value of careful data cleaning.

When Data Can Make a Difference. Inform students about case studies in which poor data collection or poor communication of data led to disasters. Examples might include the Great Smog of London (1952), Armero Tragedy (Colombia 1985), the Chernobyl Disaster (Ukraine 1986), the Love Canal Disaster (1978), Hurricane Katrina (2005), and the BP Deepwater Horizon Oil Spill (US 2010).

Make a Misleading Graphic. Provide students with data sets and ask them to create visualizations such as maps, graphs, or charts to highlight patterns in the data. Discuss how their choices in data visualization can influence the perception of the information presented. Discuss the potential for creating misleading visuals from these data sets.

The critical role of maps

Geographers need to be particularly sensitive to the way maps and geospatial data are presented online. The use of an inappropriate projection, failure to normalize data, a poor choice of data classification, ineffective color choices, and many other factors can influence a map's message.

Map Critique. Share a few contemporary and historical examples of data visualizations related to geography, such as maps, charts, or infographics. Encourage students to analyze the map's context, point of view, and intended audience. Share their findings, then discuss the effective, ineffective, and potentially misleading elements of the maps. Many examples are available online by searching "misleading maps," or "award winning maps," like those found here: (https://nacis.org/, https://www.esri.com/en-us/maps-we-love/overview). Other examples are found in Monmonier (2018), Mooney and Juhász (2020), Sandford (1982), and Tufte (2001).

Read, Search, and Present. Have students read Mark Monmonier's *How to Lie with Maps* (2018). Have students focus on different chapters and find two or three new examples that illustrate Monmonier's points in each chapter but are not presented in the book. Share their findings in class and discuss the misleading qualities of the maps.

Find a Misleading Map. Ask students to find a map related to a controversial topic and explain what position the map is advocating. Consider how the cartographer's bias is expressed in the map's design: size, shape, or emphasis on certain features. Encourage students to discuss how these variations can sway readers.

Create a Misleading Map. Introduce your students to online mapping tools that allow them to create their own maps, such as Google My Maps, Google Earth, or ArcGIS Online. Then ask them to create a misleading map about a contemporary geographical or environmental issue. Students can alter scales, distort boundaries, or emphasize specific features to manipulate their maps. Afterward, have students present their manipulated map and explain the strategies they employed to deceive viewers.

Ethical issues

Many of the activities we have provided raise ethical issues about the collection, use, and misuse of information. By the nature of their research interests, geographers are often confronted by such issues, so we advocate discussing them whenever they arise in class.

Ethical Dilemmas IRL. Present real-life scenarios or case studies that highlight ethical issues related to the use, failure to use, or misuse of data, particularly those about environmental or geographical problems. Excellent examples and discussion frameworks are available from Penn State University, GISEthics.org, https://sites.psu.edu/gisethics/, Council for Big Data, Ethics, and Society, Data Ethics Case Studies, https://datasociety.net/library/data-ethics-case-studies/, Princeton Dialogues on AI and Ethics, Case Study PDFs, https://aiethics.princeton.edu/, and Computing Cases Org, http://computingcases.org/ Have students discuss the ethical principles at stake and propose solutions that address the concerns.

Create a Data Mosaic. A data mosaic is a portrait of a person or organization created by assembling pieces of publicly available information – property and tax records, voter registrations, phone numbers, email addresses. Individually, these bits of information are not very revealing, but like the small tiles of an artistic mosaic, when these bits are assembled, they can create a detailed portrait of a person or an organization. Have students create a data mosaic for a celebrity, politician, or public figure – where they live, phone numbers, email addresses, family members, and so forth. Discuss the results and the implications of so much information being publicly available.

Mosaic Yourself. Many companies – data aggregators – create and sell data mosaics and databases drawing on public and, sometimes, private information.

An example is https://claritas.com/. Users include companies wanting to find customers for their products, politicians trying to locate potential supporters, and researchers looking for particular types of participants for surveys. One company that mosaics personal information is http://intelius.com. Have students visit Intelius or a comparable web site to look up themselves or someone in their family, and then discuss what they find. Discuss how these systems collect the information, as well as the ethical aspects of the collection and display of this information. Mosaics can be created using any unique identifier like name or social security number, but street addresses – spatial location – are key in aggregating disparate sources of information.

Backyard Spying. Data mosaics can also be assembled from graphic sources, aerial and ground-level photos, satellite images, and open-source maps. Have students use Google Earth, Bing maps, Google Maps Street View, or a similar tool to find images that document where a celebrity, politician, or public figure lives or works. Have students describe what they find in as much detail as possible. Highlight the ethical considerations of having *private* data available so readily in *public*. Who has given government agencies and private companies permission to look down on our backyards or through the windows of our homes?

Conclusion

What is the place of technology in a twenty-first-century university? In a piece in *The New York Times* titled "Why Universities Should be More like Monasteries," historian Molly Worthen mentioned a course offered at the University of Pennsylvania. In Justin McDaniel's "Living Deliberately," students learn about monastic traditions from Southeast Asia. This includes handing over their cell phones and spending a month without technology. While Worthen acknowledges universities cannot be monasteries, she also warns that too much tech can distract students and undermine their learning. She suggests an optional low-tech approach for new students, in which they would engage with big ideas only through books and discussion: "a secular monastery within the modern university, with a curated set of courses that ban glowing rectangles of any kind from the classroom" (Worthen 2023). Such an experience would certainly be valuable for those with the time and privilege to abandon technology, if only for a short time. But this does not offer a practical solution to the issue that won't go away: how to effectively and critically engage with the technologies all around us.

We cannot afford to be technology-resistant Luddites, and neither can our students. Technologies like those we introduced are simply too valuable to be ignored. They can be research tools used to tackle complex problems. At the same time, we must learn to use all technologies intentionally, and not just to make a quick point. We must get the balance right. Of course, keeping up with new technologies brings its own challenges. We constantly hear about new apps, websites, and digitized resources from colleagues, administrators, and even students that promise to transform our teaching or research. Indeed, some of the exercises listed above may be outdated as soon as this chapter goes to publication. Our central aim in this chapter has been to emphasize that specific technologies are less important than the core skill of critical thinking. Indeed, this skill is more important, and longer lasting, than mastering any given technology.

Maintaining the core skill of critical thinking means not shying away from technologies. It means understanding how they work, asking who they work for, and using them as ethically as possible. Moreover, it means modeling good technology for students. We sometimes find ourselves using technology to quickly make a point or to wow our students, but we miss a teaching moment. As we have tried to show in the examples, there is value in discussing how these technologies work, the biases baked into their algorithms, their ethical implications, and more. Through this reflection we not only model critical thinking skills, but also potentially learn from students, who may perceive these technologies differently. Indeed, students have much to teach us when it comes to using technologies and understanding their implications. This is why it is so important to take time to discuss technologies with students, so they too can learn to reflect on their enabling and disruptive potential.

Technologies have profoundly altered the teaching and research practices of the twenty-first-century university. They allow the sharing and analysis of information across vast distances, foster communication and collaboration between individuals and institutes, and allow access to the most up-to-date and comprehensive data on a range of phenomena. This connectivity and accessibility gives many students the impression that, with time and the right combination of search terms, they can answer any question. Indeed, as predictive AI tools improve in quality, some may question the value of a university education itself. In this context, we must encourage students to critically review the responses they receive from AI sources and question the source of this information; look for other sources and perspectives and consider whether the responses are logical and factual.

We conclude with optimism, believing that geography and related disciplines offer a unique response to these concerns. A quick online search on a random topic may yield a nearly limitless volume of information, and AI tools can quickly retrieve information and coherently explain any number of phenomena. But we humans – our students and ourselves – are out there asking questions, observing places and social interactions, experiencing, and feeling the world like only a human can. We geographers continue to be valuable for our focus on experiencing the world and engaging in the literal groundwork that produces new knowledge.

References

Abdul Razzak, N. 2016. Strategies for effective faculty involvement in online activities aimed at promoting critical thinking and deep learning. *Education and Information Technologies* 21(4): 881–896. https://doi.org/10.1007/s10639-014-9359-z.

Allen, M. 2008. Promoting critical thinking skills in online information literacy instruction using a constructivist approach. *College & Undergraduate Libraries* 15(1–2): 21–38.

Brookfield, S. D. 2012. *Teaching for critical thinking: Tools and techniques to help students question their assumptions.* San Francisco, CA: Jossey-Bass.

Cáceres, M., M. Nussbaum, and J. Ortiz. 2020. Integrating critical thinking into the classroom: A teacher's perspective. *Thinking Skills and Creativity* 37(September): 100674. https://doi.org/10.1016/j.tsc.2020.100674

Ertmer, P. A., A. Sadaf, and D. Ertmer. 2011. Designing effective question prompts to facilitate critical thinking in online discussions. *Design Principles and Practice: An International Journal* 5(4):1–10.

Halpern, D. F. 2014. *Thought and knowledge: An introduction to critical thinking,* 5th ed. New York: Psychology Press.

Hobbs, R. 2011. *Digital and media literacy: Connecting culture and classroom.* Media Education Lab. Thousand Oaks, CA: Sage.

Kitchin, R., and D. Fuller. 2005. *The academic' guide to publishing.* London: Sage Publications.

Mackey, T. P., and T. E. Jacobson. 2014. *Metaliteracy: Reinventing information literacy to empower learners.* Chicago, IL: American Library Association.

Major, C. H. 2015. *Teaching online: A guide to theory, research, and practice.* Baltimore, MD: Johns Hopkins University Press.

McKnight, L. 2022. Eight ways to engage with AI writers in higher education. *Times Higher Education,* https://www.timeshighereducation.com/campus/eight-ways-engage-ai-writers-higher-education

Mollick, E. R., and L. Mollick. 2022. New modes of learning enabled by AI chatbots: Three methods and assignments. https://ssrn.com/abstract=4300783 or https://dx. doi.org/10.2139/ssrn.4300783

Monmonier, M. 2018. *How to lie with maps,* 5th ed. Chicago, IL: University of Chicago Press.

Mooney, P., and L. Juhász. 2020. Mapping COVID-19: How web-based maps contribute to the infodemic. *Dialogues in Human Geography* 10(2): 265–270. https://doi.org/10.1177/2043820620934926

Murphy, D., R. Walker, and G. Webb, eds. 2013. *Online learning and teaching with technology: Case studies, experience, and practice.* New York: Routledge.

Nilson, L. B., and L. A. Goodson. 2021. *Online teaching at its best: Merging instructional design with teaching and learning research.* New York: John Wiley & Sons.

Open AI. 2023. Educator considerations for ChatGPT. https://platform.openai.com/docs/chatgpt-education#annotations:group:__world__

Paul, R., and L. Elder. 2020. *Critical thinking: Tools for taking charge of your learning and your life.* Foundation for Critical Thinking. Lanham, MD: Rowman and Littlefield.

Robertson, L., and J. Scheidler-Benns. 2016. Critical media literacy as a transformative pedagogy. *Literacy Information and Computer Education Journal* 7(1): 2247–2253.

Sandford, H. A. 1982. Misleading maps of urbanization, *Science and Public Policy* 9(1): 12–15, https://doi.org/10.1093/spp/9.1.12

Schmidt, H. 2012. Media literacy education at the university level. *Journal of Effective Teaching* 12(2): 64–77.

Smith, T. E., P. S. Rama, J. R. Helms. 2018. Teaching critical thinking in a GE class: A flipped model. *Thinking Skills and Creativity* 28 (June): 73–83. https://doi.org/10.1016/j.tsc.2018.02.010

Tufte, E. 2001. *The visual display of quantitative information*, 2nd ed. Cheshire, CT: Graphics Press.

Tyrrell, N. 2023. Making use of secondary data. *Key methods in geography*, 4th ed. N. Clifford, M. Cope, and T. W. Gillespie (eds). Los Angeles, CA: Sage, pp. 565–582.

Wardani, D.K., T. Martono, L. C. Pratomo, D. S. Rusydi, and D. H. Kusuma. 2019. Online learning in higher education to encourage critical thinking skills in the 21st century. *International Journal of Educational Research Review* 4(2): 146–153. https://doi.org/10.24331/ijere.517973

Worthen, M. 2023. Why universities should be more like monasteries, *The New York Times*. https://www.nytimes.com/2023/05/25/opinion/college-students-monks-mental-health-smart-phones.html

15

TEACHING ONLINE

Strategies to promote engagement and learning

Brandi Gaertner, Adrienne Goldsberry and Beth Fletcher King

Introduction

Online learning has surged since the early 2010s and played an integral role during the COVID-19 pandemic, allowing instructors to continue to offer classes when students and faculty were not on campus. Given that more students and faculty have experience with remote learning, enrollments will likely continue to increase. The number of undergraduates enrolled in at least one online course was 97% higher in 2020 than in fall 2019 (6 million students). In fall 2020, 75% of all undergraduate students were in at least one distance education course. Even after COVID-19, online enrollment remained high, with the percentage of students enrolled in distance education courses in post-secondary institutions in fall 2022 at 53.3% (NCES 2022).

Therefore, it would not be surprising if you were asked to teach and/or develop an online course at some point in your academic career. Online teaching offers unique opportunities and challenges for keeping students (and the instructor!) engaged in the learning process. In this chapter, relevant research in the scholarship of teaching and learning for engaging students will be reviewed, specifically focusing on how these strategies apply to the online learning environment. Several case students will highlight examples from the courses taught in Penn State's Online Geospatial Education (OGE) programs.

It is important to recognize the perspective from which we (the authors of this chapter, OGE faculty) are coming, regarding online education. OGE started offering online courses in 1999, so we were and are at the forefront of online geospatial education. Our programs are at the graduate level, so most of our students are adult learners with undergraduate degrees. Online courses can take many forms, including synchronous or asynchronous learning. In

DOI: 10.4324/9781003342953-17

synchronous learning activities, the instructor and students interact with one another at the same time, often through audio or video communication tools (Penn State 2021). In asynchronous learning, there is no requirement that the instructor and students interact at the same time. OGE courses are asynchronous, as the students are enrolled around the world, and therefore in different time zones, making same-time learner engagement challenging.

We will explore various strategies for engaging students in asynchronous online courses, including problem-solving and inquiry-based learning, using discussions to build community and enhance learning, and creating assessments and course projects that build upon students' personal and professional experiences. To support our perspectives, we offer case studies that highlight examples from our OGE courses so that you may consider how these ideas could be applied to online courses with small or large enrollments, at the undergraduate or graduate level.

Learning through problem-solving

In asynchronous online classes, students can interact with the material, instructor, and other students at their own pace, which can increase interaction (Shankar et al. 2023). The COVID-19 pandemic exposed, and in some cases created, some bad practices, such as asynchronous online classes that relied heavily on long recorded lectures. A lecture-based course (online or in-person) can be a challenge for students to gain practical knowledge. Learning through solving problems can be used to address that challenge.

Successful learning emerges when students are actively involved in the learning process, from the introduction of the problem and the implementation of a solution, to the reflection of the process (Mathews and Wikle 2019). Too frequently, university graduates lack the decision-making and problem-solving skills that their future professions demand, so it is essential for instructors to consider how to help them gain these skills. These skills, also referred to as competencies, include problem-solving, decision-making, and critical and analytical thinking (Figure 15.1).

By utilizing activities closely associated with real-world problems and issues, problem-based learning (PBL) and related methods (i.e., inquiry-based learning and project-based learning addressed later in this chapter) have been shown to promote higher-order learning and critical thinking skills within both Geographic Information Systems & Technology (GIS&T) and geography (Silviariza and Handoyo 2021). Additionally, problem-based activities, based on real-world problems that can be solved with a geographic perspective, motivate students to learn more about the themes of the discipline of geography (Golightly 2018). Healy and Walshe (2020) note that it is beneficial to establish the relevance of a particular topic or concept to motivate students. Incorporating problems related to the many industries that apply

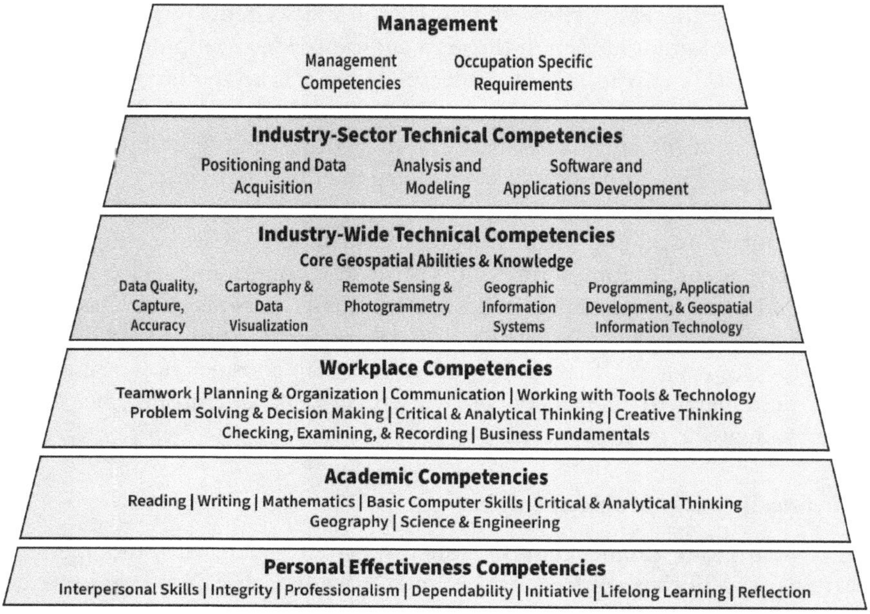

FIGURE 15.1 Geospatial Technology Competency Model. (Employment and Training Administration 2023).

geographic knowledge and employ geographers allows students to focus on academic competencies while developing workplace competencies.

Problem-solving can be particularly beneficial in online learning. Online courses are more likely to attract non-traditional learners, characterized by their wider age range, varying socioeconomic backgrounds, work experiences, and family commitments (Zamecnik et al. 2022). Learning opportunities can be used to let those students incorporate their experiences into their learning. One method for you to implement problem-solving in your online class is to focus on personally and professionally relevant problems to increase student engagement and achievement of learning outcomes because it promotes deeper discussion and critical thinking.

> For example, students in OGE's introductory GIS course employ problem-solving in their final project. They are presented with a scenario in which they are working with a local (to them) municipality. They are asked to submit a proposal for a biological reserve within the municipality that would protect lands with high species richness and habitat quality. The municipality is hiring a consulting group to perform a site selection analysis. The students' goal, as GIS analysts, is to identify land that meets the criteria defined by experts in the field of conservation biology.

The benefits of implementing this type of scenario in an online class is that students can use threaded discussions to communicate the workflows of project proposals with their peers. Additionally, this allows students to incorporate any relevant professional experience to their analysis (see also "Learning through Professional Experience").

Learning through course projects

Project-based learning (PjBL) exercises that provide interactive, application-based projects for students to learn concepts are quite feasible and effective in online courses. As a pedagogical methodology, PjBL involves students working on a project for an extended time to synthesize, apply, and analyze content (Anderson and Krathwohl 2001). The PjBL pedagogy is structured in a way that allows students to conduct meaningful and real-world projects with applicable and tangible products (Guo et al. 2020). Six important components define PjBL: (1) a driving question that outlines the problem to be solved, (2) learning goals that outline the mastery standards, (3) participation in scientific practice by learning and applying skills, (4) engaging in collaborative activities, (5) scaffolding courses to help students gradually learn new skills, and (6) creating tangible real-world products (Krajcik and Blumenfeld 2006).

Just as in a resident course, you can incorporate projects in your online course in two ways: a shorter, multi-week project or a semester-long project. The shorter, multi-week project focuses on implementing a multi-part, mini project at the end of the semester. In course design, this would entail introducing the material, methods, and concepts in the first two-thirds of the course and providing a student project in the last one-third of the course. Consider how the online learning management system (LMS) you are using for your course can influence both the design of the project assignment and the feedback you would provide. OGE uses Canvas as their LMS. For example, are student submissions visible to other students? Can students comment on other students' submissions in a discussion? Can students see your feedback? How can students use the LMS to work in groups?

Additionally, it is especially important when teaching PjBL online that both the assigned project and subsequent peer and instructor feedback are avenues for students to encourage and support each other, and to connect with their peers and instructor. Establishing a connection with others is fundamental to helping students feel motivated to learn and results in better academic output (Farrell and Brunton 2020; Ryan and Deci 2000; Shah and Cheng 2019). We will discuss this in more detail later in this chapter.

OGE's online Master of GIS (MGIS) and Master of Science in Spatial Data Science (MS SDS) degree programs implement this pedagogical method in a course titled Geographic Information Analysis, in which students learn spatial statistics and analyze spatial patterns, processes, and associations. The course culminates with a project at the end of the semester in which the students apply processes they learned in the course, conduct the analysis, and write a scientific paper. This type of project is rather open-ended and allows the students to choose their own research topic, possibly drawing from their own personal and professional experiences, and conduct the analysis individually.

The longer, semester-long project takes a more intensive approach that implements scaffolded modules, in which the concepts of each week build on those from the previous week. OGE's MGIS and MS SDS programs use scaffolding in a course titled Geospatial System Analysis and Design. In this course, students design a geospatial app, website, or enterprise system. The design process is scaffolded to guide students through the steps, which include identifying their term project topic idea, conducting a literature search, evaluating the problem statement, explaining the user and needs analysis, investigating available solutions, and outlining software architecture, data sources, and a cost-benefit analysis. The project culminates with a peer review presentation, in which students pair up with a peer based on time zones, present their project to each other, and provide suggestions for improvement. Students enjoy the opportunity to meet with their peers in this capacity. These presentations are additionally uploaded to the Canvas media space for the instructor to view and for peer comment.

Learning through research

Learning through research utilizes student-centered, inquiry-based, and discovery-based pedagogies. Conducting research allows the focus of learning to shift to the students, and the faculty advisor becomes a facilitator (Bloom 1956; Justice, Warry, and Rice 2009; Prince and Felder 2006). Research and inquiry-based pedagogy is an eight-step iterative, circular process similar to the scientific method that revolves around student self-driven learning and self-reflection (Justice et al. 2002, 2007). The inquiry process begins with developing a topic of interest and conducting background research, framing a question, deriving hypotheses and possible answers, gathering information and data, analyzing data, evaluating results, synthesizing answers, communicating new understandings, and self-reflecting on success (Santana-Vega et al. 2023).

It is important in online geospatial courses to engage students to develop a scientific understanding of spatial data. A research agenda in an online program should, on average, take around six months or more to complete. The first component should focus on developing a research topic, gaining a rapport, developing a relationship with the students through virtual meetings, and writing a research proposal including the introduction, method, and timeline. You could also consider including a 25–30-minute virtual presentation in the first semester for students to present their research topic and proposed methods to other faculty and students. An additional option could include making the presentation a peer review option, where the viewers provide constructive feedback on the student's research methodology, presentation tactics, and/or data collection methods to enhance their research process. Many LMSs provide options for peer review to be embedded into the design of the research project assessment (or any assessment in a course). In an online environment, the students will complete the analysis individually, though it is helpful to meet with the student virtually for guidance. To build a course community, you might invite students to submit questions about data sources, analysis techniques, and other aspects of the research process in a discussion forum.

While the final component of the research process in an online course would include the usual components of a capstone project or paper (i.e., abstract, methodology, literature review, results), how this product is submitted and/or formally presented is where research in an online course can be a unique learning opportunity (and sometimes a challenge!). For example, having online students attend a national or regional conference where they can present their research and where you can meet in person is a valuable learning opportunity. Meeting in person is ideal (but need not be required) in an online program because it provides a personalized feel to the otherwise remote course work. By guiding your students systematically through this research process, you can ensure they have personalized research experience and a successful research project, even if it is conducted remotely.

Learning through professional experience

Students generally respond positively to lessons that are related to the profession they are targeting (Adler and Stringer 2018). Universities are striving to develop more relevant experiences for students seeking careers outside of academia, and for those who may already have some professional experience (Hill et al. 2011; Riley et al. 2021). Given the ever-increasing need for students to develop a wide range of workplace competencies, instructors should consider the numerous ways in which professional experience can be incorporated into online course design and content, including incorporating students' professional experiences into course discussion and projects, faculty and adjunct industry and research experience when designing courses, and guest lectures from professionals in industry.

Online classes may include students who are established working professionals, which provides the opportunity to leverage students' industry experience in discussions and projects. Due to the diverse nature of the geospatial industry, these students (usually adult learners in an online course) may bring expertise to the class from industries as varied as conservation, education, business, defense, energy, health, and government, to name just a few. Therefore, offering them the opportunity to share their experiences with the class, through discussions and/or course projects and research, enables them to apply their own skill sets and provides them with the opportunity to develop their professional identities (Blanford et al. 2020), and, at the same time, benefits others in the class. For example, discussion boards could focus on students' experience with new and emerging geospatial technologies, projects could incorporate flexibility that allow students to complete career-related projects as term projects, and students can have the opportunity to contribute to "water cooler discussions," where they can post about personal experiences, such as navigating the geospatial industry, applying and interviewing for jobs, and potential networking opportunities to help their peers.

You can also draw upon your own professional experience when designing an online course. Most faculty members either have extensive industry experience prior to coming to academia or extensive research experience that leverages the application of geospatial technology in academic writing. Therefore, online courses can be designed to incorporate industry-based projects that focus on application development, remote sensing, emergency management, and/or location intelligence for business, among others. Faculty with extensive research experience can leverage their knowledge to build courses focusing on scaffolded term projects that culminate in scientifically based manuscripts. Another option is to rely on industry experience of adjunct faculty members. For example, adjunct faculty working at the National Geospatial-Intelligence Agency can provide firsthand industry expertise on geospatial intelligence, while software programmers can provide expertise on web application development and advanced Python programming.

Online courses offer endless opportunities to add the perspectives of industry experts as guests in a class, because they do not rely on having to bring the professionals to a physical classroom, while also providing the option to record the meeting and make it available in future course offerings. Given the varied nature of the geospatial industry, guest industry experts can include faculty from other universities working with new and emerging technology such as machine learning and artificial intelligence and interviews with industry experts that can provide expertise on available job opportunities, networking opportunities, and career-related projects. In your class, consider consulting with outside professionals to ensure content and methodologies taught are rooted in real-world experiences and prepare students for current job opportunities.

Learning through discussion

So much of planning and teaching an online course requires thinking about if and how the online learning environment can (or should) simulate the in-person classroom environment. This is, perhaps, most evident in the use of threaded discussions. Just as in a traditional classroom, students' interactions with the instructor and fellow classmates impact their attitudes about the class and their enthusiasm for participating in it, ultimately impacting their learning.

Before moving forward, let us first clarify what we mean by "discussion." Here are three forms that discussions in the online learning environment can take (among many others) (Griffin and Roy 2022; Muir et al. 2020):

1. A structured post with a question posed by the instructor to which the students should respond,
2. A question asking for clarification of a course concept posted by a student, to which fellow students and/or the instructor respond,
3. An online chat, most likely applied in a synchronous online course, where students can respond in real-time.

As in the traditional in-classroom environment, discussions in an online setting can occur between many or just a few participants; they can be informal, or posts can be counted toward the final grade; they can be synchronous or asynchronous (Gillett-Swan 2017; Lin and Gao 2020). All forms are important elements for helping an instructor build a course community. In short, discussion forums are the key element of online learning that facilitates interactions between teachers and students, and between students themselves, providing a structure for interactive learning and communication and an essential means for enhancing student participation (Kebritchi et al. 2017; Onyema et al. 2019; Song and McNary 2011). The advantage that online courses have is that discussions can span hours or days, well beyond the amount of time in a typical resident lecture or meeting.

Students who have a sense of belonging, created through frequent and positive interactions with others in their learning community (like discussions), have better academic achievement, increased retention, and enhanced learning experiences (Farrell and Brunton 2020; Shah and Cheng 2019). Creating this "sense of belonging" to improve learning is particularly relevant to online courses – the frequently asynchronous nature of online communication and resulting potential for feelings of isolation can negatively impact the learning environment (Blau et al. 2020; Pilcher 2016). Trimble and Fan (2023) comment on the loneliness of online learning, noting that connections with other students can mitigate the loneliness and make online learning more enjoyable.

Discussion forums also enhance learning by offering an opportunity for students to think critically about course content (Ahmad et al. 2022; Beckmann and Weber 2016) and are especially effective when students can assist each other (Baglione and Nastanski 2007; Dengler 2008). Online discussions can be more substantive than those in the classroom because students have more time to research their responses, can reflect upon other students' contributions, and can participate more equally (it is more challenging when one or a few students dominate the discussion) (Baglione and Nastanski 2007; Ahmad et al. 2022).

You may wonder what factors should be considered when designing online discussions to best develop a sense of course community and enhance student learning. Woods and Bliss (2016) offer a summary of research-based suggestions for creating effective discussions that encourage and facilitate student participation while also enhancing student learning. For example, discussion questions should relate to students' personal experiences when applicable (see earlier mention of using students' professional experiences), students should have choices about which questions to answer, and topics should relate to real-world issues to make a connection between course concepts and real practice (Woods and Bliss 2016). Additionally, numerous universities offer their own research-based guides for designing effective online discussions – do not hesitate to use them!

The most important factor for ensuring discussions' effectiveness toward improving learning is instructor engagement, which happens through moderating, answering questions, and asking probing questions to encourage further discussion (Ebrahimi et al. 2017; Muir et al. 2020; Parks-Stamm et al. 2017; Kember and Hicks 2023). Kember and Hicks (2023) found that where discussion forums positively contributed to helping students, there were higher levels of teacher–student and student–student exchanges, generating a feeling of a learning community in the class.

OGE courses offer several types of discussions. For initial community-building, all of our courses offer an introductory discussion forum where students post information about themselves. Almost all classes incorporate discussions into at least some part of the grade. Our introductory class, titled Making Maps that Matter with GIS, is highly centered around the use of discussion forums to build a repository of knowledge. Students submit two graded discussion forum posts for each of the seven course lessons. In the discussions, students investigate assigned topics independently and share findings with the class to collaboratively construct understandings of these topics. Quiz questions are taken from this student-constructed knowledge base (which has been vetted by the instructor). Grading rubrics are used to provide consistent feedback and clearly defined expectations of students' participation (Giacumo and Savenye 2020; McKinney 2018; Rovai 2007). Figure 15.2 illustrates an example of a grading rubric used in this introductory OGE course.

The success of an online course is highly dependent on the success of the discussions in the course, which create a dynamic learning opportunity that fosters community among students and the teacher (Maddix 2012). Engagement with the course materials and engagement in discussion forums are

Assignment Rubric Details ×

Discussion Rubric

Criteria	Ratings			Pts
Correct and relevant The post addresses one or more questions posed in the discussion assignment. It provides accurate information that advances the class's understanding of the topic.	2 pts Full Marks	1 pts Partial Marks	0 pts No Marks	2 pts
Distinct The post is substantively different from others. It adds new sources, facts, and perspective to the discussion. If your chosen topic has already been discussed, find a different source or sources. Acknowledge classmates' similar posts (by name), and compare what your source(s) has to say.	1 pts Full Marks	0 pts No Marks		1 pts
Authoritative sources The post is paraphrased from trustworthy sources. Questions about the accuracy or timeliness of the source, if any, are disclosed. Circumstances of posts based on personal experience are explained.	1 pts Full Marks	0 pts No Marks		1 pts
Properly paraphrased Sources are paraphrased in students' own words. Quotations are limited to a small portion of the post.	1 pts Full Marks	0 pts No Marks		1 pts
Source(s) acknowledged Online, open-access sources are linked directly from the post. Other (offline) sources are cited and referenced as in the Style Guide in the course introduction.	1 pts Full Marks	0 pts No Marks		1 pts
				Total Points: 6

FIGURE 15.2 Example of a grading rubric for a discussion forum in an introductory OGE course (Geog 482: Making Maps that Matter with GIS).

found to be predictors of course completion (Soffer 2019). As you design or teach your online course, consider how you can be an active facilitator of and participant in these discussions, and how you can design them to encourage and support student participation.

Summary and recommendations for early career faculty

In this chapter, we reviewed research-based strategies for engaging students in online courses (many of which can be used in the traditional classroom environment too). Creating and teaching online courses that encourage student engagement requires deliberate efforts in course design. Quality engagement occurs when instructors create student-centered assignments, projects, and assessments that promote critical thinking and in which students can build upon their backgrounds and/or professional and personal experiences. Additionally, instructors should create safe and open-communication course environments that encourage community among students, where students feel comfortable asking questions, engaging in collaborative activities, and responding to effective feedback.

Here are a few specific recommendations taken from discussions throughout this chapter that we hope will help you with developing and teaching your own online course.

Assignment/project design & implementation

Create assignments and projects using PjBL principles that allow students to develop higher-order thinking and learning skills, which are important components for increasing student motivation and learning. Suggestions for designing these assessments include the following:

1. Create scaffolded courses in which each online course module actively incorporates writing, presentation, and communication skills. Students can meet virtually in peer groups to present their projects and receive constructive feedback. Additionally, online courses can implement scaffolded term projects, in which each module expertly guides the student through writing one section at a time, which provides the opportunity for students to invest more time in writing thorough and detailed projects while also providing the instructor the opportunity to provide iterative thorough feedback.
2. Allow students to gather, analyze, and synthesize information by solving problems, completing projects, and performing research. Online courses can be scaffolded to allow students to propose term project ideas, collect data, and culminate in a science-based manuscript.
3. Incorporate the faculty and students' professional experiences. Some options may include creating discussion posts that allow the students to explain professional projects, stories, or applications of geospatial technology, creating courses that leverage faculty industry or research experience, and/or inviting industry experts as guest lecturers to present virtually and answer questions.
4. Develop grading rubrics with clear criteria and the opportunity for thorough instructor feedback. Provide rubrics early and often.

While these suggestions can also be applied to traditional classroom environments, explore the tools of the learning management system your university provides for online courses to implement them. For example, (1) employ asynchronous discussions to provide a space for students to help and provide feedback to each other on projects, (2) use online grading rubrics that provide space for written feedback, (3) allow voice recordings or video submissions for certain assessments as appropriate.

Course communication and discussion

Effective course communication from the instructor to student, between students, and between the students and instructor is essential to building course community and to motivate students. Consider these tips as you plan your course:

1. Post frequently to an announcement area with videos and create a welcome message at the beginning of the course.
2. Make regular contact with students through email, assignment/rubric comments, or a chat/discussion space.
3. Create a discussion forum for "hallway discussions," introductions, and so forth.
4. Provide discussion prompts, facilitate discussion responses, and model good discussion etiquette in your responses.
5. Provide prompt, specific, and constructive feedback on all student work.
6. Sequence feedback. For example, include a positive statement about specific deliverables where the student performed well, a corrective suggestion about the deliverables the student could change, and an overall positive statement (Henley and DiGennaro Reed 2015; James and Shephard 2001).

As the instructor of an online course, you have the important responsibility of creating a learning environment that challenges students with information that engages them, builds upon their interests and experiences, and one where they feel connected to you and their fellow students. You can then share your experiences and expertise with your peers. We wish you well!

References

Adler, R., and C. Stringer. 2018. Practitioner mentoring of undergraduate accounting students: Helping prepare students to become accounting professionals. *Accounting & Finance* 58 (4): 939–963. https://doi.org/10.1111/acfi.12249

Ahmad, M., K. Junus, and H. B. Santoso. 2022. Automatic content analysis of asynchronous discussion forum transcripts: A systematic literature review. *Education and Information Technologies* 27 (8): 11355–11410. https://doi.org/10.1007/s10639-022-11065-w

Anderson, L., and Krathwohl, D. A. 2001. *Taxonomy for learning, teaching and assessing: A revision of Bloom's Taxonomy of Educational Objectives.* New York: Longman.

Baglione, S., and M. Nastanski. 2007. The superiority of online discussion: faculty perceptions. *The Quarterly Review of Distance Education* 8: 139–150.

Beckmann J., and P. Weber. 2016. Cognitive presence in virtual collaborative learning: Assessing and improving critical thinking in online discussion forums. *Interactive Technology and Smart Education* 13 (1): 52–70.

Blanford, J., P. Kennelly, B. King, D. Miller, and T. Bracken. 2020. Merits of capstone projects in an online graduate program for working professionals. *Journal of Geography in Higher Education* 44 (1): 45–69. https://do.org/10.1080/03098265.2019.1694874

Bloom, B. S. 1956. *Taxonomy of Educational Objectives, Handbook: The Cognitive Domain*. David McKay, New York.

Blau, I., S. Tamar, and O. Avdiel. 2020. How does the pedagogical design of a technology-enhanced collaborative academic course promote digital literacies, self-regulation, and perceived learning of students? *The Internet and Higher Education* 45: 100722. 10.1016/j.iheduc.2019.100722.

Dengler, M. 2008. Classroom active learning complemented by an online discussion forum to teach sustainability. *Journal of Geography in Higher Education* 32 (3): 481–494. https://doi.org/10.1080/03098260701514108

Ebrahimi, A., E. Faghih, and M. Dabir-Moghaddam. 2017. Student perceptions of effective discussion in online forums: A case study of pre-service teachers. *Innovations in Education and Teaching International* 54 (5): 467–475. DOI:10.1080/14703297.2016.1143858

Employment and Training Administration (ETA). 2023. Geospatial Technology Competency Model. Retrieved from https://www.careeronestop.org/CompetencyModel/competency-models/geospatial-technology.aspx

Farrell, O., and J. Brunton. 2020. A balancing act: A window into online student engagement experiences. *International Journal of Educational Technology in Higher Education* 17 (1): 1–19. https://doi-org.ezaccess.libraries.psu.edu/10.1186/s41239-020-00199-x

Giacumo, L. A., and W. Savenye. 2020. Asynchronous discussion forum design to support cognition: effects of rubrics and instructor prompts on learner's critical thinking, achievement, and satisfaction. *Education Technology Research and Development* 68: 37–66. https://doi-org.ezaccess.libraries.psu.edu/10.1007/s11423-019-09664-5

Gillett-Swan, J. 2017. The challenges of online learning: Supporting and engaging the isolated learner. *Journal of Learning Design* 10 (1): 20–30. https://doi-org.ezaccess.libraries.psu.edu/ https://doi-org.ezaccess.libraries.psu.edu/10.5204/jld.v9i3.293

Golightly, A. 2018. The influence of an integrated PBL format on geography students' perceptions of their self-directedness in learning. *Journal of Geography in Higher Education* 42 (3): 460–478. https://doi.org/10.1080/03098265.2018.1463974

Griffin, L., and J. Roy. 2022. A great resource that should be utilised more, but also a place of anxiety: Student perspectives on using an online discussion forum. *Open Learning: The Journal of Open and Distance Learning* 37 (3): 235–250. https://doi.org/10.1080/02680513.2019.1644159

Guo, P., N. Saab, L. S. Post, and W. Admiraal. 2020. A review of project-based learning in higher education: Student outcomes and measures. *International Journal of Educational Research* 102: 101586.

Healy, G., and N. Walshe. 2020. Real-world geographers and geography students using GIS: Relevance, everyday applications, and the development of geographical knowledge, *International Research in Geographical and Environmental Education* 29 (2): 178–196.

Henley, A. J., and F. D. DiGennaro Reed. 2015. Should you order the feedback sandwich? Efficacy of feedback sequence and timing. *Journal of Organizational Behavior Management* 35: 321–335.

Hill, J., P. Kneale, D. Nicholson, S. Waddington, and W. Ray. 2011. Re-framing the geography dissertation: A consideration of alternative, innovative and creative approaches. *Journal of Geography in Higher Education* 35: 331–349.

James, T., and D. Shephard. 2001. *Presenting Magically: Transforming Your Stage Presence with NLP.* Crown House, Carmarthen, Wales.

Justice, C., W. Warry, and J. Rice. 2009. Academic skill development-inquiry seminars can make a difference: Evidence from a quasi-experimental study. *International Journal for the Scholarship of Teaching and Learning* 3: Article 9.

Justice, C., J. Rice, W. Warry, S. Inglis, S. Miller, and S. Sammon. 2007. Inquiry in higher education: Reflections and directions on course design and teaching methods. *Innovative Higher Education* 31: 201–214.

Justice, C., W. Warry, C. Cuneo, S. Inglis, S. Miller, J. Rice, and S. Sammon. 2002. A grammar for inquiry: Linking goals and methods in a collaboratively taught social sciences inquiry course. *The Alan Blizzard award paper: The award-winning papers.*

Kebritchi, M., A. Lipschuetz, and L. Santiague. 2017. Issues and challenges for teaching successful online courses in higher education: A literature review. *Journal of Educational Technology Systems* 46 (1): 4–29. https://doi.org/10.1177/0047239516661713.

Kember, D., and D. Hicks. 2023. Modelling the way teachers can support the retention and success of online students. In: Kember, D., Ellis, R. A., Fan, S., Trimble, A. (eds) *Adapting to Online and Blended Learning in Higher Education.* Springer, Singapore. https://doi-org.ezaccess.libraries.psu.edu/10.1007/978-981-99-0898-1_14

Krajcik, J. S., and P. C. Blumenfeld. 2006. Project-based learning. In: Sawyer, R. (ed.) *The Cambridge Handbook of the Learning Sciences*, 275–297. Cambridge University Press, Cambridge.

Lin, X., and L. Gao. 2020. Students' sense of community and perspectives of taking synchronous and asynchronous online courses. *Asian Journal of Distance Education* 15 (1): 169–179. https://www.asianjde.com/ojs/index.php/AsianJDE/article/view/448

Maddix, M. 2012. Generating and facilitating effective online learning through discussion. *Christian Education Journal* 9 (2): 372–385.

Mathews, A., and T. Wikle. 2019. GIS&T pedagogies and instructional challenges in higher education: A survey of educators. *Transactions in GIS* 23: 892–907. https://doi.org/10.1111/tgis.12534

McKinney, B. 2018. The impact of program-wide discussion board grading rubrics on students and faculty satisfaction. *Online Learning* 22 (2): 289–299.

Muir, T., T. Douglas, and A. Trimble. 2020. Facilitation strategies for enhancing the learning and engagement of online students. *Journal of University Teaching & Learning Practice* 17 (3): 8. https://doi-org.ezaccess.libraries.psu.edu/10.53761/1.17.3.8

National Center for Education Statistics (NCES). 2022. Postbaccalaureate Enrollment. *Condition of Education.* U.S. Department of Education, Institute of Education

Sciences. Accessed February 2023. https://nces.ed.gov/programs/coe/indicator/chb.

Onyema, E. M., E. C. Deborah, A. O. Alsayed, Q. Noorulhasan, and S. Sanober. 2019. Online discussion forum as a tool for interactive learning and communication. *International Journal of Recent Technology and Engineering* 8 (4): 4852–4859.

Parks-Stamm, E. J., M. Zafonte, and S. M. Palenque. 2017. The effects of instructor participation and class size on student participation in an online class discussion forum. *British Journal of Educational Technology* 48 (6): 1250–1259. doi:10.1111/bjet.12512

Penn State University. 2021. Strategies for creating engaging synchronous and asynchronous learning environments. Web Learning @ Penn State. Accessed February 2022. https://weblearning.psu.edu/strategies-for-creating-engaging-synchronous-and-asynchronous-learning-environments/

Pilcher, A. J. 2016. Establishing community in online courses: A literature review. *College Student Affairs Leadership* 3 (1): Article 6.

Prince, M. J., and R. M. Felder. 2006. Inductive teaching and learning methods: Definitions, comparisons, and research bases. *Journal of Engineering Education* 95: 123–138.

Riley, J., S. Eldridge, and R. Venkatesh. 2021. Engaging the experts to embrace the future: An interactive professional learning experience. *Issues in Accounting Education* 36 (4): 41–58.

Rovai, Alfred. 2007. Facilitating online discussions effectively. *The Internet and Higher Education* 10: 77–88.

Ryan, R. M., and E. L. Deci. 2000. Self-determination theory and the facilitation of intrinsic motivation, social development, and well-being. *American Psychologist* 55 (1): 6878. https://doi.org/10.1037/0003-066X.55.1.68

Santana-Vega, L. E., L. Feliciano-García, and A. Suárez-Perdomo. 2023. Inquiry-based learning in the university context: A systematic review. *Revista Espanola De Pedagogia* 78 (277): 9.

Shah, M. and M. Cheng. 2019. Exploring factors impacting student engagement in open access courses. *Open Learning: The Journal of Open, Distance and e-Learning* 34 (2): 187–202, https://doi.org/10.1080/02680513.2018.1508337

Shankar, K., P. Arora, and M. C. Binz-Scharf. 2023. Evidence on Online Higher Education: The Promise of COVID-19 Pandemic Data. Management and Labour Studies, 48(2), 242-249. https://doi.org/10.1177/0258042X211064783

Silviariza, W. Y., and B. Handoyo. 2021. Improving critical thinking skills of geography students with spatial-problem based learning (SPBL). *International Journal of Instruction* 14 (3): 133–152.

Soffer, T., and A. Cohen. 2019. Students' engagement characteristics predict success and completion of online courses. *Journal of Computer Assisted Learning* 35: 378–389. https://doi-org.ezaccess.libraries.psu.edu/10.1111/jcal.12340

Song, L., and S. W. McNary. 2011. Understanding students' online interaction: Analysis of discussion board postings. *Journal of Interactive Online Learning* 10 (1): 1–14. www.ncolr.org/jiol ISSN: 1541–4914.

Trimble, A., and S. Fan. 2023. Peer student support and the formation of learning communities. In: Kember, D., Ellis, R. A., Fan, S., Trimble, A. (eds) *Adapting to Online and Blended Learning in Higher Education*. Springer, Singapore. https://doi-org.ezaccess.libraries.psu.edu/10.1007/978-981-99-0898-1_17

Woods, K., and K. Bliss. 2016. Facilitating successful online discussions. *Journal of Effective Teaching* 16 (2): 76–92.

Zamecnik, A., V. Kovanović, S. Joksimović, and L. Liu. 2022. Exploring non-traditional learner motivations and characteristics in online learning: A learner profile study. *Computers and Education: Artificial Intelligence* 3: 100051. https://doi.org/10.1016/j.caeai.2022.10051

16

INTERNATIONALIZING THE CURRICULUM

Gil Latz, Robin Matross Helms and Keiko Ikeda

Introduction

Universities today increasingly recognize that they have a responsibility to prepare all graduates to live and work in a global society, an interconnected world where the local and the global have become ever more intertwined (Nussbaum 2002; Altbach and Knight 2007; Latz, Sutton, and Hill 2015). Curriculum internationalization is one key to exploring and exercising this responsibility.

Recognition that the curriculum is a defining element of the effort to internationalize higher education is well articulated by Leask (2015, i): "As the point of connection between broader institutional strategies and the student experience, the curriculum plays a key role in the success or failure of [a campus's] internationalization agenda."

The broad outlines of comprehensive internationalization can be captured as follows: "A strategic, coordinated framework that seeks to align and integrate policies, programs, initiatives, and individuals to position colleges and universities to be more globally oriented and internationally connected" (American Council on Education 2021).

This chapter is organized in three parts: a brief overview of curriculum internationalization; discussion of ways in which early career faculty can successfully incorporate global and international themes into their teaching portfolio; and a profile of an exemplary Japanese approach to virtual exchange/collaborative online international learning (VE/COIL) as an emergent and effective modality for addressing global learning opportunities.

DOI: 10.4324/9781003342953-18

The value of internationalizing the curriculum

Leading thinkers on internationalization and the curriculum argue that for

> the United States to have a truly world-class higher education system, colleges and universities must be globally engaged and prepare students to be citizens of a multicultural community both at home and in a globalized world. Institutions accomplish this by having a multi-dimensional, comprehensive strategy that includes internationalization at home and engagement with global issues and partners.
>
> *(American Council on Education 2022)*

The component parts of internationalization are captured in Figure 16.1, the American Council on Education's Model for Comprehensive

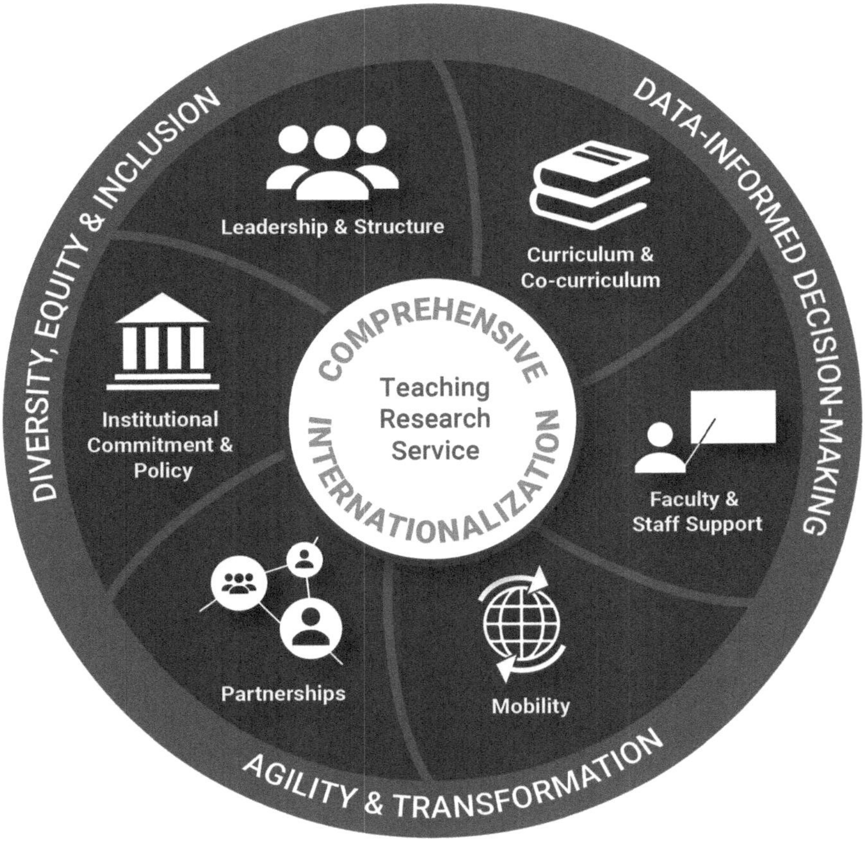

FIGURE 16.1 ACE model for comprehensive internationalization (American Council on Education 2021).

Internationalization, which calls out "curriculum and co-curriculum" as one of six interconnected target areas that require strategic attention and resources in order to advance higher education's teaching, research, and service missions.

The ACE model organizes curriculum and co-curriculum internationalization by

- Placing student learning as a central feature of internationalization;
- Identifying the curriculum as the principal pathway to learning for all students;
- Establishing that curriculum internationalization, regardless of study focus, exposes all students to international perspectives they can build on to acquire global and intercultural competence;
- Ensuring that workforce-ready global competencies are included in institution- or system-wide learning outcomes and assessments; and
- Recognizing that co-curricular programs and activities deliver high-quality learning experiences that complement course-based instruction and align with competencies and skills for working in a diverse postgraduate environment.

Several specific curriculum and co-curriculum internationalization examples are highlighted below.

EXAMPLES OF CURRICULUM AND CO-CURRICULUM INTERNATIONALIZATION

Undergraduate general education/first-degree compulsory curricula require focus on foreign language, regional studies, global issues, and intentional opportunities for self-reflection, intercultural interaction, and identity exploration. All students engage with global and national issues of historical and contemporary racism, colonialism, and systemic injustice.

Courses in each major, program of study, discipline, or research area are internationalized by incorporating international perspectives and highlighting global issues in the field. They provide students with a global and historical context as well as resources and scholarship.

Co-curriculum programs and activities address global issues, reinforce the international and intercultural elements of the curriculum, facilitate discussion and interaction among learners of different backgrounds, and support the integration and success of diverse/international students, faculty, and staff. Learners have opportunities to engage with culturally diverse individuals and organizations in the local community through projects and partnerships built on principles of just reciprocity and collaborative development.

Technology is used in innovative ways to enhance global learning, communication and social skills, research, and collaboration through interactions with students, faculty, and staff across borders. Interactions might be facilitated through collaborative international online learning (COIL), research partnerships, virtual exchange, guest speakers, or administrative collaboration. Adapted from American Council on Education's Comprehensive Internationalization Framework (2021).

Graduating students with an understanding of global dynamics and issues is central to addressing the problem of global warming and climate change, a complex phenomenon that extends beyond the purview of any one discipline. Such recognition reinforces the need to encourage interdisciplinary strategies for global learning. New approaches to the study of complex global occurrences, it should be underscored, do not reject disciplinary-based methodologies. In fact, the goal is to promote disciplinary integration into a diverse and interdisciplinary curriculum. For example, innovative research and practice at Leiden University College in The Netherlands establishes geography as central to the interdisciplinarity found in an emerging international liberal arts program, due to its adaptability and depth (Hudson and Hinman 2017).

Individual academic disciplines have contributed to curriculum and co-curriculum internationalization. Acknowledging that "the disciplines are the 'intellectual homes' for faculty," and "disciplinary associations can and should lead the way in promoting internationalization," four disciplines can be singled out for their initiatives, including the American Association of Geographers (AAG), American Historical Association (AHA), American Political Science Association (APSA), and the American Psychological Association (APA) (Helms and Ward 2014). In the case of geography, for example, the discipline's self-scrutiny has led to calls for new perspectives on complex issues associated with internationalization of the curriculum through surveys and research that ascertain what motivates faculty to pursue international collaboration in their work and incorporate global learning outcomes into the undergraduate curriculum (Solem and Ray 2005; Ray and Solem 2009); new approaches to geography and spatial thinking in the social sciences (Murphy and Latz 1992; Latz 2012); and by questioning the field's over-reliance on traditional world regional geography courses as a means to achieve global learning, resulting in a limited approach to rigorous analysis of global dynamics and issues (Solem 2009, 111).

The American Association of Colleges and Universities (AAC&U) has developed a set of 16 VALUE (Valid Assessment of Learning in Undergraduate Education) rubrics, including four that define the fundamental criteria to identify and evaluate the knowledge, attitudes, and skills associated with curriculum and co-curriculum internationalization (Association of American Colleges & Universities 2009b): global learning, intercultural knowledge and competence, civic engagement – local and global, and ethical reasoning. Each includes

> performance descriptors demonstrating progressively more sophisticated levels of attainment.... [Their] utility...is to position learning at all undergraduate levels within a basic framework of expectations such that evidence of learning can by [*sic*] shared nationally [and internationally] through a common dialog and understanding of student success.
>
> *(Association of American Colleges & Universities 2009a)*

The AAC&U's global learning assessment learning rubric, for example, utilizes the following criteria: global self-awareness, perspective taking, cultural diversity, personal and social responsibility, understanding global systems, and applying knowledge to understanding global systems (Association of American Colleges & Universities 2009b). As will be discussed later in the chapter, VE/COIL is identified in the global learning and other AAC&U rubrics as an example of a faculty-driven, accessible, and effective means to prepare students in various national settings with the knowledge, attitudes, and skills "to understand, proactively engage in, and work collaboratively to solve global issues and problems of importance at local, national, and international levels after graduation" (Association of American Colleges & Universities 2009a).

The strategy and tactics appropriate to meeting curriculum internationalization needs of early career faculty are explored in detail in the following section.

Getting started and making an impact

As described throughout this chapter, global competence has become a critical element of student success; the examples cited illustrate that institutions and organizations have recognized this and are developing curricular programs, initiatives, assessments, and opportunities to ensure students acquire the global skills and knowledge they need. As the linchpins of teaching and learning, faculty are essential to these efforts. But in an era of competing priorities and increasing demands on faculty time and energy, how does incorporating international elements into the curriculum play out on a day-to-day level for early career faculty, without it feeling like an added obligation, or something that might detract attention from activities that are most crucial to faculty evaluation?

Assess and amplify what you are already doing

Rather than starting with a blank page and creating an entirely new course, take a look at existing course syllabi and consider incorporating global perspectives through the following content and activities:

- Globally focused content modules or discussions of topics such as cross-border, regional, or global trends in the field; various national historical, political, and cultural perspectives on course material; challenges for developing countries; intercultural issues in professional practice; international and national laws, standards, and customs pertaining to professional practice in different national settings; nuances of field-specific terminology in different linguistic and cultural contexts.

- Materials such as books and articles written by scholars from other countries; texts that explicitly include an international perspective on subject matter (e.g., in the topics presented, through non-US-based examples, and as part of problem sets and other assignments); case studies that are set in non-US countries or that explore international themes and challenges; articles from newspapers and international journals in the field; foreign films, television, and radio broadcasts; international and non-US-based websites; international data sets.
- Activities and assignments that center on perspective-sharing, including guest lectures by scholars and practitioners with international background and experience (in person or virtually); field trips to local offices of internationally owned companies and other organizations engaged in international activities; participation in campus and local events with an international component (e.g., talks by visiting faculty, museum exhibits, concerts, and festivals); simulations, role-plays, and debates to approach issues from different cultural perspectives; analysis and interpretation of media reports from other countries; student presentations to a real or simulated international/cross-cultural audience; interviews with international students or professionals who have worked internationally; group projects or assignments involving teams comprising both domestic and international students collaborating in person or virtually (see American Council on Education 2013 for additional ideas).

Leverage campus resources

Many of the campus offices and resources that are designed to facilitate faculty development and student success broadly are well-positioned to support faculty members' curriculum internationalization efforts. Consider seeking guidance, expertise, and resources from the following:

- The global programs office. While the office for global programs may be most recognized for working with international students and facilitating study abroad programs, the "senior international officer" (or equivalent) and their staff build and maintain broad networks throughout and beyond campus. With an institution-wide view and scope of activity, they are an excellent starting point for assessing administrative support for curriculum internationalization, and may be aware of faculty in other departments with global expertise who might be tapped for interdisciplinary, internationally focused curriculum components. They are often able to introduce faculty to locally based experts with global expertise who might be invited to provide guest lectures in person, as well as international colleagues to contribute virtual content.

- The academic support unit, such as a center for teaching and learning. Experts in course design are available to review syllabi and serve as a sounding board for what activities will likely resonate with students and lead to global learning. Through their work with international students and other student populations, they are often well-positioned to advise on how to create effective group activities and assignments that facilitate mutual learning and sharing of perspectives without "othering" students or inappropriately expecting them to represent a particular culture or background.
- Grants offices. At institutions where comprehensive internationalization has been set forth as a strategic priority, there may be specific funds for which faculty can apply in order to acquire international course materials, pay for their own travel or that of a visiting lecturer, and so forth. Indeed, as illustrated in Figure 16.2, American Council on Education data show that the proportion of institutions offering such funding increased between 2000 and 2020. When dedicated internationalization funds are not available, it is still worth a conversation with a grants officer about what internal or external monetary

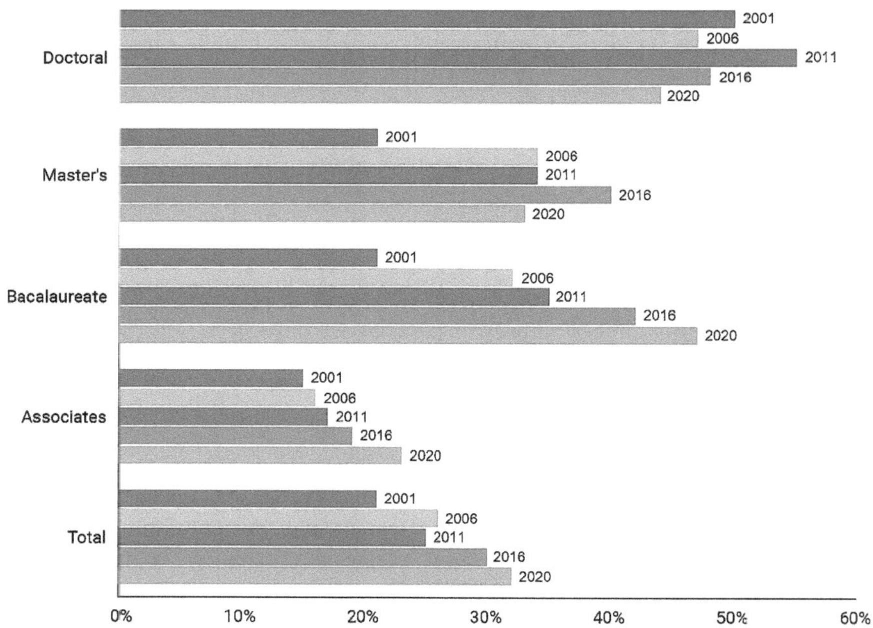

FIGURE 16.2 Percent of institutions providing funding to faculty for course internationalization. Adapted from *Mapping Internationalization on U.S. Campuses*, American Council on Education (2022).

sources are close enough in purpose that faculty could make the case for their application to curriculum internationalization projects.

- Technology support structures. While all faculty quickly became acquainted with virtual learning during the pandemic, campus IT professionals have a depth of expertise and experience with creative solutions that can help even technologically seasoned faculty maximize the effectiveness of virtual activities such as guest lectures, globally shared course content, and broader technology-enabled collaborations such as the VE/COIL courses and programs described in the next section – from which IT platforms to employ to effective pedagogy in a virtual environment. They may also control the scheduling of "smart" classrooms and other technology that supports global learning endeavors, and can direct individual faculty to lesser-known equipment and facilities on campus.
- Cross-campus faculty networks. While the international programs office may be the best facilitator of individual introductions to colleagues with specific global or geographic expertise or interest, for faculty just starting to think about global ties and internationalizing the curriculum, casting a wide net and networking broadly with other faculty in a variety of settings can spark ideas and highlight opportunities and potential directions of interest. Attending book talks and lectures focused on international topics, for example, can generate ideas for content to incorporate into course syllabi, as well as identify potential guest speakers. Globally oriented faculty affinity groups can also be a source of connections; many universities, for example, have Fulbright program alumni groups that host events not only for their members but also to generate interest among the broader campus community.

DON'T FORGET YOUR DISCIPLINARY ASSOCIATIONS

In addition to campus-based resources, faculty disciplinary associations are often an important source of guidance and support for curriculum internationalization. Not only can they point to field-specific issues and examples to incorporate into course content, but they may, in fact, require it.

- The APA's *Guidelines for the undergraduate psychology major,* for example, includes a section on "internationalizing our courses," and specific student learning outcomes such as "Apply psychological principles to address issues of global concern (e.g., poverty, health, migration, human rights, international conflict, sustainability)" (2023, 14).
- The American Chemical Society's *Guidelines and evaluation procedures for bachelor's degree programs* states, "The curriculum should expose students

to the role of chemistry in contemporary societal and global issues, including areas such as sustainability and green chemistry" (2015, 19).

Larger professional organizations may have sections or tracks that focus on global issues, and in some cases global chapters and meetings. A discussion of guidelines (or lack thereof, in some cases) set forth by relevant disciplinary associations around global learning at a campus departmental meeting can be an excellent starting point for developing or refining curricular initiatives within the department, and in individual courses as well as in majors, minors, and certificate programs.

Develop your value narrative and communicate your accomplishments

Whether preparing for a formal, high-stakes evaluation such as a tenure and promotion review in the United States, or more broadly as early career faculty establish themselves at their institutions, articulating how curriculum internationalization efforts and accomplishments demonstrate faculty success and value will help them leverage this work for their career development and progress.

Institutional strategy and policy documents are an excellent starting point in developing this narrative. If the institutional strategic plan, for example, explicitly references global learning goals for students, then a faculty member's narrative for the value of their curriculum internationalization work might focus on furthering these as a critical institutional goal. If there are no such explicit references, however, the onus may be on the faculty member to make the case for global learning as a key component of student success – which is almost always a key pillar of institutional strategic plans. When institutions have articulated campus-wide student learning outcomes, these also present an opportunity to link global learning to broader institutional tenets.

Though faculty work is often by nature very independent, sharing curriculum internationalization efforts and achievements with departmental colleagues and administrators – formally and informally – will help highlight successes, and allow faculty to test and refine their value narrative. Documenting global activities and connections, as well as concrete student achievements and recognition of faculty efforts through campus teaching awards, grants received, and so forth, will help early career faculty tell a compelling story of how their curriculum internationalization efforts demonstrate their broader contributions to their institutions, their fields, and the global academic enterprise. National associations whose missions center on international education can be found throughout the world and provide another networking opportunity for faculty to share with and learn from through participation in annual conferences, accessing publications, and identifying regional affinity groups. Examples include NAFSA: Association of International Educators, the "world's largest and most diverse nonprofit association dedicated to international education and exchange"; the Association of Public and Land-Grant Universities (APLU); Institute of International Education (IIE); Association of International Education Administrators (AIEA); Forum on Education Abroad (FORUMEA);[1] American Council on Education (ACE); Fulbright Program (Fulbright); and a number of other organizations throughout the world.

KNOW THE NUANCES OF YOUR INSTITUTION'S TENURE PROCESS

A 2015 study by the American Council on Education examined nearly 100 tenure and promotion policies from US institutions, and analyzed how they address faculty international experience. The overall conclusion: it's complicated.

There are a variety of different types of polices, and different areas in which international experience may be relevant. In developing a value narrative, faculty should not only reference policy documents, but should also examine and understand the organizational culture that impacts how policies are operationalized. Discussions with the department chair, other faculty mentors, and tenured faculty members who have engaged internationally about how best to position their international experience for maximum value will help junior faculty focus their efforts on global work that will positively impact their evaluations and develop the strongest possible value narrative (American Council on Education 2015).

In the final section, VE/COIL is reviewed as an emergent, innovative, and effective modality that enhances global learning opportunities.

Program spotlight: virtual exchange (VE) and collaborative online international learning (COIL) in Japan

VE/COIL is an innovative pedagogy involving collaborative teaching and learning between classrooms and institutions in two or more countries facilitated by online communication. Delivering global learning and intercultural experiences to greater numbers of domestic and international students, irrespective of their financial circumstances, is a priority of this pedagogy. Because the VE/COIL approach is much more affordable, accessible, and scalable in comparison to physical study abroad, a wider range of students have been able to participate in this endeavor and experience internationalization without leaving their home country (Ikeda and Ahmad 2022). Indeed, VE/COIL enables students to connect with peers and gain insights into cultures from multiple countries, concurrently or sequentially.

Japan is an international leader in the implementation of VE/COIL. Kansai University (KU), a large, private university located in Osaka, is an exemplary case study, illustrating both how individual faculty members have internationalized their course curricula, and how such efforts have been supported and sustained administratively and as institution-wide initiatives (Onorevole 2021; Fujii and Ikeda 2021). KU's VE/COIL engagement began in 2014, with a small cadre of faculty implementing VE/COIL practices in their classrooms.

KU rapidly scaled up its COIL offerings from 2014. By the 2018–2019 academic year, in conjunction with the receipt of a five-year Ministry of Education, Culture, Sports, Science and Technology (MEXT) Inter-University Exchange Project grant, KU created the Institute for Innovative Global Education (IIGE). IIGE was established to help faculty design COIL courses that equip students to be "future ready," not just in terms of their global employability but also their ability to help solve future global issues. One of the main goals of the MEXT grant was to promote VE/COIL partnerships between colleges and universities in Japan and the United States.[2] From 2018 to 2020, IIGE facilitated COIL partnerships between KU faculty and counterparts at 12 universities in the United States as well as with other countries around the world. Concurrently, a growing number of colleges and universities initiated development of COIL courses on their campuses. A new university consortium, the JPN [JAPAN]–COIL Association, was launched in 2018, and as of 2023, a total of 57 institutions have joined the network and meet regularly, sharing exemplary practices and exploring new horizons to realize their international education goals through use of digital technology (JPN-COIL Association 2023).

Student learning goals and outcomes

The VE/COIL global learning experience has had profound outcomes for KU students. Such encounters, enabled by the digitally enhanced pedagogy to connect and learn with non-Japanese students, have provided a positive and measurable impact by nurturing a deeper awareness of global affairs. This awareness is accomplished by designing the KU VE/COIL curriculum to help students learn how to (1) become informed, open-minded, and responsible with respect to differing national perspectives; (2) understand how their actions affect both local and global communities; and (3) address the world's most pressing and enduring issues collaboratively and equitably. Integrating

VE/COIL into course design guides the participating students on global or international matters, and the interdisciplinary nature of VE/COIL practices provides students with a significant opportunity to grapple with contemporary global challenges (Ikeda and Ahmad 2022).

In the authors' view, the most valuable benefit resulting from KU's VE/COIL international partnerships has been to realize global learning across the breadth and diversity of colleges and universities. For instance, KU has had the experience of bringing together 140 students from 11 countries in one VE course, generating a small group project opportunity in each that is comprised of students from at least four or five different national backgrounds (IIGE 2021). KU benefited from having students from Myanmar (Yangon University) in this course amidst the country's coup d'état in 2021, and anecdotal evidence suggested the same occurred for all students in the course. These reports are now being verified by faculty at each institution. In another COIL course on diversity and inclusion that took place after the Russian invasion of Ukraine, Russian students overcame ongoing domestic turmoil to seek learning opportunities with international peers.

Starting simple, scaling support

The KU example highlights multiple dimensions of curriculum internationalization. As a starting point, engaging in the VE/COIL pedagogy can be done simply through an individual faculty member's own initiative. If a faculty member knows someone overseas and they decide to arrange for their classes to meet online using whatever technology is available and preferred in pursuit of collaborative learning, the technical elements of a VE/COIL course are established, and curriculum internationalization is underway at the classroom level.

While faculty can begin utilizing the VE/COIL global learning methodology as classroom-by-classroom, on-the-spot intervention for internationalization, the higher education institution's leadership might consider more comprehensive approaches that best support and prepare faculty throughout the institution to be leaders in imparting global learning to their students. One example is that institutional leadership can help faculty understand the importance of formulating a Memorandum of Understanding (MOU) or Memorandum of Agreement (MOA) with a partner university. The MOU/MOA framework clarifies how each institution can commit to supporting a learning ecosystem capable of nurturing, codifying, and measuring the outcomes of inter-institutional VE/COIL projects.

A second example is to recognize that in order to advance campus-wide transformation through internationalization, institutional commitment can play a key role through establishment of a teaching and learning center on campus that could assist in identifying exemplary practices from the VE/COIL learning modality. Faculty, especially early career faculty, may not have sufficient readiness to implement courses with globally embedded themes on their own. Even if they have been abroad (e.g., study abroad; career abroad), the scope of such "international experiences" may be limited when considered from the perspective of the number and variety of travel destinations. Whether early career faculty have had the necessary opportunities to "sublimate" their knowledge and experience, allowing them to impart global knowledge to their students is a key question worth posing. Capacity-building tactics to intentionally enhance the intercultural competence of faculty is a cornerstone of successful internationalization of the curriculum, as illustrated by the establishment of the IIGE at KU in 2018. This is an important insight from Japan that has bearing on faculty in other countries as well.

Notes

1 https://www.forumea.org/.
2 For more on the US–Japan COIL initiative, see https://www.acenet.edu/Programs-Services/Pages/Communities/US-Japan-COIL-Initiative.aspx.

References

American Chemical Society. 2015. *Undergraduate professional education in Chemistry: Guidelines and evaluation procedures for bachelor's degree programs.* Washington, DC: American Chemical Society. https://www.acs.org/content/dam/acsorg/about/governance/committees/training/2015-acs-guidelines-for-bachelors-degree-programs.pdf

American Council on Education. 2022. Making the case for internationalization. Accessed September 11, 2023. https://www.acenet.edu/Research-Insights/Pages/Internationalization/Making-the-Case-for-Internationalization.aspx

American Council on Education. 2021. Comprehensive Internationalization Framework. Accessed September 11, 2023. https://www.acenet.edu/Research-Insights/Pages/Internationalization/CIGE-Model-for-Comprehensive-Internationalization.aspx

American Psychological Association Board of Educational Affairs. 2023. APA guidelines for the undergraduate psychology major, Version 3.0: Empowering people to make a difference in their lives and communities. Washington, DC: American Psychological Association. https://www.apa.org/about/policy/undergraduate-psychology-major.pdf.

Altbach, P. and J. Knight. 2007. The internationalization of higher education: Motivations and realities. *Journal of Studies in International Education 11*(3–4): 290–305.

Association of American Colleges and Universities. 2009a. Global Learning VALUE Rubric. Accessed September 11, 2023. https://www.aacu.org/initiatives/value-initiative/value-rubrics/value-rubrics-global-learning

Association of American Colleges and Universities. 2009b. VALUE Rubrics. Accessed September 11, 2023. https://www.aacu.org/initiatives/value-initiative/value-rubrics

Fujii, M., and K. Ikeda. 2021. Case study: Virtual exchange/COIL beyond the classroom: Impact on Kansai University's U.S. partnerships. In *U.S.-Japan Higher Education Engagement Study*, V. M. Onorevole, ed., 43–45. Washington, DC: American Council on Education.

Helms, R. M. 2015. *Internationalizing the tenure code: Policies to promote a globally focused faculty*. Washington, DC: American Council on Education, Center for Internationalization and Global Engagement. https://www.acenet.edu/Documents/Internationalizing-the-Tenure-Code-Policies-to-Promote-a-Globally-Focused-Faculty.pdf

Helms, R. M., and M. Tukibayeva. 2013. Internationalization in action: Internationalizing the curriculum, Part 1 – Individual courses. American Council on Education. Accessed September 11, 2023. https://www.acenet.edu/Research-Insights/Pages/Internationalization/Intlz-in-Action-2013-December.aspx

Helms, R. M., and H. H. Ward. 2014. Internationalization in action: Internationalizing the curriculum, Part 4 – Disciplines. American Council on Education. Accessed September 11, 2023. https://www.acenet.edu/Research-Insights/Pages/Internationalization/Intlz-in-Action-June-2014.aspx

Hudson, P. F., and S. E. Hinman. 2017. The integration of geography in a curriculum focused to internationalization: An interdisciplinary liberal arts perspective from the Netherlands. *Journal of Geography in Higher Education 41*(4): 549–561. doi: 10.1080/03098265.2017.1337089.

IIGE. 2021. UMAP-COIL Joint Program. Accessed July 15, 2023. https://kuiige.wixsite.com/umapcoil/2020-online-program

Ikeda, K., and A. L. Ahmad. 2022. COIL/Virtual Exchange as a driver for high-performance international partnership building beyond the response to COVID-19. In *Higher Education in Southeast Asia and Beyond (HESB)*, ed. L. K. Yeong, 23–26. Report No. 12. Singapore: The HEAD Foundation.

JPN-COIL Association. 2023. JPN-COIL Association news. Accessed September 11, 2023. https://www.kansai-u.ac.jp/Kokusai/IIGE/JPN-COIL/

Latz, G. 2012. Asia: Transnational organizations. In *21st Century Geography: A Reference Handbook*, J. P. Stoltman, ed., 585–600. Los Angeles: Sage Publications, Inc.

Latz, G., S. Sutton, and B. Hill. 2015. An internationalized stewardship of urban places. *Metropolitan Universities Journal 25*(3): 83–98.

Leask, B. 2015. *Internationalizing the Curriculum*. London: Routledge.

Murphy, A., and G. Latz. 1992. Geography and International Studies. In *Revealing the World: An Interdisciplinary Reader for International Studies*, D. Lieberman and M. Gurtov, eds., 23–51. Dubuque, IA: Kendall-Hunt.

Nussbaum, M. 2002. Education for citizenship in an era of global connection. *Studies in Philosophy and Education 21:* 289–303.

Onorevole, V. (ed). 2021. *U.S.-Japan Higher Education Engagement Study*. Washington, DC: American Council on Education.

Ray, W., and M. Solem. 2009. Gauging disciplinary engagement with internationalization: A survey of geographers in the United States. *Journal of Geography in Higher Education 33*(1): 103–121. doi: 10.1080/03098260802276425

Solem, M. 2009. Geography and global learning. In *Teaching College Geography: A Practical Guide for Graduate Students and Early Career Faculty*, M. Solem and K. Foote, eds., 111–120. Upper Saddle River, NJ: Pearson Prentice Hall.

Solem, M., and W. Ray. 2005. Gauging disciplinary support for internationalization: A survey of geographers. *AAG Report to the American Council on Education.* Upublished manuscript. Accessed September 11, 2023. https://citeseerx.ist.psu.edu/document?repid=rep1&type=pdf&doi=8aca513abf02679aa826ce9c24ef4d4a5004de45

Soler, M.C., J. H. Kim, and B. G. Cecil. 2022. *Mapping internationalization on U.S. campuses.* Washington, DC: American Council on Education. https://www.acenet.edu/Documents/Mapping-Internationalization-2022.pdf

SECTION III

Leading in research and scholarship

Michael Solem

One of the major attractions of an academic career is the opportunity to advance knowledge through research. The research process can be exciting, and there is real joy to be found in discovering meaningful patterns and information contained in quantitative and qualitative datasets. Research appeals to many faculty for its inherent creativity and the potential for the results of research projects to improve human welfare and environmental conditions from the local to the global.

In her chapter, Antoinette WinklerPrins discusses the importance of having a vision for your research – an agenda that encourages you to think about the contributions you hope to make over the short term and further into the future. Much of this will depend on the nature of your institution, as expectations for research vary across teaching-intensive colleges and universities known for high levels of research productivity. Antoinette differentiates between basic and applied research while noting the importance of understanding the broader impacts of both types. As in many aspects of academic life, researchers benefit from having a robust professional network from which productive, long-lasting collaborations can emerge.

Getting research off the ground can be a challenge especially for complex projects requiring external funding support. While some research doesn't require funding, many faculty will seek support that enables them to hire research assistants and pay for instrumentation, data licenses, and other costs. Patricia Solís offers a set of recommendations that, taken together, can result in a highly competitive research proposal for a sponsor. By following Patricia's advice, you can craft a proposal that is marked by strong coherency – a careful alignment of all the elements and components that compose

DOI: 10.4324/9781003342953-19

an application, from the problem statement and research questions to the methods, dissemination plan, and budget.

Let's say you successfully funded or otherwise implemented your research ideas. Great! Soon you will begin the scholarship portion of the research activity. By that, we mean taking your work public through the publication process. Just like the process of investigating questions and collecting and analyzing data, publishing research results can come with a lot of satisfaction. There are, however, many issues to be mindful of, from selecting an appropriate journal to understanding what goes on behind the scenes as editors consider manuscript submissions and send them out for peer review. Scholars as diverse as those you will find in geography and related sciences have a wide array of potential outlets where they can possibly get their work in print (and online). The chapter by Lynn Resler, May Yuan, and Josh Inwood guides you through the nuts and bolts and procedural considerations of publishing.

Let's stick with the optimistic outlook and suppose your manuscript is accepted and published in a desired journal. Congratulations! Yet as Derek Alderman explains in his chapter, publication is not the end of the research story. Derek advocates for all faculty to promote their hard work beyond the readership of academic journals. This is important not only for raising general awareness of the existence of one's research but also for communicating the work's significance to broader publics. Derek recommends a variety of strategies for maximizing the exposure of research publications via social media, the press, conference presentations, editorials, and other forms of outreach.

The editors of this book wish you well on your research journey and hope the experience is enjoyable and impactful.

17

PLANNING A RESEARCH CAREER

Antoinette M.G.A. WinklerPrins

Introduction

The goal of this chapter is to help you think comprehensively about the arc of your research career. For many of you this may feel strange, as you are or have been so focused on completing your dissertation, or finding a job, or just surviving as a post-doc, a contingent faculty member, or an assistant professor. But it is important at this early stage of your career for you to actively think about the arc of your career – being proactive about it. Just taking things day to day and being reactive will not be adequate or satisfactory for your career in the long run. Most academic careers last decades, 30, 40, even 50 years. The work of a scholar or scientist is one that can have a long arc, and it is worth contemplating where you might end up. What do you want to be known for? What is the legacy you wish to leave behind as a scholar, scientist, teacher, or as an academic administrator?

Why consider a possible path for the future, now, early on in your career? Because strategic thinking about your goals and objectives will set you up better for success. Many benchmarks for academic careers necessitate a demonstration of a coherent *corpus* of work, so you want to be strategic about how the various pursuits "hang together" as it were. Also, many activities you do as a scholar-scientist involve some planning, for example, which topics to pursue, which publication outlets to target, which collaborations to pursue and which to decline, how to integrate teaching and research, or how to effectively integrate service into a research career. One area that takes a lot of strategic planning is finding financial support for your research – writing and succeeding at grant applications. Thinking about and setting targets is an iterative process that will help you achieve your career goals, even if these are not yet fully formed.

DOI: 10.4324/9781003342953-20

Specific objectives for this chapter are to help you:

1 Appraise your process of research topic selection,
2 Identify the many considerations that impinge on and enable your research career, and
3 Recognize different types of science and scholarship.

How do you identify research topics?

Take a moment and reflect on how you settled on your dissertation topic. Was it a topic you became interested in early as an undergraduate and you had since been determined to find the answers to certain questions (perhaps related to personal experience)? Did the topic for your dissertation come from identifying a *lacuna* in the literature you were exposed to during seminars in graduate school? Perhaps you were a graduate assistant in a lab and were assigned a particular component of a larger project your advisor or lab director was pursuing? Or was your topic motivated by the needs of a community you identify with? Likely it was a combination of interests and opportunities – yours and your advisor's.

Other considerations can influence the selection of a dissertation topic, and these may well continue to be important factors in your career. Could the topic be pursued given the timeframe of your degree? Was it practical to pursue given your personal circumstances and the time available? Did you have the financial support to conduct the study? Did you have the professional connections, networks, and skills to conduct the fieldwork or necessary analysis? Did you have access to the necessary data?

Pursuing a dissertation topic has traditionally been internally driven – it was what you wanted to do, a deep dive into a singularly focused area of inquiry. This is changing as more students in geography, environmental science, and related fields work in labs and come from communities that have expectations from them. What you will also find is that, as you pursue research foci beyond your dissertation as a junior scholar, external factors, such as what others (department chairs, deans, community leaders) expect you to do, may form and direct your research more strongly.

One aspect of planning the arc of your career is that serendipity plays an important role. So yes, you need to plan and set objectives and targets for your career, but you also need to set yourself up for unplanned opportunities that might come your way as the result of chance encounters. You do so by networking on your campus and in your community; being widely engaged in your field, not just in your narrow focus area; participating in a variety of conferences small and large, local and international; and accepting invitations to speak (including volunteering to speak!) formally or informally on areas you are developing expertise in.

Factors that influence the type of research you will be doing

Many factors will influence the research you will do throughout your career. Some of these factors you can control, whereas others you cannot. It is important to work with what you have available and what is possible. Embrace where you are located to pursue local or regional work and seek opportunity from turns of events that might be unanticipated.

The type of position and the type of institution you are employed at will significantly influence the arc of your career. Familiarize yourself with the research expectations of (and support provided by) different types of institutions (in the US, see the *Carnegie Classification of Institutions of Higher Education*). If you are on a limited-term appointment with a heavy teaching load, the time available for research may be limited. But if remaining an active scholar is important to you and essential for securing your next position, then you will need to be very deliberate about setting aside dedicated time for research. Learn to incorporate research into your teaching – not only will that assist you with your career, but you will be exposing a new generation to research as well. This is important not only in an institution that is focused on teaching, but also in an institution where you are asked to teach and conduct research. What will also factor into your research decisions is that there may be connections with the agendas of your department chairs, college dean, or institution. You want to maintain your own identity, but being open to and working with the opportunities that present themselves via institutional agendas can be advantageous. And what is critical for your success is that you are well aware of the expectations of what gets you to the next level (tenure, promotion to Associate or Full Professor, or other career ladder steps). This should be an ongoing dialogue between you and your chair, supervisor, and/or mentor(s) – be active and deliberate in these conversations.

There is no denying that your personal situation factors into the arc of your career. Time for partners, children, elders, other family, pets, and community obligations vary in their demands in your life, circumstances are not always predictable, and will change over the course of your career. It is important to bring your whole self to the arc of your career, not deny your life outside work, and to see the possible stages and situations as opportunities and not impediments to pursuing your research.

Realistically there are better times for certain types of research than others, and you want to have a research agenda that is flexible enough to accommodate. If getting to a far-away location for research becomes difficult, dig deeper into previously collected data or secondary data or find a meta-analysis project that requires no travel. Also, consider pursuing local/regionally based research, leveraging what your place has to offer. Not only will this be much more convenient, but this approach has the additional benefits of being able to include students more easily. Also, fully embrace technology to assist your work and become familiar with the electronic resources available to access from the convenience of your office – the digital revolution has truly changed access to data and archives. The creativity we saw during the COVID-19 pandemic of research pivots (e.g., changing a research site from those far away to one closer by; a switch to using virtual meeting software to conduct interviews) and creative approaches demonstrated that novel circumstances can be accommodated.

While we are decolonizing the curriculum (Chapter 10), research too needs to be decolonized. There are inherent power structures in the process of conducting research, and these need to at minimum be made explicit, but better yet is to frame the research process differently. One way this is being done is by "co-producing" research. More broadly, this is termed "engaged" research, defined as "research conducted via meaningful collaboration among scientist and non-scientist actors that explicitly recognizes that scientific expertise alone is not always sufficient to pose effective research questions, enable new discoveries, and rapidly translate scientific discoveries to address society's grand challenges" (National Science Foundation 2022, 5). Many of you are embedded in communities and may be motivated by questions that your community wants you to study. If you are not from the community being studied, it is essential that you be an engaged researcher, developing relationships with the target community early in the process, and co-producing the research agenda, the research itself, and the research outcomes, together with participants and other stakeholders (e.g., Lemos et al. 2018).

If your research necessitates funding at any level, then a very important factor in influencing your research agenda is the priorities of the funders. The organizations that financially support research do not operate in a vacuum, and this includes public funding, and you need to realize that you need to work with their agendas, which may change over time, molding your research to fit those demands. Funders set priorities and have missions – become familiar with them, and set up your research to respond to those priorities and missions. More on this in the following sections, and in Chapter 18.

Types of science

One of the considerations is the type of science or scholarship that you participate in. Geography and related disciplines span across a broad range, from humanities-focused research to positivist science. Where do you sit? Which frameworks have relevance to you – and, more importantly, which do you wish to be engaged with? Within science, there is a range from basic (fundamental) to applied research, with differential considerations of use versus the quest for fundamental understanding (Stokes, 1997).

Pure basic (Bohr's quadrant) = fundamental science, curiosity drive, theoretically motivated

Use-inspired (Pasteur's quadrant) = fundamental but motivated by use, some theory

Pure applied (Edison's quadrant) = problem-solving, no theory

It is important to understand what motivates you (more theory or more application), and what motivates the institution that employs you. Become familiar with the mission of the institution or organization that employs you and embrace how that can motivate and direct your type of science and scholarship. It is also important to realize that understanding the range of basic to applied research relates to where you seek funding to support your research and scholarly activity. Different funders have different priorities, for example, in the US, the core mission of the US National Science Foundation (NSF – a government agency) is to support basic research (Bohr's quadrant). However, it requires societal relevance of the research it supports, and is increasingly interested in supporting use-inspired research (Pasteur's quadrant). Mission-focused government agencies (in the US, for example, NASA, NOAA, USDA) and most non-governmental organizations or foundations focus much more on use-inspired research or are completely focused on problem-solving (Edison's quadrant). It is important to align your research interests with those of possible funders (see Chapter 18 for a detailed discussion about writing a competitive grant proposal).

Types of scholarship

Another consideration in planning for the arc of your career is the type of scholarship you want to be known for. This overlaps with the types of science discussed above but adds some additional considerations. Ernest Boyer, in his influential book, *Scholarship Reconsidered*, provides a framework for types of scholarship (Boyer 1990). First, the **scholarship of discovery** is basic research, knowledge for knowledge's sake, curiosity-driven, akin to the aforementioned "pure" research. When most people hear the term "research," this is what they think of. This type of research "contributes not only to the stock of human knowledge but also to the intellectual climate of a college or university" (Boyer 1990, 17).

Second is the **scholarship of integration**. This is the scholarship of "making connections across disciplines, placing the specialties in larger context, illuminating data in a revealing way, [and] often educating non-specialists too" (Boyer 1990, 18). Geographers are particularly well suited to this.

Third is the **scholarship of application**, putting new knowledge to use in some sort of application, the scholarship of putting knowledge to societal benefit. This is akin to the aforementioned applied research, as it asks the question, "How can knowledge be applied to consequential problems?" (Boyer 1990, 21).

Fourth is the **scholarship of teaching**, focusing research on the teaching process. Boyer notes that "teaching is often viewed as a routine function," but "the work of a professor becomes consequential only as it is understood by others" (Boyer 1990, 23). He builds on Aristotle's famous statement that "teaching is the highest form of understanding," and this form of scholarship is very important for any of you working in the academy. It is important to see teaching as an opportunity to inform your research, as some of the best questions come from students. This may not have been part of your graduate school experience but may evolve from your work as a professor.

In 1996, Boyer added a fifth type to his scholarship framework, the **scholarship of engagement** (Boyer 1996). This type of scholarship focuses on "connecting … the university to … [the] most pressing social, civic, and ethical problems" (Boyer 1996, 32),

> ultimately, the scholarship of engagement means creating a special climate in which the academic and civic cultures communicate more continuously and more creatively [with] each other, helping to enlarge … the universe of human discourse and enriching quality of life for all of.
>
> *(Boyer 1996, 33)*

Boyer set in motion an important conversation about engaged research that matured into the recommendations in the report cited earlier (US National Science Foundation 2022).

The type of scholarship you practice is likely to be a combination of types, and the type or combination will likely change over the course of your career. A dissertation is often a work of discovery, a piece of newly generated knowledge, where original data were collected and analyzed. For many of you, primary, original data collection and analysis may become the dominant form of your scholarship throughout the arc of your career. A dissertation can also be a work of integration, bringing together secondary data sets and deriving new knowledge from that integration. Research and scholarship as meta-analysis is becoming increasingly common and central in many fields because of the rise and availability of big data, and ever richer pools of secondary data that have become available, in part, because of the requirement of better and more consistent, streamlined, and available data sets. Many research grants, especially those supported by public (government) funding, require data management and sharing plans (DMSPs), and this has led to changes in how primary data is collected and archived in accessible repositories, and made available for others to use.

Science and scholarship can consist of individual or collaborative work. This varies by discipline, but in geography and related fields, researchers operate across the full continuum from individual to team science. Where do you fit in? Your dissertation experience may have been conducted in a lab, and you may believe that is how research is done and how you want to work. Alternatively, you may have been advised that you must produce single-authored research in order to be taken seriously. How you pursue this will depend on your experience, comfort, the specific topic, and the expectations and constraints of the position you have – be sure to ask explicitly about these expectations from your department chair or supervisor. It is healthy for most researchers to produce a mix of products demonstrating your capacity to conduct research on your own while also indicating your capacity to collaborate in small and large teams. If you do work in collaborative teams, ensure you participate actively and complete your work in a timely manner. Similarly, seek out collaborators who finish their work.

TEAM SCIENCE

Participating in research generated by a team of collaborators ("Team Science") is quite common in geography and related sciences, especially if you are working on inter- or transdisciplinary topics, and regardless of whether you are comfortable working in this space or not, it is likely you will need to engage in it. It is important to recognize that team research requires active work in the collaboration process itself; it does not happen automatically. This is especially important if you have team members from different disciplines and from outside academia (e.g., NGOs, government). There is an emerging literature on Team Science, and it is worth becoming familiar with it if you are working in or need to form a team (e.g., Lyall and Fletcher 2013; National Research Council 2015; Palmer et al. 2016). The following are a few key points: (1) *Communication is essential*. All aspects of the project need to be communicated explicitly: project expectations, theoretical frameworks, priorities, timeline, semantics, methods, products, member roles, and so forth, all need to be discussed. Take time to set these clearly, and use facilitation if you can. (2) *Avoid predetermination*. If you are the leader of the project, ensure that you are open to perspectives that team members bring. In other words, do not predetermine or be rigid in the way the project should unfold. (3) *Take adequate time for ideation and iteration*. Working with others will be slower because of the time needed to handle the research process, but in the end, this time invested with likely result in a stronger project.

For many scholars today, the desire to conduct research collaboratively with or on behalf of a community is becoming much more commonplace. But it is important to understand the difference between activist research and engaged research. Activist research includes the active promotion and targeting of certain goals or endpoints. This type of work can be incredibly rewarding, but it is important that you understand the degree to which this type of research and scholarship is valued given the position you have. Activism is, by definition, not objective, and some funders do not support this type of research, while others do so explicitly.

Funding sources

Understanding the landscape of funding opportunities is critical to support your research career, and grants writing is an essential component of a successful scholarly career. But obtaining funding is hard work and requires persistence. Chapter 18 in this book provides more detail about funding your research and writing competitive grant proposals, but in the context of planning your career, there are several matters to be aware of.

Understanding what funders want to support and aligning your work with those priorities is essential. Creating the right fit is key to success. This means that you need to learn a lot about the universe of funders, and what they are looking for, and you need to be willing to accommodate your research interests to fit specific opportunities.

Read acknowledgments in papers to learn how researchers you admire were supported. Then read up and ask about those funders, and attend relevant workshops and outreach seminars given by them, as well as those organized by professional organizations at your own university.

If you work at an organization with a sponsored research office, establish connections with it and utilize the resources they provide. Talk with colleagues who have been successful. Recognize that for every grant an investigator obtains, there are likely many declined applications. Persistence is essential. Also be sure to leverage internal/start-up funds from your organization to pilot research ideas and collect preliminary data. Demonstrated capacity is essential for success in obtaining many external grants.

The path ahead

As you embark on developing the arc of your scholarly career, there are many considerations. Over time, your skill and knowledge will deepen, leading to a career that is multilayered and multifaceted, and it is best that the career arc contains a unifying thread that defines your *corpus* of new knowledge creation. Careers develop in phases, and what is central later in a career may be different from what was of central importance early on. Here are some key takeaways:

1. It is important to **think actively about your career**, which you are doing by reading this chapter, and by doing some of the suggested follow-up activities, but be sure to keep this up!
2. Identify a **unifying theme for your work**, even while working on multiple projects at the same time. When you apply for positions, and when your

work is evaluated (for tenure, for promotion), those reviewing applications and promotion packets will want to see a coherent trajectory of your research. You want to clearly project a direction of where you are going with your research – demonstrating that you are planning a research career. So be careful with what could be perceived of as "one-offs." Perhaps you are asked to provide methodological support on a project by a colleague and you need to say yes. Then ensure that the skill advanced in that work is also applicable to your core work. Also, learn to build on previous work; sometimes, a direction that you were heading to does not work out, requiring you to redirect or to pivot, but there may be threads you may return to years later. Some examples of a unifying theme are response and recovery of geomorphic systems to stress and change; changing smallholder livelihoods in the Global South; territorial arrangements and how these link to identity and a sense of place; migration and transnational networks; urban spatial dynamics in a post-pandemic world; rural–urban migration and socioeconomic mobility.

3 Be **deliberate and strategic, but be open to serendipity and opportunity**. Have a plan but be flexible, and if you pivot, remain coherent in your overarching theme. Have a list of goals and objectives, and revisit these periodically. Maintain several active projects at once.
4 Actively seek and find **mentors and collaborators** – within your institution and beyond. To be successful, you need to network broadly. To do this, be sure to present papers and give presentations frequently, on all dimensions of your work (theory, methods, and analysis), specifics and the generalizable findings, and in many settings. Include brown bags and seminars in your own department and institution and attend small and large conferences – those within your specific field, and those beyond (think about who you wish to disseminate your findings to). You never know who might be in the room – that individual may need you as a collaborator. Also, do not be afraid to reach out directly to people whose work you appreciate, and ask them specifically about their path. Participate in formal mentoring programs and seek additional informal arrangements. Seek out and attend workshops about writing grants, especially those that provide feedback on drafts of proposals. Seek feedback on proposal drafts from various colleagues, those both close to your research and those further away from it. The type of feedback they provide is different; both are needed.
5 Actively and constantly be on the **lookout for potential sources of support** for your research. Support for your research should be for both funds and time. Look for small and large pots of money to tap into, fellowship opportunities, buy-outs, internal and external grants, public and private funding; one can lead to the next. Also, be conscientious about seeking research assistance, time being a valuable commodity. Pursue opportunities to engage students, graduate, undergraduate, as well as post-docs, to work for you as

research assistants. This liberates you from doing all the work while helping to develop others. Engaging with students incorporates an educational component into your research career, alleviates you from having to do all the tasks, and helps you learn from their skills. This approach comes with time needed to delegate and supervise, but if done well, can be rewarding.

6 **Remain curious**. It is important to stay interested and engaged in your research – be strategic but not at the expense of what excites you.

Conclusions

Setting a course for your career is an important part of your early career development and requires active work. Being deliberate about your pursuits can reap rewards in the long term. Hopefully, this chapter has influenced how you think about the work required for shaping your career while also preparing you for serendipitous opportunities. Be cognizant of the long game, have a story to tell, and have something to say – an overarching theme for which you want to be known.

Acknowledgements

The writing of this chapter was supported by the US National Science Foundation while the author worked at the agency. Any opinion, findings, conclusions, or recommendations expressed in this material are those of the author and do not necessarily reflect the views of the US National Science Foundation.

I thank the geographers and environmental scientists who shared the stories of their scholarly career arcs with me during informal conversations.

References

Boyer, E. L. 1990. *Scholarship Reconsidered: Priorities of the Professoriate*. Jossey-Bass: San Francisco, CA.

Boyer, E. L. 1996. The Scholarship of Engagement. *Bulletin of the American Academy of Arts and Sciences* 49 (7): 18–33.

Carnegie Classification of Institutes of Higher Education. https://carnegieclassifications.acenet.edu/index.php [last accessed 8/26/2023].

Lemos, M. C., J. Arnott, N. M. Ardoin, K. Baja, A. Bednarek, A. Dewulf, C. Fieseler, K. Goodrich, K. Jagannathan, N. Klenk, K. J. Mach, A. M. Meadow, R. Meyer, R. Moss, L. Nichols, K. D. Sjostrom, M. Stults, E. Turnhout, C. Vaughan, G. Wong-Parodi and C. Wyborn. 2018. To co-produce or not to co-produce. *Nature Sustainability* 1: 722–724.

Lyall, C., I. Fletcher. 2013. Experiments in Interdisciplinary Capacity-building: The Successes and Challenges of Large-Scale Interdisciplinary Investments. *Science and Public Policy* 40: 1–7.

National Research Council. 2015. *Enhancing the Effectiveness of Team Science*. The National Academies Press: Washington, DC.

Palmer, M. A., J. G. Kramer, J. Boyd, and D. Hawthorne. 2016. Practices for Facilitating Interdisciplinary Research: The National Socio-Environmental Synthesis Center (SESYNC). *Current Opinion in Environmental Sustainability* 19: 111–122.

Stokes, D. E. 1997. *Pasteur's Quadrant: Basic Science and Technological Innovation.* Brookings Institution: Washington, DC.

US National Science Foundation (NSF) Advisory Committee for Environmental Research and Education (AC-ERE). 2022. *Engaged Research for Environmental Grand Challenges: Accelerating Discovery and Innovation for Societal Impacts.* US National Science Foundation: Alexandria, VA.

18

THE COHERENCE FRAMEWORK

Ten actions for crafting competitive research grant proposals[1]

Patricia Solís

Beyond the academic duties of teaching, research, publishing, and service, expectations that faculty should generate external funding are increasing – and are increasingly important. Tenure and promotion evaluations of faculty may or may not explicitly review dollars won, but they are clearly influenced by successful grant-getting as a measure of contribution to one's field, especially for peer-reviewed proposals. In nearly every type of academic institution, from small liberal arts colleges to state public universities, the pressures of shrinking state budget allocations, coupled with increasing proportions of organizational revenue coming from outside sources (Lee and Clery 2004; Musselin 2018; Flannery 2022), make grantsmanship an ever-critical aspect of professional life for tenure-track professors, research faculty, lecturers, and part-time faculty alike. Graduate students, as well, seek outside monies to conduct their research activities, not only to garner additional financial support, but also to demonstrate their abilities in obtaining funds and gain early grants experience that will serve them well in their budding careers.

Even if a grant is not awarded, the process of writing research grant proposals can help you solidify your research ideas, and make them tangible, realistic, and programmatic. Proposals, funded or unfunded, can themselves become sources for publishing journal articles. Writing to external sponsors can aid you in thinking through your rationale or finding ways to connect with others in the scientific community or with the broader society – whether you end up with the money or not.

This chapter aims to give you some advice for how to do that successfully. Admittedly, there have been a great many articles and books written on the

DOI: 10.4324/9781003342953-21

subject of grant proposal preparation. These resources adeptly cover issues such as choosing research topics, mechanics of writing, necessary elements of proposals, hallmarks of successful proposals, obstacles to overcome, strategies for locating funding sources, perspectives of sponsor agencies, and effective research design. In this chapter, I include a focus not only on practical recommendations that will help you prepare competitive research grant proposals, but also on habits of mind and action that will prepare YOU as a researcher who is successful at generating external funding.

Here, then, are **Ten Actions You Should Take**. These of course do not represent everything you need to know, but they will help you develop your own strategies and tactics for success to incorporate into your professional practice:

1. START WITH A GOOD PROBLEM.
2. CREATE THE RIGHT FIT.
3. ASSEMBLE A WINNING TEAM.
4. DESIGN TO DELIVER.
5. BE PERFECTLY PERSUASIVE.
6. MAKE IT REAL.
7. DEMONSTRATE YOUR UNIQUE VALUE.
8. GO THE EXTRA MILE.
9. ACHIEVE AND COMMUNICATE COHERENCE.
10. LIVE IT.

The remainder of this chapter explains each of these ten imperatives and includes some suggested concrete actions you can implement either as a novice to begin preparing winning proposals or, perhaps, as a seasoned grant writer trying out some new ideas and tools to refine your skills – ultimately to become a more effective research grant proposal writer.

Ten actions you should take

Start with a good problem

No amount of writing skill will compensate for a poorly developed research problem. A good problem is more than just a topic; it encapsulates some kind of tension, contradiction, unresolved issue, challenge, question, or even a mystery of sorts. A particular problem makes for a good proposal if it is very specific, precise, and focused. You should be able to clearly express what this research problem is in no more than three to five sentences.

Good problems lead to good questions, which underlie the scientific enterprise, be it theoretical or applied, quantitative or qualitative. The objective of a research project is always to answer those questions. This sequence of problem-question-purpose should be explicit as your thesis. Clarity, lucidness and a sharp thesis will allow you to develop a sound project and a convincing proposal. It will leave your reviewers with something to remember after reading the many other competing proposals. Make sure this appears in your first paragraph. Like first impressions, reviewers will likely make up their minds quickly about whether your proposal is competitive or not.

A good problem is properly contextualized. A critical element of grant proposals is a summary of the current state of knowledge that is supported by references to relevant scholarly publications. You should carefully and explicitly position your work relative to this context. This entails a different approach to writing than would a traditional "literature review," especially as you consider that most likely not all of your reviewers will be familiar with your specialized literature. A good rule of thumb is to precisely answer this series of questions: "What is already known? What don't we know that is worth finding out? What will be learned from this research?" "What [have others] done to address the problem and why wasn't that sufficient?" (Gerin 2006, 73). This could also be conceptualized as a narrative path, moving from a generalized idea to a specific research problem to be solved, identifying the "objective and the rationale for why it fills a gap in knowledge," according to Wentz (2013, 7).

To communicate this contextualized problem, it is helpful to consider three general models of what research is supposed to accomplish with respect to existing knowledge. These include that research might (1) advance an existing line of research; (2) resolve a contradiction; or (3) develop a new line of inquiry. There are certainly other functions of research, but these three are common models to begin thinking about how your project contributes to the academic enterprise of knowledge production.

Create the right fit

Like people, institutions have their own personalities, and so do funding institutions. Whether they are private, federal, state, foundation, or other kinds of organizations, each has their specific mission, goals, and program objectives. Each sponsor organization also has its own preferences on operational protocol, including whether they invite unsolicited proposals or not, whether they issue calls for proposals that are clearly or vaguely defined, whether they prefer formal or informal inquiries, whether program officers should be contacted prior to deadlines, and other tacit protocol. You should make it an integral part of your research proposal preparation process to find out about the individual characteristics of potential resource providers. In other words, do your homework before you write. Many organizations have extensive and easily accessible information online where you can research past funding history and current priorities. If information about previously funded projects is available, study it. If only the contact information of their previous awardees is listed, consider talking to funded principal investigators (PIs) about their projects and experiences with the sponsor organization. Talk to program officers if possible, but have a clear objective for your conversation.

There are also general guides available about how the funding process works with sponsors from different sectors. Government agencies are very important sponsors of research, yet private enterprise donors are increasing in importance for researchers, and resource requests to businesses call for a specialized approach that understands the sector's perspective and timing (Schot and Steinmueller 2018). Similarly, foundations often require more insider knowledge, and they conduct calls that are unlike the competitive proposal processes of public institutions (Geever 2012).

The purpose of knowing your potential funding organization well is to be able to write to the right audience and create what I call the "right fit" – finding the overlap between your own research agenda and the goals and mission of the funding organization. You might think of a good fit as a temporary alignment in the same general direction to advance goals together. This does not mean that your research agenda should be wholly subservient to the aims of the sponsor. Certainly, external funding from industry and foundations is known to influence overall research agendas for aims beyond academia (Fabbri et al 2018; Giridharadas 2018). Thinking of your ideas in terms of the "right fit" means you will need to consider how to "package" your research idea, which, as Michael Watts (2001, p. 2) notes, is not the same as to "compromise" it.

The way one goes about searching for potential funding opportunities is itself an opportunity to ensure the right fit. One approach is to follow the path of leaders in your field. The CVs of experienced researchers in your area are a gold mine of potential funding sources for your own research. Many scholars publish their full resumes online, or at least post some kind of list of recent projects. Look up the name of a senior scholar in your subfield, find their homepage or LinkedIn profile, ResearchGate, or other tool, or simply find them through their department's faculty page if you know their institution. Locate their CV or resume, and make a list of all of the names of organizations where their research has been funded or where they have conducted projects in collaboration. Odds are that your research will be eligible for funding from those same organizations. They are also more likely to be a better fit than less focused searches.

Organizing and systematically prioritizing which funding opportunities align best with your work – rather than jumping at the nearest deadline – will help you focus on the best fit and, by extension, the highest probability of success. Create a spreadsheet of possibilities from a variety of sources, listing the potential sponsor organizations' name, address, website, contact person, email, telephone, published funding program priorities, grant funding ranges (minimum to maximum amount of awards), eligibility criteria, deadlines, and so on. Annotate each record with a reference to a particular research problem or question (not just a topic) that might be developed into a proposal. Rate the likelihood you believe you have in receiving funds from each potential source, using a percentage scale. This is a bit of an instinctive process, weighing a multiplicity of factors such as the level of competitiveness, how good the "fit" is, how amenable the timeframe is to your research agenda, how well you match the eligibility profile, and whether the request limits will meet your resource needs. In this way, you can prioritize the list of possibilities according to the likelihood of success and your own preferences for research questions.

For discovering and creating the "right fit" as you design and write the proposal, you might find common ground in one or more of the following areas:

Organizational mission: There may be some shared vision between the goals of the organization and your respective research goals. These may not align perfectly, so the objective is to discover where they overlap and exploit those connections fully.

Targeted beneficiaries and research subjects: Pay special attention to shared constituencies and their needs (e.g. both you and the funder are interested in youth populations) or relationships between respective constituencies (e.g. your interest in youth and their interest in retired professionals possibly connecting through mentoring relationships).

Operational Approach: The means of executing your projects may coincide with those that sponsors prefer or promote (e.g. the use of geographic technologies, mapping, communication technologies, community participation, industry competencies, or shared markets).

Mutually Beneficial Deliverables: What might your project produce that serves both you and the funding organization? (e.g. a tourism brochure for them that includes a community map you seek to create).

Global Objectives: What global or international targets might you both care about? (e.g. investigating progress toward the U.N.'s Sustainable Development Goals).

Communicating in the language of the sponsor is a critical part of creating the "right fit" for your proposal. Scholars should recognize that the style of writing for a grant proposal differs sharply from that of an academic journal article, for instance, and different modes of expression are called for. Even proposals to scientific funding agencies like the U.S. National Science Foundation should avoid scientific jargon and use an appropriate style that reflects the manner in which the funding agency communicates and in which panel reviewers are situated – typically an interdisciplinary context. When key terms are used in the proposal announcement, integrate those same terms in your proposal. Don't leave reviewers guessing if your research meets their specific objectives; tell them exactly, in *their* own words, how it does. To practice, read through an entire proposal announcement carefully and identify "buzzwords" that, in the language of the sponsor, express important key concepts significant to their organizational mission or the goals of the granting program. These might be words, word combinations, or phrases. List them, and then define each of them using the kinds of language that your research would normally use to express ideas to create a kind of glossary, a handy reference when you are writing and revising your proposal.

Beyond the horizon of one particular grant proposal, long-term success in generating resources for your research rests upon your ability to build and maintain solid mutually beneficial relationships with sponsors. You should be willing to invest in developing these relationships over the long term. Start planting seeds with a good understanding of who the sponsor is and with what language they communicate. Resource requests, including everything from informal inquiries to formal proposals, often both draw from and generate relationships, whether they ultimately are short or long term, whether intimate or formal or casual. Doing it right may make the difference not only for a positive response to your current proposal, but also for the long-term success of your research agenda.

Assemble a winning team

The saying goes that "no one is an island." Grant preparation is no different. Assemble a winning team on your side to increase your chances of success. This advice holds for experienced faculty, early career faculty, and graduate students alike.

There are at least three kinds of help you might need in preparing your research grant proposals. First, you should early on ask for help from your university's sponsored research office (SRO), where they can provide a wealth (and often literally a library) of information, including granting agency directories, model proposals, writing services, and human subjects advice. In most cases, consulting with them from the start can also help you get through university review more easily and quickly.

University review is often required because most grants are contracts between the granting agency and the university (or a unit within the university), not between the agency and the researcher. A typical university review may consist of a series of forms and procedures, including some kind of researcher information form, a proposal or abstract form, budgeting forms (including your final budget with justification), human subjects approval, and other information. Generally, a litany of signatures is required, from you as the PI, your department chair, your dean, the university office of grants and contracts, and finally the SRO. If your university does not specify the time frame by which you must submit all materials prior to the granting agency deadline, it is recommended that you provide your complete written proposal (and all related documents) at least three to five business days before, but at some universities a much longer approval period is needed. You should check with your office well ahead of time to know what their internal deadlines and requirements are. Some sponsors limit the number of proposals from any single institution, in which case there will be a Limited Submissions team that conducts an internal process to determine who will represent the institution. Any research that involves human subjects must also eventually undergo a review by your university's Institutional Review Board, for which you should leave at least a couple of weeks' time if the granting organization requires final approval prior to submission. Finally, a common required element of the research grant narrative is some general information about the applicant institution (i.e., your university). Your SRO should be able to assist you with standard responses to those requirements.

Secondly, you might seek out the aid of a seasoned grant writer that can help you keep on task and on time and cover all the bases. This person can play the role of practice reviewer for later drafts of your proposal. If willing and available, scholars experienced in proposal writing can serve as critical mentors as you develop your skills.

Thirdly, when appropriate, you might invite these same scholars to team up with you on the research project as a co-principal investigator. Nothing breeds success like success, and if you can develop collaborations and partnerships with other successful researchers, you will see your own success expanding. By building a team, you may also be able to be eligible and competitive for larger funding programs than you otherwise might be. Be careful to keep your team to a manageable number appropriate to the needs of the project and the allowed budget size.

If you are not quite ready yet to call a team to task around a particular research proposal, you could simply make an appointment with a scholar or two in your subfield who are experienced in writing and getting grants. Interview them, share your ideas and questions with them, invite them to mentor you on specific aspects of the process, or generally consider developing joint proposals with them. If you do not find an experienced colleague, you could also benefit greatly from doing the same even with a less-experienced peer who also wishes to gain greater experience.

Finally, some proposals require or are improved by partnering with non-academic or community stakeholders. This aspect of engaged scholarship should be taken seriously, with great care, thought, and insight. Acknowledging underlying perspectives such as community geography (Shannon et al. 2020) as well as practical realities of community-based research (Hinrichs and Solís 2021) is an important dimension of planning a collaboration that will itself be successful, let alone generating a successful proposal.

Design to deliver

The actual research activities and methodologies that you will undertake should be very carefully designed and clearly organized to instill confidence in your reviewer that you can deliver on your promises. As Krathwohl (1988, 15) advises, "while projects typically start with an idea, sponsors fund activities, not ideas." Focus on what you will *do*. Keep your scope realistic and doable within the granted time. Think carefully through your entire plan, step-by-step, from start to finish, to ensure that you will be able to administer the research project on time and on budget. Practice good time management, not only in your writing process but also in the design of the project; determine how long and at what point in time each of the set of activities will need to occur.

To be clear about what you will do when, how, and with what resources, you should, as a matter of practice, always include a separate section called "project design" where you identify the specific *activities* needed to carry out your research methodology, a *timeline*, and a bulleted list of "*deliverables*" or products or outcomes expected – even if the call for proposals does not specifically request these items.

Designing to deliver also means that you will be able to realize your project with the requested resources. Above and beyond serving as your formal request for a particular dollar amount, your budget section should demonstrate your resource plan. The single most important key to designing a good budget you can live with and get approved is to be very clear, first, about what you will ***do***, and then ask for the resources you will need according to each activity. Ask yourself for each activity, what do I need to perform this task? Convert how much time it will take into dollars, since most budgets also pay for salary or wages. Doing this early on in the process can help you refine the intellectual scope of the research, so you do not promise more than can be delivered.

You may ultimately have to negotiate a final budget with some sponsors, so be sure that each item in the budget is fully justified and necessary, or you may find that it gets cut. Most novice grant writers need some guidance on deciding how much to ask for and how to ask for it. A good rule of thumb is to use the guidelines and limits in the call for proposals, remembering that you do not have to ask for the full amount available, but you should ask for the amount you need and can appropriately justify. For travel expenses, look up current federal government guidelines for per diem rates; for equipment expenses, get some quotes from vendors to provide evidence of costs; for indirect rates that differ widely among universities, be sure to check with your SRO for proper procedure and documentation.

In short, basic criteria for designing a budget that delivers include the following:

a Is each item eligible for funding according to donor rules?
b Is each item necessary for the project and linked to a particular activity?

c Is each amount properly justified?
d How much "bang for the buck" can your project deliver? Put your request in perspective according to what impact your research will have.

For more advice on budgeting fundamentals that are valid across sponsor types, see Henson (2003) and Pain (2017).

What you ultimately propose to deliver should strive to meet the expectations of the granting organizations, or at least of your reviewers. These, of course, differ by field, with research in the humanities, social science, and physical science having their own unwritten codes of expectations of their communities of scholars. Conform your proposal to the appropriate context. If you don't know much about that set of unspoken assumptions, talk to a mentor or prior awardee. In any case, this usually means that you should deliver some kind of impact – be it impact on practitioners of the field, impact on our knowledge about the problem, or impact on a community affected by the issue. Clearly specify what that impact is and show how you will achieve it.

Be perfectly persuasive

Grant writing uses persuasive communication, aiming to convince the reader that the project is worth investing in. This means that you should use carefully crafted arguments backed up by evidence, not unfounded assertions that your reviewers might doubt or, worse, challenge. It also means that the language should exude confidence, eschewing grammatical formulations that are overly reliant on "might" or "could," opting instead for simple present or future tenses as often as appropriate, like "is" and "will be." Be sure to avoid passive voice unless completely necessary, to convey the sense of action that you will perform as you carry out your project.

Excellent persuasive proposals distinguish themselves by anticipating opposing arguments or identifying points of contention in your research plan. By preparing a positive response to defend likely criticisms, rather than avoiding them, you demonstrate thoroughness in your thought process and can convince reviewers that might otherwise remain skeptical about how your research will contribute to the field. Mentors and colleagues can help identify opposing arguments. Or consider reviewing your proposed work in a seminar where the discussion might raise ideas about possible responses to your work.

Being persuasive is not only about what you say, but also how you say it. Be sure that your document is perfect, error-free, and looks great. First and foremost, conform *exactly* to the formatting specifications published in the call for proposals. Use the very same headings and subheadings for proposal narrative elements in the same order as requested (e.g., Introduction, Significance, Methodology). Don't cheat line spacing or margins; use standard, black color and readable fonts and font-sizes (never less than ten points) and stick with the same choice throughout all proposal documents. Leave plenty of white space, and break up long paragraphs of text. Use, but don't overuse, bullets, tables, and graphics to draw attention to particular elements of your proposal. Don't rely on automatic spell check in word processing programs to catch all spelling or grammar errors. Don't allow widow or orphan lines, always number your pages, and use a common paragraph justification for the whole document.

When you finish, step back from your document to check for possible formatting problems by using the "zoom" feature of your word processing program – scroll through to look for inconsistencies. Print out a hard copy to see what it looks like, even when submitting electronically. Leave enough time in your writing schedule to let your proposal gather a little dust and reread it fresh again, or ask a colleague to proofread for you.

In short, a good rule of thumb is to avoid any kind of presentation problems that might irritate your reviewer or distract from your content. Similarly, follow all submission guidelines precisely and on time. Leave extra time for using online submission systems, especially if it is your first time preparing an electronic proposal for that sponsor. By paying attention to these details, you not only produce a polished proposal, but you will instill confidence in your reviewer that you can deliver a well-conducted research project.

Make it real

Related to the idea of persuasion is the strategy of making your project real. The discipline of geography, for example, has a distinct intrinsic advantage of being closely connected to the actual phenomena that we study, phenomena that generally intersect with our everyday lives in a real way, whether it be about income distribution patterns in a major city, or linkages between rural land use and groundwater quality, or statistical evaluations of global climate change models. Even theoretical research often draws from and has implications for very real everyday matters. Take advantage of this perspective to connect to your reviewers by providing them with real-life examples and connections that avoid jargon. Infuse your narrative with *one or two* (don't overdo it!) carefully selected and highly relevant maps or photographs that validate the adage of a picture being worth a thousand words. Writing with passion (but not hype) also makes it real for your reviewers, as legendary geographer Abler (1989) so elegantly enjoins, offering good advice that stands the test of time. Flip through news and magazines for inspiration on compelling examples. Explain your unique geographic work in a way that will implant an image into the minds of your reviewers to remember you as they wade through the other scores of proposals they still have to finish reviewing this afternoon.

Demonstrate your unique value

The proposals that are successful are the ones that stand out from the crowd. Explicitly state what is special about your work. Usually, every call for proposals requires a special section to describe the "significance" of the project, where you identify what is unique about the proposal. Brainstorming answers to the following questions may help you pinpoint what aspect or combinations of characteristics are valuable in the work you plan to do:

- What is special about the problem I have chosen to address?
- Is my work conceptually innovative?
- Does the field know less about my case than other similar cases elsewhere?
- Have there been any recent critical breakthroughs in the field that my work draws from or seeks to build upon?
- What about my methodology might represent a contribution to the field?
- How is my approach different from the way others have looked at the issue?
- What's the expected impact?
- What is new about what my expected research results?

Certainly, as Przeworski and Salomon (1995) point out,

> disciplinary norms and personal tastes in justifying research activities differ greatly: Some scholars are swayed by the statement that it has not been studied (e.g., an historian may argue that no book has been written about a particular event, and therefore one is needed), while other scholars sometimes reflect that there may be a good reason why not.

In any case, justifications should be based on evidence and argumentation rather than on assertions or opinions about what is appealing about the subject matter. Sometimes that unique value you offer is what you are able to create beyond the creation of new knowledge from your research. Beyond outputs or outcomes, impact captures a value to society that could arise from knowledge being put into action. Funding agencies increasingly ask for impact statements, which should neither be exaggerated, nor should they be perfunctory (see Wilbert and Tedjasaputra 2023).

Go the extra mile

Turn up the notch on your competitiveness by including one or more strategic elements that may not be required but can help round out your proposal. While some of these ideas apply better to faculty proposals, including others in student proposals may demonstrate early professional maturity. All should be applied to enhance your work, not distract from it. Be sure to check that your target sponsor agency allows these items in proposals; some sponsors have very strict guidelines. At best, incorporating these strategies regularly can help you transform individual research projects into a solid research program.

An *advisory board* is a group of people who can help you ensure quality and provide a sounding board for ideas and problems. If you can identify and convince experienced scholars to agree to serve on such a board (pending acceptance of your proposal), referring to their tentative agreement to participate in your proposal also lends a measure of credibility to your efforts. Advisors usually serve on a volunteer basis, but if it is allowed by your sponsor agency, you might consider writing in a budget line item to cover a small amount of basic communication or travel expenses. A cost-effective strategy is to plan your advisory board meeting(s) during a national conference where you and your advisors are already likely to be in attendance. If you are a student, form your board now, invite scholars from inside and outside of your home university, and establish a track record to support your career development goals.

Letters of support are also good means of demonstrating the importance of your proposed research. Even if not required, a key endorsement from a leading scholar, a partnering agency, or other group can help your proposal stand out. To get these letters, be prepared to write each letter draft yourself, and tailor every single one to each individual supporter. Do not use a form letter (unless required), or you will receive five letters back with exactly the same wording. Make sure that the letter refers to the exact, correct title of your project and is addressed to the right person or department at the funding organization, or generally to the "Review Committee." Alternatively, have the letters addressed to you. Check that your sponsor does not prohibit such letters.

A *research agenda* is a common way to demonstrate that you anticipate how this project contributes to a larger and broader knowledge generating enterprise. By proposing to develop one during the course of your research project (and even better, in conjunction with your advisory board), you set yourself up for possible future funding opportunities. Later, you can then point to the research agenda as the source for new questions you wish to raise in subsequent proposals.

A *sustainability plan* is a tool to allow you to continue making progress beyond the grant period. Many granting agencies like to see that you are thinking about how to leverage their resources as a way of launching new, long-term efforts. Say a few words in your proposal about how you envision sustaining your work after the funds are gone. Sustaining the results of a project does not always mean continuing in the same way with the same funding needs. There are many ways to transition a project at the end stages to lead to a productive conclusion that sustains what was valuable about your work (see Hinrichs and Solís 2021, 36).

Matching funds or *in-kind contributions* are another way to demonstrate to funders that you will be using their resources efficiently and stretching them to their highest impact. Even if not required, if acknowledgment of outside contributions is permitted, it is valuable to explicitly name what other resources you or your institution will bring to bear upon the effort, if nothing more than pointing to the use of facilities, extra faculty time spent working or writing on the project, and so forth.

References to your own publications in your proposal narrative and bibliography can help demonstrate your competence in the field and position yourself relative to other scholars. Having publications that relate to the work you propose certainly reinforces credentials in the eyes of proposal reviewers, but be sure that the references are relevant and appropriate to the project. Furthermore, even unfunded proposals themselves may later become a source of future articles, creating a synergy between your grant writing and scholarly publication writing activities.

Memorable project titles or *acronyms* can help reviewers recall your proposal over others, particularly in discussions among review committees. Proposal titles make the first impression, so be careful to choose wisely, or the memory reviewers have may not be a positive one.

Dissemination activities include ways that you will ensure that your research results are broadly known. Donors are often interested in investing in work that will be widely recognized and continue to have further impact on the scientific community or the public. It is easy to add a few strategic dissemination activities like publishing your work in a high-impact journal, writing a newspaper op-ed about the results, holding a workshop or seminar or a panel at a conference to consider the results, or creating a website or blog for new audiences.

Evaluation processes are often required for major research projects, but even if yours is not mandatory, it is advisable to incorporate some kinds of evaluation procedures into your work. This ensures quality and effectiveness and will allow you to map outcomes to original goals. It also provides excellent material for future proposals that build upon the same line of research. You may only need to include a few statements regarding how you plan to assess success. Or you may be required to contract with a professional evaluator that is external to the project. Formal, external evaluations generally should be budgeted at about 10% of the overall budget (before overhead percentages are applied). Be sure that the evaluation utilizes a valid and reliable evaluation design, referring to methods that are appropriate to the kind of funding agency you are seeking support from.

Achieve and communicate coherence

The whole is truly more than the sum of its parts. While writing each section of the narrative, completing the budget, putting together your biographical sketch, and all of the other parts of the proposal, you should strive for coherence among all of the elements. While you will not be able to eliminate all possible contradictions from any text, the idea is to create a complete package that minimizes a sense of the disjoined. This concept underlies all of the advice given, and serves as a framework for your writing. Systematically structure the connections across the components of your proposal, using a matrix or other tool that allows you to interrogate where there may be opportunities to amplify your research aims by presenting an elegant and coherent proposal.

Live it

Preparing competitive research grant proposals is just as much about preparing yourself for a successful career as it is about getting resources for a particular project. Writing grants successfully is a continual process that is best practiced as integral to your professional life – not just parallel to it. Here, I highlight a few simple organizational tools you can employ to create a constant state of readiness that will allow you to respond to funding opportunities as a matter of professional practice, rather than approaching it as a project (or chore!) in itself.

- Keep files / folders (either digital or hard copy) on possible research questions or problems that interest you. A research idea diary can help you focus on lines of work that you would like to elaborate, without having to flesh them out completely in the moment. You will be surprised in reviewing past entries just how your thoughts develop over time. This can then be a source for responding quickly to new funding opportunities with short deadlines.
- Keep similar but separate files / folders on possible sponsors amenable to your work. Organize this information, as you gather it, into a spreadsheet, so that you can quickly find what you are looking for and you can see relationships between opportunities. Information on these funding agencies can come from a variety of sources, including your university's sponsored programs office. You can also simply use any search engine on the Internet combining key words for your research and the word "funding" or "grant" to dig up potential opportunities. Or mine the CVs of scholars in your field, as suggested above. Be sure the spreadsheet includes a column for deadlines, plan ahead, and prioritize.
- When you have identified a match between a sponsor and one of your research questions, or when you have a sufficiently developed research question arising in response to a particular call for proposals, you should launch a special writing operation. Start with a proposal writing timeline for each proposal, working backward from the deadline, and identify key dates for finishing certain benchmarks.
- Write for 15 minutes a day on any proposal. Whether or not you have a particular funding opportunity, call for proposals, or research sponsor in mind,

your writing skills to address this special audience will develop by frequent practice. Successful grant seekers almost always have one or more proposals in development, with draft documents saved on their hard drives specifically designed to capture ideas, arguments, and thoughts as they occur. This reduces the need to create the research proposal completely from scratch once you find a suitable sponsor organization.

Finally, there are clear ethical considerations that must be taken into account when writing grants. These are familiar refrains against plagiarism, integrity, honesty, informed consent, reciprocity, conflicts of interest, using AI tools, and other stances appropriate for aspiring scholars to live up to (see Locke et al. 2014; Chapin 2004).

Now what?

It can be an exhilarating moment to finally submit a grant proposal for consideration after all the hard work involved in preparing it. However, soon the realization hits that you will have to wait sometimes as long as six months for a decision. At these moments, turn to the next proposal and do not continually ask for updates from the program officers. Meanwhile, prepare yourself realistically for what to expect next: actual funding rates are indeed quite low, and even experienced grant writers can expect success rates on the order of 30% or even far less, depending on the sponsor. Instead, focus on the other benefits that the writing process brought you. Reflect on what you might do better next time, regardless of whether you receive a positive or negative response for funding.

After proposals are submitted, they enter some kind of formal review process. Often this includes peer review, but for some foundations, the staff or a board of directors considers the applications. In any case, once you receive notification of a decision, you will often also receive copies of the feedback from reviewers. If you do not, you should certainly request them. In the event of a negative review (a likelihood given the competitiveness of many programs), looking carefully through the feedback will help you understand that the process itself is instructive. A rejection of a proposal does not mean your ideas are not worthy of future consideration – it may simply indicate temporarily limited funding or unusually heavy competition. Read all reviews and use them to revise and resubmit rejected proposals, or to reformulate the next one. Even the reviews of accepted proposals should be carefully considered, since they often include very helpful comments that will improve your funded project. As you gain greater experience with the cycles of the grant preparation process, your ability to coherently develop and present new ideas will mature.

In conclusion, the entire process of writing research grant proposals – from developing ideas to writing convincing text to considering reviewers' feedback— can be conducted in a way that solidifies your research ideas and shapes your academic career in positive directions. Integrating these practical recommendations and habits of mind and action through a coherent framework into your professional life can help you prepare yourself as a successful researcher *and* grant winner.

Note

1 This chapter is edited from an original manuscript that appeared in the volume *Aspiring Academics* (Solem, Foote, and Monk 2009).

Additional resources

Abler, Ronald. 1989. How to win extramural research awards. In *On becoming a professional geographer*, ed. M. Kenzer, 170–82. Caldwell, NJ: Blackburn Press.

Although written some time ago, Dr. Abler's advice is salient and timeless. Coming from this renowned geographer, in an elegant and witty style, the text itself is a good example of clear and persuasive writing. A short and easy piece to cover a good deal of conceptual ground.

Geever, Jane C. 2012. *The Foundation Center's guide to proposal writing*, 3rd ed. New York: The Foundation Center.

A widely read go-to guide with multiple updated editions and indispensable general resource for proposal writing, especially for applied projects, but also appropriate for studies. The perspective is particularly valuable for researchers seeking to submit proposals to private foundations or donors that are non-traditional in terms of academic research proposals.

Locke, Lawrence F., Waneen Wyrick Spirduso, and Stephen J. Silverman. 2014. *Proposals that work: A guide for planning dissertations and grant proposals.* Thousand Oaks, CA: Sage Publications.

For graduate students, this book not only reliably advises on key aspects of developing a research thesis, but also navigates through the dissertation proposal process itself, including how to form a dissertation committee and make the oral presentation. Its organization is particularly creative around nine functions that proposals must perform. The sections on writing a literature review, designing budgets, and discussing qualitative research will make this a valuable reference work in any academician's library.

Hall, Mary, and Susan Howlett. 2016. *Getting funded: The complete guide to writing grant proposals.* Portland, OR: Portland State University Press.

Although written for a much broader audience than just academic researchers, readers will find plenty of practical value in this volume. This resource includes a section for instructors teaching proposal writing, including a sample syllabus for courses on grant writing, and suggested assignments related to each chapter.

Wentz, E.A. 2013. *How to design, write, and present a successful dissertation proposal*, 1st ed. Thousand Oaks, CA: Sage Publications.

Geared toward graduate students in the social or behavioral sciences using quantitative or mixed methods research, this book is a valuable start-to-finish guide. Topics include ethical scientific research, curriculum vitae, effective reading, literature reviews, oral presentations, and next steps after the proposal.

Sternberg, R. J. (Editor). 2014. *Writing Successful Grant proposals from the top down and bottom up*. Los Angeles, CA: Sage Publications.

A terrific edited volume, this book presents chapters authored by many successful grant winning scientists and scholars, dispensing relevant general writing advice to be competitive across many disciplines. One section is dedicated to navigating specific federal agency calls for proposals, with additional advice tailored to building effective research collaborations that are often necessary for larger successful efforts. A final word about reviewers offers a valuable insider perspective.

References

Abler, R. 1989. How to win extramural research awards. In *On becoming a professional geographer*, ed. M. Kenzer, 170–182. Caldwell, NJ: Blackburn Press.

Chapin, P. 2004. *Research projects and research proposals: A guide for scientists seeking funding*. Cambridge, MA: Cambridge University Press.

Fabbri, A., A. Lai, Q. Grundy, and L.A. Bero. 2018. The influence of industry sponsorship on the research agenda: A scoping review. *American Journal of Public Health* 108 (11): e9–e16. DOI: 10.2105/AJPH.2018.304677.

Flannery, M.E. 2022. State Funding for Higher Education Still Lagging. *neaToday* 25 October. https://www.nea.org/nea-today/all-news-articles/state-funding-higher-education-still-lagging (last accessed 27 December 2023).

Geever, J. 2012. *The Foundation Center's guide to proposal writing*, 6th ed. New York: The Foundation Center.

Gerin, W. 2006. *Writing the NIH grant proposal: A step–by-step guide*. Thousand Oaks, CA: Sage Publications.

Giridharadas, A. 2018. *Winners take all: The elite charade of changing the world*. New York: Knopf Doubleday.

Henson, K. 2003. *Grant writing in higher education: A step-by-step guide*. Boston, MA: Allyn & Bacon.

Hinrichs, Margaret M., and Patricia Solís, eds. 2021. *A knowledge exchange playbook to build resilience*. Tempe: Knowledge Exchange for Resilience, Arizona State University; Washington, DC: Global Council for Science and the Environment. https://hdl.handle.net/2286/R.2.N.160839

Krathwohl, D. 1988. *How to prepare a research proposal: Guidelines for funding and dissertations in the social and behavioral sciences*, 3rd ed. Syracuse, NY: Syracuse University Press.

Lee, J., and S. Clery. 2004. Key trends in higher education. *American Academic* 1 (1): 21–36.

Locke, L., W. Spirduso, and S. Silverman. 2014. *Proposals that work: A guide for planning dissertations and grant proposals*. Thousand Oaks, CA: Sage Publications.

Musselin, C. 2018. New forms of competition in higher education. *Socio-Economic Review* 16 (3): 657–683. DOI: 10.1093/ser/mwy033.

Pain, E. 2017. How to budget your grant proposal. *Science* 27 September. https://www.science.org/content/article/how-budget-your-grant-proposal (last accessed 27 December 2023).

Przeworski, A., and F. Salomon. 1995 rev. 1998. The art of writing proposals: Some candid suggestions for applicants to Social Science Research Council competitions https://www.ssrc.org/publications/the-art-of-writing-proposals/ (last accessed 14 September 2024).

Schot, J. W. and E. Steinmueller, 2018. Three frames for innovation policy: R&D, systems of innovation and transformative change. *Research Policy* 47 (9): 1554–1567. DOI: 10.1016/j.respol.2018.08.011.

Shannon, J., K. B. Hankins, T. Shelton, A. J. Bosse, D. Scott, D. Block, H. Fischer, L. E. Eaves, J. -K Jung, J. Robinson, P. Solís, H. Pearsall, A. Rees, and A. Nicolas. 2020. Community geography: Toward a disciplinary framework. *Progress in Human Geography* 45 (5):1147–1168. DOI: 10.1177/0309132520961468.

Watts, M. 2001. The Holy Grail: In pursuit of the dissertation proposal. https://iis.berkeley.edu/sites/default/files/general/inpursuitofphd.pdf (last accessed 14 September 2024).

Wentz, E.A. 2013. *How to design, write, and present a successful dissertation proposal*, 1st ed. Thousand Oaks, CA: Sage Publications.

Wilbert, T., and V. Tedjasaputra. 2023. *NSF 101: Five tips for your Broader Impacts statement. Science Matters.* Washington DC: National Science Foundation. https://new.nsf.gov/science-matters/nsf-101-five-tips-your-broader-impacts-statement (last accessed 27 December 2023).

19

THE PUBLISHING PROCESS AND GETTING YOUR WORK INTO PRINT

Lynn M. Resler, May Yuan and Joshua Inwood

How has publishing changed over the last decades?

The publishing process, expectations, and how the academy views publishing are dynamic. Throughout the authors' professional careers (over the last ~20 years), some of the changes we have experienced include the following:

1. The proliferation of publishing-related metrics and increased requirements for official reporting of these metrics in promotion dossiers and annual reports.
2. Increased attention to publishing ethics and responsible authorship. For example, more journals now require statements of individual author contributions to justify authorship, sharing data and codes for reproducibility and replicability, and affirming ethical conduct.
3. Obsolescence of print publishing formats, including open-access (e.g., Forgues and Liarte 2013) and digital-only publications, obsolescent print publishing formats, and migration to online-only submissions.
4. Many peer review practices. Double-blinded reviews (anonymizing both authors and reviewers), single-blinded reviews (disclosing authors but anonymizing reviewers), open peer reviews (Ross-Hellauer 2017), and, the latest, no rejection with reviews (Else 2022).

Some, if not all, of the aforementioned changes could impact how you strategize publishing throughout your career, how you present your promotion case, and the journals you target. For example, since publishing is fully online, open-access journals typically result in broader access to publication once an article is published. As a result, the "bar" for publishing expectations may be formally or informally

DOI: 10.4324/9781003342953-22

raised according to the quality of journals where your papers are published, especially for job applicants. Each university differs in its requirements for publishing scholarly output. The expansion of corporate publishing also means you should be extra cautious of predatory journals (e.g., Cobey et al. 2018).

Why publish?

One aspect of academic publishing that hasn't changed over the last 20 years is the expectation that we share our ideas with the broader scientific community. As incentive-based budget models have expanded across US universities, we suspect that publishing expectations at many institutions have also increased. Furthermore, current academia awards promotion and tenure to those who contribute to advancing knowledge through evidence of excellence in scholarship. Scholarship, a bulk of which is likely to include refereed journal articles, is an essential criterion for promotion with tenure, especially for graduate-degree granting institutions. However, the "big-picture" may be helpful to put publishing in perspective. Publishing establishes us as scholars in our field. It advances and diversifies thought within our discipline by providing an avenue for new voices and changing traditional mindsets about knowledge production. We thus advocate strongly for all voices in social, physical/environmental, and technology fields to publish – regularly and consistently – to diversify scientific contributions and change the dynamic of the productivity puzzle (e.g., Rigg et al. 2012). Publishing also benefits us personally. As Brunn (2009) noted, many academics love their work, and a sense of excitement accompanies the opportunity to share it. By sharing our hard-earned knowledge and experience, we have ultimately helped create new knowledge and experience. Publishing is both powerful and empowering.

What is the publishing process?

The publishing process includes the collective steps from initial manuscript preparation to final acceptance and copyediting. These steps typically include the following:

1 Identifying a target journal that is a good "fit" for your scholarship, and formatting your manuscript to adhere to journal requirements;
2 Submission, involving a cover letter and authorship disclosures;
3 The peer review process, which occurs after submission;
4 Proofreading and editing the galley proofs.

Identifying a target journal

We recommend that you identify a target journal early in manuscript preparation. Journal articles should be framed and formatted to a specific outlet.

Targeting a journal before you put the work into preparing the manuscript will save you time and effort later on (Butler 1989).

The task of selecting a target journal is an important one, but it may seem overwhelming. A wide variety of journals exist, some of which attract a general audience, such as *Science, Annals of the American Association of Geographers,* and *American Scientist.* Others are intended for specialist communities such as *Geomorphology, Remote Sensing of Environment, Arctic, Alpine, and Antarctic Research, Progress in Human Geography,* or *International Journal of Geographical Information Science.* Journals with a regional focus, such as *Southeastern Geographer* or *Castanea,* may be suited for scholarship emphasizing regional or applied work, including case studies. The match between your manuscript and the journal's aim and scope (or "fit") is essential. Well-conceived scholarship could be rejected if it is a poor match for a journal. Fit will depend upon several factors, including your audience, the angle you choose for your topic, and what kind of article you have written, such as a review paper or an original research article (Table 19.1). The journal's website is the best place to consult when determining if the scope and aim of the journal are a suitable match. It is also useful to look through the journal's editorial board to get a sense of the kinds of scholars the journal relies on to set the publications agenda and who might review your manuscript. Browsing recent publications in a journal can help you grasp the scope and writing styles of the journal and can also help you respond to recent research that the journal has highlighted.

Many people focus on impact factors when selecting a target journal. There is discussion and even disagreement about both impact factors and ranking of journals. Some universities adhere to journal rankings, whereas others do not. Journal quality is typically ascertained through qualitative attributes such as journal tradition and recognition within the discipline, as well as through metrics such as impact factors. A journal impact factor (JIF) is calculated by a private company Clarivate and reflects the average number of citations in a given year for the journal. Often a higher-impact score can indicate a journal that is well read and well respected, though for some specialties JIFs can be deceiving. Impact factors may be used (or potentially misused) to estimate the influence of a journal within a field. Additionally, at some institutions, JIFs are considered in decisions related to faculty advancement (McKiernan et al. 2019). We believe journal quality should not equate solely to its impact factor. However, the impact factor, despite its drawbacks (see Meadows et al. 2016, for perspective on the increased reliance on publishing metrics), is one way your work may be evaluated, especially by those unfamiliar with a journal. Impact factors vary substantially both within and among fields, and they change on a regular basis.

Furthermore, not all quality journals have impact factors. Consider that solid, quality research is published in first-, second-, and third-tier journals (Figure 19.1). Ultimately, we believe it is best for early career scholars to consider their audience, and then publish in outlets that would have several readers with similar interests. Potential peer evaluators would likely recognize these journals for promotion and tenure.

BOX 19.1 PREPARING FOR A RESEARCH CAREER: A LIFE COURSE IN LEARNING AND RELEARNING

FIGURE 19.1 Dr. Stanley D. Brunn in the exhibit hall at the AAG 2023 Annual Meeting in Denver, Colorado, USA. Photo taken by L. Resler on March 25, 2023.

1 Writing is a lifelong experience; it should be constant learning and relearning. Consider if you enjoy writing and want it to be part of your career.
2 There are many more acceptable quality journals to publish in than there were 20–30 years ago.
3 Provide a strong, informative, and detailed letter to the editor so she/he will know your background and experience related to the research. Ensure it is original research, and you have not submitted it elsewhere.

4 Prepare papers with a title that addresses the research focus and relate to some appropriate conceptual/theoretical framework in geography or related fields.
5 Select a journal with great care; examine previous issues if the research topic, content, methodology and theoretical framework suit the journal. Be sure the submission "fits" the journal's focus.
6 Quality research is published and cited in first-, second- and third-tier journals. There is always a good place for innovative, timely, and quality research.
7 Hope that the manuscript receives an "R&R" (revise and resubmit) recommendation from the reviewers and editor. When submitting revisions, respond directly to points made by reviewers and editors, even if you disagree.
8 Departments and universities need to exercise flexibility in publishing in major interdisciplinary as well as disciplinary journals. Weigh journal rankings with some scrutiny in submissions and career advances.
9 Even very successful junior and senior scholars in their careers have experienced rejections and R&R recommendations, sometimes multiple times.
10 Invite present and past editors to visit universities and hold "learning" sessions about publishing at regional and annual meetings; listen to their advice and experiences as gatekeepers.

Stanley D. Brunn, University of Kentucky, Lexington, KY 40506-0027 Brunn@uky.edu.

When you are an early career academic or a doctoral student, your dissertation (or, in some cases, your master's thesis) is likely to be a primary source of publications. It is thus logical and advisable to discuss possible outlets with your close colleagues, academic advisor, and committee members. Consider the journals you read when conceptualizing your research. Try browsing through a few potential journals to get a feel for article layout and framing. Doing so will help you determine if your scholarship is similar in scope to those articles already published in the journal. Using a recent example from your target journal as a guide may be helpful as you prepare your work for submission. If you receive an invitation to submit your work to a special journal issue, *go for it*! Editors of special issues seek articles with targeted topics, and the editorial process may be streamlined.

We suggest identifying two to three potential target journals that could be a good match for your work. If your first choice doesn't work, you already have a backup, which may help to alleviate some stress. Remember that "Plan B" may not necessarily need to be a lower-tiered journal – it could simply be a journal with a slightly different aim or scope.

TABLE 19.1 Questions to Consider to When Selecting Target Journals

Who do I want to read my work? *Consider* not just the intended audience of a journal. More importantly, ask yourself, "who do I want to reach?" *Examples:* Broad disciplinary audience; international audience; specialized audience (e.g., biogeographers, remote sensors); regional or topical audience (e.g., geographers within the Southeast Regional Division of the AAG, tropical ecologists).
What type of article am I writing? *Consider* that scholarship can assume different forms in journals. Consider your many options. *Examples:* Review article; original research; editorial/commentary; progress report; technical note; short communication
What is the nature/scope/scale of my work? *Consider* the extent to which the research area will be explored in the work (boundaries), the approach used, and the delimitations within which the study will be operating (i.e., a specific location, or population, or sample size of interest). *Examples:* Local case study framed within a broader context; global or continental scale analysis; empirical study; methodological study; conceptual framework
What is the impact of my work? *Consider*: How generalizable are my results? How widely applicable is my framework? Does my work have global or local implications? Do the results transform the way we think about ways of knowing or doing? Do my results mainly support the findings of other studies, or are they novel?

A final tip to consider when selecting a target journal is cost. Article processing charges (APCs) can be prohibitive, especially for open-access journals. *Please* investigate article processing charges well before submission if, like most academics, you are on a budget. Imagine your disappointment and frustration if you had an article accepted but could not fund its publication. Additional resources through your university (i.e., subvention funds) or waivers through the publisher (typically for addressing hardship in underserved geographic regions or research areas) may make publishing in open-access journals affordable. Your university librarian can likely answer questions about the availability of subvention funds. Additionally, remember that many quality journals in geography do not have APCs. Submitting your work to a professional society journal, such as *The Professional Geographer*, *The Annals of the American Association of Geographers*, and the regional *Southeastern Geographer*, is a way to support non-profit societies and will ensure journal recognition by other members.

Tips for a smooth submission

The recipe for smooth journal article submission begins with careful adherence to the author's instructions. The author instructions, typically posted on the journal's website, tell you how the journal editors expect to see your manuscript formatted upon submission. The instructions will outline the required elements (e.g., keywords, cover page with names and affiliations removed), citation and referencing style, word count, statement of author contributions, abstract formatting (e.g., bulleted formatting or paragraph), preparation of figures and tables, and rules for the use of third-party materials. Some author instructions also include article templates for preparing your paper. You will make an impression on the editors and reviewers with how polished (or unpolished) your work is, so careful attention to detail is vital.

Finally, when you submit your manuscript, you engage in an agreement with the publisher that your research has been conducted and presented ethically. Be mindful that you will likely need to present documentation of IRB approval and permission to use third-party materials not in the public domain. In your cover letter, reflect on how your work fits within the scope and aims of the journal, and also include a statement to guarantee your manuscript is not currently under consideration elsewhere. Some journals now require an online acknowledgment from each co-author that they have read and approved the submitted version of the manuscript. Don't ever surprise a co-author by submitting a manuscript they had no knowledge of! Although written for a medical audience, Shewan and Coats (2010) outline some best practices on ethics in authorship that are widely applicable.

What to expect from the editorial process

Now that your manuscript has been submitted, it is officially part of the editorial process. At the initial stage, an editor will conduct a preliminary review to ensure the journal is an appropriate match, that the manuscript meets the journal's quality standards, that the work is original, and that the work generally adheres to the formatting guidelines. Upon passing this initial review, the editor will seek peer reviews. In some instances, perhaps due to a perceived lack of fit between your manuscript and the journal aims and scope, an editor may choose to "desk reject" the manuscript before it is sent out for review. With the increasing number of journals and submissions, editors face growing challenges in seeking reviewers and securing quality reviews. Reviewers are more likely to accept invitations to review manuscripts with expressive titles and well-written abstracts. Make extra efforts to ensure the title and abstract attractive, self-explanatory, and informative of the presented research in the manuscript.

The purpose of peer review is to ensure the validity and reliability of research scholarship, which helps to maintain and advance the integrity of a discipline. At its best, peer review challenges your ideas and helps strengthen your argument for your work. Manuscripts in peer review typically are assessed by a ***minimum*** of two peer scholars (chosen by the editor) using a single-blind or double-blind review approach. Be willing to make changes as requested by the reviewers and the editor. When receiving multiple reviews for one manuscript, you may receive divergent reviews and opposing suggestions. In the case of opposing feedback, the role of the editor becomes key. Ideally, the editor's guidance will direct you how to focus your revisions. If necessary, seek additional guidance. Regardless of the outcome, we suggest that you try to see the value in each review, especially if it is highly critical. Most reviewers who agree to review a paper do so because they find the topic compelling, and will thus provide legitimate and constructive criticism. Keep in mind that it is the writer's responsibility to communicate research ideas and findings clearly and unambiguously to avoid possibilities of misinterpretations by readers, including reviewers.

An editor's decision to allow you to revise a manuscript under any circumstance means that publishing your scholarship is solidly possible. Minor revisions typically involve some clarification of terms and ideas, or minor reworking of the manuscript without substantial reframing or additional analysis. A decision of "major revisions" or "revise and resubmit" may indicate that the work has merit; however, there is a need for reanalysis, or substantial reframing, before it can be considered suitable for publication. These revisions can involve substantial time and effort, as well as multiple iterations of reviews. The payoff (in addition to publication) will be a stronger presentation of your work. The revision process can also lead to new insights and directions for future work. Make sure to respond professionally, thoroughly, and methodically to each reviewer's concerns, including details of the corrections made and any well-justified rebuttals. Submit this response as a separate document with your revision.

What if your paper is rejected? Rejection of a manuscript is likely to happen during an academic career, and the key is not to let a rejection or a hard review keep you from moving forward with the writing process and pushing your work forward. Rejection can be a sign that you are taking some risks with the target journal. For example, maybe your manuscript wasn't quite ready for *Science*, but *The Journal of Biogeography* is an excellent match. An article rejection can also mean that your scholarship needs more reflection, that you must communicate more clearly, or that your methodology has a flaw (which is always good to know before publication!). Another possibility is that your work is not framed well for the target journal. For example, a study that applies geospatial methods to model urban land use may be more suited for journals in urban studies than for GIScience journals. Alternatively, a study that develops a new geospatial method to model urban land use will fit GIScience journals well. Rejection never feels good, but remember that even the most prolific researchers have had papers rejected. The rejection of a journal manuscript is not personal. Give yourself some time to respond and reflect, and then get busy! Use the reviewers' comments to improve your manuscript for submission to another journal. Persistence usually leads to success in academic publishing.

Author reflections: publishing strategies across geographic sub-disciplines for holistic and balanced career practice and planning

Traditions vary across disciplines; a successful publishing career may thus follow one of several trajectories. For example, physical, environmental, and GIScientists will often co-author articles with a team of students and faculty collaborators from an array of cognate disciplines. Publications in journals from allied fields, such as landscape/plant ecology, earth sciences, econometrics, social science, urban and regional studies, environmental science, civil engineering, and computer science, are also common. Alternatively, a career oriented toward human, the social sciences, or the humanities may be characterized primarily by solo-authored scholarship and different publishing modes, including books (Brunn 2009). Regardless, each career is singular. In this section, we share thoughts and tips from our publishing strategies to help you establish your own path.

Publishing in the social sciences and the humanities: Joshua Inwood

Publishing in the social sciences/humanities (specifically human geography) is a rewarding, if at times, challenging experience. Work in human geography is diverse, ranging from theoretical pieces aimed at changing paradigms to studies that focus on specific locations and show how social processes occur. Sometimes, this work results from individual efforts and is solo-authored, while at other times, it emerges from collaborative efforts. At its most basic level, publishing in human geography is about advancing our understanding of the human condition – broadly defined – and helping us to understand how humans create space and place. This work is sometimes political, and some journals have specific political orientations that contribute to advancing the cause of social justice. Other journals are more eclectic and are geared toward broader work that advances knowledge in ways that are sometimes open to wider audiences. In either case, publishing in human geography necessarily means navigating a diverse landscape. Being open to the broad array of journals and outlets is essential in getting your work into print.

Publishing in the social sciences and humanities necessarily means finding the right outlet for your work, and there are many ways you can do this. First, as we mentioned previously, you should ask yourself what authors you drew on in your work and where did they publish? This can help you understand the kinds of outlets you might publish with and help you identify your work's target audience. Finding the right journal is more than just finding an outlet for your work; it also is a chance to think through who you want to read your work. This is important not only for driving citations and downloads but also for those scholars on the tenure track, as it helps reach potential external reviewers and others who might comment on your work at a later stage of your career. Being strategic in selecting a possible audience is sometimes as important as choosing the outlet where you want your work to be published. Other ways to identify journals are to check out the kinds of paper sessions they sponsor at national meetings and the people they invite to speak and highlight. Major journals like *Antipode* and *Journal of Economic Geography* often support plenary sessions at major conferences, and these sessions are often agenda-setting. They can help you understand if a journal is a right fit for your work.

Finally, publishing in the social sciences and humanities is always collaborative because as you send out your work to be reviewed, you will be asked to review other manuscripts. This labor is essential, as the review process primarily runs through an understanding that as others review your work, you will be working to provide high-quality feedback for other manuscripts that editors will send out to you to review. In my experience, the relationship between sending out your manuscripts for review, receiving timely feedback, and providing timely feedback for others is challenging. We work in academic communities, and as your career progresses, you are often asked to work for others almost as much as you work for yourself. This includes being asked to review manuscripts, which is a critically important piece of finding your voice in the discipline.

Perspectives on publishing in GIScience: May Yuan

Generally, there are two broad categories of GIScience papers: fundamental research and use-inspired research. Fundamental GIScience papers inquire about conceptual and computational foundations of geographic information and knowledge. Use-inspired GIScience papers address a targeted geographic problem with novel geospatial solutions or perspectives, including critical GIScience. I work to sustain contributions to both research directions for two reasons: (1) to keep a steady stream of manuscripts, and (2) to cross-validate and stimulate conceptual and empirical investigations.

Manuscripts culminate research projects. GIScience is well suited for multidisciplinary collaboration with issues across humanity, social science, and physical science. Research challenges of representation, integration, and computation vary with the thematic, spatial, and temporal considerations. Hence, multidisciplinary collaborative projects offer fertile grounds for GIScience researchers to develop novel methods or workflows that delve into data for geographic patterns, relationships, and predictions. Growing diverse geospatial data and algorithms available from online sources accelerate use-inspired GIScience research. Geotagged Twitter data, for example, have facilitated thousands of use-inspired GIScience papers, ranging from risk communication and population estimation to sentiment mapping. With targeted research questions, empirical data, and novel solutions, use-inspired papers populate most GIScience journals. Early GIScience scholars may find use-inspired research easier to write and get published than fundamental GIScience research.

In contrast to problem-driven, use-inspired research, fundamental GIScience research examines basic properties of spatial information and develops conceptual and computational foundations that transform spatial data into geographic information and prediction. From the bottom-up, fundamental research draws common data abstractions and generalizable functions out of use-inspired empirical solutions. From the top-down, GIScience fundamental research provides conceptual and computational frameworks to ensure epistemological and methodological bases for use-inspired research. Fundamental GIScience papers often take a long time to write and are subject to intensive scrutiny and criticism because of their broad-based claims as the fundamentals and their agenda-setting intent. As such, fundamental GIScience manuscripts commonly receive harsh reviews and require additional rounds of revisions. Junior researchers, under the pressure of the tenure clock, may better focus on use-inspired papers instead, even though publishing a fundamental paper gives us great satisfaction from intellectual contributions that shape GIScience.

Perspectives on publishing in the physical/environmental sciences: Lynn Resler

Increasingly, the interdisciplinary problems we explore as physical/environmental scientists necessitates the integration of theory and method from the natural sciences, physical sciences, GIScience, and social sciences. As such, it is difficult for one person to possess the required excellence in all the skills required to conduct and publish cutting-edge integrative earth and environmental research. It is okay, and even expected, that you seek out collaborators from cognate fields. The benefits of collaborative work are immense. Many physical/environment scientists rely on primary data collection through fieldwork, a process that often necessitates collaboration to maximize efficiency and enhance safety. Collaboration also enhances creativity and helps one develop new skills, test new ideas, and expand networks. For me, working in research teams has aided my productivity through shared resources and expertise, but has also been an avenue for career enjoyment; working with students and other faculty has helped me to establish a greater sense of purpose as a physical geographer. It is exciting for me to reflect upon how the complementary strengths of each contributing author mesh into a final manuscript. I enjoy learning more about my field each time I submit or review a paper.

Meaningful collaborations are reciprocal; as a meaningful collaborator, you assume both leading and supporting roles to achieve scholarly output. Achieving a publication record that reflects a balance of authorship roles (including the occasional solo-authored piece) is a good strategy for promotion as a physical scientist because it demonstrates your ability to support *and* lead research. These collaborations could include those you develop with students and faculty from both within your field and from allied disciplines. Consider also seeking out new collaborators to expand your work into new directions and to diversify your publishing portfolio to include, for example, a systematic literature review paper, or a manuscript that demonstrates a teaching application of a STEM topic. As an early career faculty member, especially consider the importance of new collaborations to demonstrate work that has diverged from your doctoral advisor's mentorship.

Publishing in popular outlets

A more recent change in academic publishing has been the growth of popular outlets for scholarly publishing. This includes everything – from curated websites like *theconversation.com* to posting threads on Twitter, Mastodon, or other social media outlets. In addition, publishing in the popular press can consist of newspaper editorials and blog posts that different outlets host. There are challenges to this kind of work but also benefits. Much like academic publishing, the ultimate goal remains the same: to get your work to audiences who will find your work exciting and help in shaping broad debates in society. In our experience, there are a few things that you need to keep in mind as you put your work out into the world.

First, publishing in popular outlets differs from publishing in academic journals. Because popular outlets are geared toward a broader audience, it is sometimes difficult to switch from using the specialized languages and vocabulary used in academic work to a wide-appeal speech. Working closely with editors and being open to critique can make this transition easier. On the bright side, learning to write in a way that has a wider appeal can open your work to broader audiences. In our experience, pieces written for popular press outlets can receive hundreds of thousands of downloads and open your work to audiences who might not otherwise know your work. This can lead to productive conversations, invitations to appear on other media platforms, and invitations from potential collaborators and even the general public, who will reach out with ideas and resources that can expand your research profile.

Along these lines, it is also important to realize that putting your work into popular outlets can sometimes lead to blowback from the broader public, which are threatened by your ideas. We are living in a political moment in which there is an effort by politicians and others to politicize work that threatens powerful special interests. Climate change, work on critical race theory, and other "controversial" topics often elicit powerful reactions from the broader public that can sometimes lead to threats and unwanted public scrutiny. As you contemplate writing for a broader audience, consider the implications of how your work will be received. It might be necessary to take some precautions to protect yourself, including locking your social media accounts. While it is rare to receive a threatening message, this sometimes happens, and awareness of this reality is essential.

Overall, our experience shows that writing for popular outlets can have benefits that outweigh the risks, especially in terms of reaching general audiences. Even policymakers can and does have broad benefits. This includes helping shape important debates in society and focusing on the powerful role that geography plays in a society that does not fully appreciate what geography has to offer. This includes helping to keep geography's relevance central to debates about its role in educating a broader public desperately in need of geographic perspectives and knowledge.

Closing remarks

Each publishing experience is unique, but if you follow the tips and guidance provided here, you will be on the right track. Publishing is a career-long learning (and relearning) process (Box 19.1), and should be approached as such. To be successful at publishing, we advocate keeping an open mindset. Be prepared to learn something new with each item you write, submit, revise, re-revise, and publish. Consider the many number of publishing outlets available and the types of scholarship geographers can pursue to achieve balanced and holistic career success. We advocate for the points raised in this chapter to be discussed openly within departments, at division meetings, and at the national level.

Acknowledgement

Yuan's contribution to the material is, in part, based on work while serving at the US National Science Foundation.

References

Brunn, S. D. 2009. Academic publishing. In *Aspiring academics,* eds. M. Solem, F. Foote, and J. Monk, 179–189. Upper Saddle River, NJ: Pearson Education.

Butler, D. R. 1989. Conducting research and writing an article in physical geography. In *On becoming a professional geographer*, ed. M. S. Kenzer, 88–99. Columbus: Merrill.

Cobey, K. D., M. M. Lalu, B. Skidmore, N. Ahmadzai, A. Grudniewicz, and D. Moherr. 2018. What is a predatory journal? A scoping review. *F1000Research*, 7:1001. doi: 10.12688/f1000research.

Else, H. 2022. eLife won't reject papers once they are under review — what researchers think. Nature News. 03 November 2022. doi: https://doi.org/10.1038/d41586-022-03534-6

Forgues, B., and S. Liarte. 2013. Academic publishing: Past and Future. *M@n@gement*, 16(5): 735–756.
McKiernan, E. C., L. A. Schimanski, C. Muñoz Nieves, L. Matthias, M. T. Niles, and J. P. Alperin. 2019. Use of the journal impact factor in academic review, promotion, and tenure evaluations. *Elife*, 8: e47338. doi: 10.7554/eLife.47338.
Meadows, M., T. Dietz, and C. Vandermotten. 2016. A perspective on problems and prospects for academic publishing in Geography. *Geo: Geography and Environment*, 3(1): e00016.
Ross-Hellauer, T. 2017. What is open peer review? A systematic review [version 2; peer review: 4 approved]. *F1000Research*, 6: 588.
Rigg, L., S. McCarragher, and A. Krmenec. 2012. Authorship, collaboration, and gender: fifteen years of publication productivity in selected geography journals. *The Professional Geographer*, 64(4):491–502. doi: 10.1080/00330124.2011.611434
Shewan, L. G. and A. Coats 2010. Ethics in the authorship and publishing of scientific articles. *International Journal of Cardiology*, 144: 1–2.

20

PROMOTING YOUR WORK

A professional responsibility[1]

Derek H. Alderman

Why promotion matters

Many of us can vividly remember the first time we published something and the sense of exhilaration and accomplishment that came with seeing one's name on an article, chapter, or book. We had good reason to feel that way. The publication was the culmination of long hours spent analyzing data, reviewing literature, developing theory, writing up results and interpretations, and responding to what were sometimes humbling peer reviews. Given the labor and emotion poured into bringing one's ideas to fruition, it is perhaps natural for many of us to see the final published manuscript or completed project as the end of the research process. However, this is really just the start of the hard work of scholarship.

Communicating and promoting our work to wider scholarly, public, and policy audiences constitute a core professional responsibility crucial for achieving individual career security and fulfillment, advancing the relevance and the overall health of one's field, and making interventions in socially and environmentally consequential scientific and civic debates.

The academy has long viewed the publicizing of scholarship as simply a matter of marketing rather than the "real" work of scientists. Yet, has science ever been about just the work in the laboratory, in the field, or in the archives? Very seldom do we carry out science merely for the purposes of testing hypotheses, formulating new social theory, or mapping patterns. We undertake our scholarship to affect change, whether that means changing the conversation about a scientific issue or resisting what we see as an injustice within society. Promoting the broader impact of research is now a clear requirement for grant proposals submitted to funding agencies and foundations.

DOI: 10.4324/9781003342953-23

Disseminating and translating one's work to wider audiences is central to the scientific process because it legitimizes the social relevance and broader implications of the science. Good science makes those findings understandable, relatable, and useful to a larger world. The term "translational science" has been coined to capture the importance of making research findings accessible to the public and converting them into real-world interventions that improve community well-being (Wethington and Dunifon 2012). Translational science has thus far been framed in the context of the medical sciences, but it has broader applicability as any field considers the social efficacy of promoting its scholarship.

Higher education increasingly faces questions about its relevance, particularly in the face of tightening budgets, program eliminations, and increased public suspicion about the practical value of university education (Salhorta 2022). A program or department is not built solely on promotion, but increasing our work's visibility can help in filling classes, recruiting majors, and securing resources from upper administrators. Greater communication from and about your discipline within scholarly and public circles is crucial to making the case for its survival.

Promoting our scholarship is more than a matter of public relations or marketing – it also involves a critical set of story-telling practices. Because stories are core to how we as humans think and bring meaning to our lives (McKittrick 2020), they are essential to making our ideas and findings socially important. Promotion, when framed in this context of story-telling, is about creating shared understanding of one's work and transforming scholarship into the bodies of actionable knowledge necessary for shaping how policy decision-makers and other public groups think and talk about and act upon pressing global problems. Such actionable knowledge is key to challenging the current "post-truth" environment characterized by the devaluing and suppression of scientific evidence, the political re-scripting of curriculum, and widespread denials of climate change and the lived realities of racism (Alderman and Inwood 2019).

Given these current challenges, communicating the merits of scholarship is key to developing a generation of public intellectuals within universities and professional workplaces. Public intellectualism, as understood here, pushes back on the traditional notion of a lone elite, celebrity-commentator, but recognizes the multiple roles and settings in which scholars can use their work to make a difference while also fitting their promotion and communication to their skills, identities, and work–life contexts. A more democratized public intellectual promotes their work not to monopolize truth but to contribute to ongoing debates, mixing their work in creative, dialogic ways that inform and connect with diverse public groups (Kurtz 2019).

Graduate schools prepare students to formulate, conduct, and write research, but not enough programs provide formal training in outreach and advocacy, the intricacies of doing interviews with journalists (see Box 20.1), or on how to bring basic scholarship into public and policy arenas. When one's career begins, it is uncertain whether or to what extent universities, research labs, or other professional workplaces will value or support efforts to promote and translate scholarship to broader audiences. Promotion can come naturally to some, but for most it requires intentionality and additional skill development.

BOX 20.1 TIPS FOR DOING INTERVIEWS WITH JOURNALISTS

- Prepare to drop what you are doing to talk to a journalist. Journalists often have tight deadlines, and news stories develop rapidly.
- Avoid an interview request for a story outside of your expertise. Recommend more qualified colleagues in your field, especially scholars from underrepresented groups.
- Research the journalist and read some of their work to get a sense of how they cover issues and their ideological point of view as well as that of their outlet or publication.
- Ask for questions the journalist plans to ask and prepare for those questions in advance. Yet, remember interviews develop organically and take new directions once they start.
- Prepare for questions not provided by the journalist but that you know should be addressed. Most journalists end interviews asking, "is there anything else you want me to know."
- Collect key facts and figures from one's scholarship for the interview, confirm their accuracy, and have them ready for quick reference or commit to memory.
- Don't assume the journalist has time to read all of your publications. Communicate a compelling, understandable summary of your work and why it matters publicly.
- If it's a radio or television interview, ask whether it will be broadcast live or recorded and edited. Knowing this will affect your delivery and how much time you have to talk.
- If doing live radio or TV, there is only enough time to make a few key points rather than show all that you know. Reinforce those points throughout the interview.
- If talking with a print journalist or doing a pre-interview with a producer, take the opportunity to collaborate with them to shape and refine the story's direction.
- Pay close attention to what is asked in an interview. Think carefully about your response. Resist over-generalizing about complex issues and people's lives.

- If you use jargon in an interview, do so sparingly and always define terms for journalist.
- Never assume anything you say is "off the record." Even light-hearted comments can end up quoted. If you have doubts about a point you might make, then don't make it.
- When encountering a "gotcha" question for which you don't have an answer, don't make something up. Offer to get back with an answer later. You may not have a specific answer, but you can share what you do know at a more general level.
- Manage the feeling of "imposter syndrome." You may not know everything (and really who does?), but you have plenty of great ideas to offer wider audiences. Expect some criticism from the public but also praise and appreciation.
- Ensure the journalist has correct information on your affiliation, rank, and discipline along with the spelling of your name. Ask the journalist for a timeline on publication.
- Review published results of the interview. See if you have been misquoted or taken out of context. Egregious cases can be brought to the attention of supervising editors.
- Prepare for possible additional media interview requests. If you are comfortable, share with your home institution, disciplinary societies, and colleagues through social media.

Source: Derek Alderman, University of Tennessee, and Joshua Inwood, Penn State University.

Broadening and caring for citation profiles

Our publications are significant personal and professional investments, and it makes sense to expand who knows of and engages our work. There is no greater feeling of frustration than to find one's scholarship not cited in papers and books that address the same issues we do. To use a dated pop cultural reference from the 1980s movie *Field of Dreams*, we often labor under the impression that if we "build it, they will come" or, more specific to our case, "if we publish scholarship, they will surely read, reference, and use it." Yes, good research published in prominent places can and does attract attention on its own, but there is no guarantee that this will always happen, and even for some of our most gifted colleagues, it does not happen.

Self-promotion of one's publications can feel uncomfortable to many professionals who fear that such promotion is unbecoming of a scholar. Yet, the reality is that the advocacy of scholarship has become a necessity in our current political economy of proliferating journals and the increased pressure placed on faculty– even at teaching and service intensive institutions – to publish and have a record of impact and citation. At the neoliberal university, faculty members – especially non-tenured ones – must convincingly demonstrate that their work is cited among other scientists and scholars. As the then AAG President Mona Domosh (2015, np) noted, universities increasingly use citation counts and other metrics to measure research productivity and impact. She says that we "must always act strategically to be in the 'game' (of increasing and counting citations)," while not allowing this "audit culture" to hurt our interdisciplinary richness and collegiality.

A lack of recognition of our research has important ramifications for not just our welfare as individual scholar-teachers and professionals but also the perceived relevance of our field and the vitality of the outlets within which we publish. Academic journals – interested in enhancing their own citation impact profile – now provide advice, software platforms, and communication channels to assist scholars in getting the word out about their publications. Well-heeled scholars, grant teams, and universities can pay to have publications open access, and bringing our work out from behind paywalls is increasingly required of research institutions, research funders, and governments. Publishers sometimes provide authors a limited number of free article downloads to send to colleagues. These complimentary downloads and open access pieces can be combined with the well-worn practice of sending copies of one's work to scholars who would be most interested in reading and likely citing it as well as to those authors who are unaware of the work you do. Active sharing of scholarship is also important to building professional networks, especially with researchers and educators who have a limited working knowledge of one's discipline. A robust citation profile, when combined with this networking, can lead to invitations to contribute to special journal issues, edited books, conference sessions, and funded grant projects.

Maximizing the influence of our publications does not mean we ignore the dangers of focusing so intently on growing a citation profile. There is a difference between "citation" and "impact," recognizing that an impactful publication does not necessarily have to be heavily cited in the academic literature since it may be actively referenced and used by public groups and policymakers outside of the academy. Conversely, scholarship might be widely cited by colleagues, but its language and focus are so academically specialized that it is less usable and thus less visible to wider audiences. Regrettably, universities have long relied heavily upon citation statistics as shorthand indicators of the value of scholar's research. A growing number of universities are incorporating greater public outreach expectations into performance review guidelines, but traditional metrics remain entrenched. Scholars – especially early career faculty scrapping to be tenured and promoted – are well served to balance their time and energy in promoting their work in terms of both citation and societal impact. Attracting fellow scholars or broader public groups to one's publications requires additional labor, which we may or may not be able to contribute. It is important for scholars to take on research promotion efforts that make sense in light of their many other duties at work and home. However, this is not to suggest that one does nothing. Even the smallest effort, if properly targeted and conceived, can have a positive impact.

Finally, academic promotion and citation, while it may appear to be based on individual merit, is actually a collective enterprise shaped by social practices that facilitate or inhibit the participation and recognition of wider communities of scholars and thus affect disciplinary health and equity. It is important to expand our focus beyond maximizing the influence and citation footprint of our own individual publications, but also consider what active role we can play as stewards or caretakers of the promotional process for others in the discipline. Carrie Mott and Daniel Cockayne (2017) direct our attention to the politics of citing published research. Citation practices, more than innocent metrics for measuring scholarly impact, are complicit in the reproduction of racialized and gendered power hierarchies in the academy and the lack of acknowledgment given to the intellectual contributions of women, people of color, students, and non-academics. These unequal footprints affect what is considered valuable knowledge and impact the professional opportunities and job security of underrepresented scholars. Mott and Cockayne argue for academicians to rethink citation as a form of conscientious engagement, examine who is cited (or not) within research, and encourage giving visibility to a greater range of voices in publications even as we promote our own ideas.

Adopting and adapting to new technologies

Promoting one's work requires staying current with and taking advantage of a wide array of communication and media platforms as well as being nimble, responsive, and strategic in not being confined to the traditional reach and pace of academic discourse, dissemination, and collaboration. Adopting and adapting to new technologies is important to building what Brunn and O'Lear (1999) call the "invisible college" of scholars from different disciplines and in different locations using electronic networks to share ideas and benefit from the wider contributions of others researching similar topics. An important tool in this invisible college is Google Scholar, which allows researchers to create, free of charge, a profile to help boost the visibility and accessibility of their scholarship in online searches for literature. A Google Scholar profile provides the Internet community with a researcher's full list of academic productions, citation counts, and an h-index (which measures both the productivity and citation impact of their publications). A profile in Google Scholar, however, does not manage itself, and it is imperative that you regularly review it for accuracy, ensuring that the research attributed to you as an author is indeed yours. You should also periodically check who is citing your scholarship, given that this can help identify further avenues of networking and promotion. Be careful of investing too much into comparing your Google Scholar profile to that of others given that citation footprints can vary widely by field of study, research trends and timing, extent of co-authorship, type and visibility of publication outlet, and the degree and length of time one has been actively promoting their work.

Google Scholar is noted for providing reliable and consistent counts of citations and arguably measures the impact of one's work more comprehensively than other citation databases. Yet, Google Scholar has also been criticized for exacerbating certain social inequalities in the citation process, disadvantaging certain underrepresented groups, and devaluing publications in languages other than in English (Jensenius et al. 2018; Rovira et al. 2021). The promotional effectiveness of Google Scholar can only be fully realized if it is combined with other strategies for publicizing research and generating citations. For example, many scholars have developed profiles at academic sharing and network sites (e.g. ResearchGate, Academia.edu, Mendeley) and posted copies of journal articles, book chapters, conference papers, and other creative products on these sites.

Scholars increasingly supplement their articles with extra data and graphics, highlights written in non-specialist language, animated video abstracts, infographics, recorded author interviews, and blogs and podcasts. In some instances, publishers have resources to facilitate these promotions. Cambridge University Press partners with Kudos, an online platform with services to help scholars tell stories about their research and grow exposure and audience for their publications. The journal *Tourism Geographies* hosts a podcast series highlighting the authors of recent articles. *Antipode* has a companion website (Antipodeonline.com) that provides access to video abstracts of papers along with "intervention" commentaries and teaching resources built upon published content in the journal.

There are companies that work with researchers on a fee basis to produce podcasts and magazine pieces to raise the visibility and social media impact of publications; but it is crucial to make sure that these are not predatory enterprises before committing any funds or time. Having a "reference check" conversation with other scholars at your institution or within your field who have used these companies in the past is highly advisable. These commercial services expose the reality that there is often unequal access to funding needed to support promotional efforts. Disseminating one's work and hence garnering greater scholarly and public recognition is a social practice not open to everyone to the same degree, and there is a need for institutional and disciplinary systems of mutual support to work against these inequities.

Some scholars pursue the development of supplementary promotional materials and communication tools on their own. This is happening in the midst of a larger creative turn among researchers as they begin to make films, produce art works, curate museum exhibits, write poetry, and perform theater and music – not just to share scholarly insights in novel ways but as research methodologies and forms of knowledge production in their own right. Faculty – especially those beginning their careers – are encouraged to hold conversations with department chairs and senior evaluating colleagues about how this creative outreach work of both supporting and moving beyond traditional publications is weighed as scholarly productivity in the annual review and tenure process.

Authors increasingly use social media (e.g., X/Twitter, Facebook, LinkedIn, Instagram, and, more recently, Threads) to lead readers and viewers to their scholarship. Social media promotion requires a thoughtful weighing of opportunities and challenges. It is important to research any social media channel you will use, asking questions such as: What company or owner is behind the platform? How will they use and control my profile and information? What kind of inclusive or exclusive places are they trying to make online through content policies and fee structures? Who will be granted access (or not) to my scholarship, and what political agenda are my social media promotions (un) knowingly supporting?

Over the years, social media-active colleagues have shared that their online interactions have led to "real-world" introductions at conferences, and later to collaborations, requests for interviews from journalists, and citations by scholars from other disciplines. Social media is a potentially important vehicle for greater visibility of one's work; however, social media is not merely an academic tool. The openness of its platforms can facilitate a culture of public outrage and even abuse. Academicians are easy targets, and our employing institutions and agencies may provide little assistance in these tense moments as they are averse to stoking controversy and concerned with protecting their own brand or reputation. One of my colleagues suggests a "10-minute rule" when one sees something that is disagreeable. If you have to react, then draft an angry post, save it, and then come back in ten minutes. The issue of time is also salient in that social media is just one of many of our other duties. Some of the more successful users/promoters limit their online time to specific times of the day or to a set number of minutes a day.

Promoting one's work through social media does not always provide immediate payoff. For example, in the case of X/Twitter, which has traditionally been a popular promotional and communication platform for scholars, gaining followers can be a slow process that requires frequent posting, searching for and following others, tagging highly regarded people and organizations in posts, using hashtags # that are trending, including X/Twitter handles on presentations and publications, and generating meaningful content and opinion. According to one consulting colleague, "people who share good information and provide thoughtful perspectives do tend to build a following over time." It is also easier to create compelling content and build a following when you think strategically about what you want to accomplish through social media. Simply joining the social media bandwagon without a goal will probably leave one unsatisfied and one's discipline poorly served. Yet, with meaningful direction and a spirit of engagement, investments of time and energy in social media can be a productive and natural extension of one's traditional academic and professional work rather than a distraction from it.

Public-facing writing: joining the conversation

An important part of any effort to promote our scholarship must address strategies for writing and speaking to wider publics. I emphasize "publics" as plural, recognizing that there is a vast number of community organizations, governmental agencies, educational institutions, media outlets, nonprofits, and private firms within which our work and perspectives can and do make a difference. Recognizing your audience heavily shapes the promotional message and medium you pursue in translating the value of scholarship to public audiences, whether by writing a blog or op-ed (see Box 20.2), preparing a press release, participating in a news broadcast or documentary, or producing educational videos, pamphlets, or zines. Zines are usually physical or digital booklets that are self-published and have small, non-commercial circulations often intended to trigger wider community discussions of a topic of interest or a call for social action – thus expanding our scholarly understandings of how and where knowledge is produced, taught, and mobilized politically (FLOCK 2021).

BOX 20.2 TIPS FOR WRITING OP-EDS

- Be timely, write quickly for the news cycle, and respond to breaking issues or stories.
- Write on issues in which you have expertise. Connect your scholarship to current events.
- Don't treat an op-ed as merely an analysis. Embrace your informed personal perspective.
- Effective op-eds are clear and persuasive calls to action. Consider what you want the reader to do, think, or feel as a result of your piece.
- Be concise (650–750 words), use active voice, and write short sentences and paragraphs.
- Hook readers by putting the main idea at the beginning of the op-ed. Use plain but attention-holding language.
- Summarize your argument and/or offer recommendations in a strong final paragraph.

- Determine the audience of your op-ed. Are you writing for the general public or for a targeted audience? Consider which media outlet might be interested in publishing. Is there a local angle? Is it environmentally oriented? Is it politically focused? Is it industry-specific?
- Read op-ed submission instructions. Some outlets want exclusive consideration of your piece (you can't submit elsewhere). Depending on the outlet, acceptance rate can be low.
- Submit an op-ed to a media outlet with succinct, attention-grabbing email to the editor.
- Once published, share news of op-ed across your social media channels.

Source: Moseley (2010), Duke University Communicator Toolkit, Boston University Public Relations.

Recognizing that many scholars are poorly prepared to communicate outside of their academic cultures, organizations such as Compass, the Alan Alda Center for Communicating Science, the AAAS (Association for the Advancement of Science), and the National Geographic Society have established resources and workshops to help scientists move beyond dry presentations of data to tell evocative stories about themselves and their work to journalists, policy-makers, and other audiences. Some of these initiatives such as the OpEd Project work with experts from underrepresented groups to help promote their ideas and voices. It is valuable to locate outstanding scholar-communicators within one's own field and consult them for advice in public communication (see Box 20.3). In geography, for example, William Moseley (2010) has contributed numerous op-eds to major media outlets on agricultural, environmental, and development challenges facing African countries. Minelle Mahtani, an award-winning researcher and specialist in media communication, hosted a daily radio called "Sense of Place" in Canada from 2015 to 2018. Mahtani views radio not only as a tool of dissemination but as a geographic space in and of itself for creating scholar–public dialogues that are decidedly antiracist and anticolonial (Mahtani and Shafer 2018).

Marshall Shepherd, a regular contributor to *Forbes* magazine and long-time host of the television show turned podcast series *Weather Geeks*, has built a career diffusing and translating his work in atmospheric science to large segments of the public. Shepherd (2016) provides helpful tips for communicating to non-scientists, emphasizing the importance of concise and jargon-free communication tailored to one's audience. Shepherd notes that researchers and public audiences have inherently different styles of communication. While scholars frequently present considerable amounts of background information and theory before delivering results, public groups and policy-makers need and expect key findings to be provided early on and in a straightforward manner.

BOX 20.3 SELECT EXAMPLES OF MEDIA ENGAGEMENT BY SCHOLARS

Op-eds, blogs, guest analyses

- Sophia Grijalva. "How Fog Catchers Improve Water Scarcity in Urban Settlements in Peru." *London School of Economics (LSE) Blogs*. January 12, 2023. https://blogs.lse.ac.uk/latamcaribbean/2023/01/12/how-fog-catchers-improve-water-scarcity-in-urban-settlements-in-peru/
- John Rennie Short. "Settler colonialism helps explain current events in Xinjiang and Ukraine – and the history of Australia and US, too." *The Conversation*. March 14, 2022. https://theconversation.com/settler-colonialism-helps-explain-current-events-in-xinjiang-and-ukraine-and-the-history-of-australia-and-us-too-176975
- Farhana Sultana. "Opinion: Climate and COVID-19 Crises Both Need Feminism – Here's Why." *The Hill*. May 16, 2021. https://thehill.com/opinion/energy-environment/553707-climate-and-covid-19-crises-both-need-feminism-heres-why/
- Brandi T. Summers. "Op-Ed: What Black America Knows About Quarantine." *New York Times*. May 15, 2020. https://www.nytimes.com/2020/05/15/opinion/sunday/coronavirus-ahmaud-arbery-race.html

Radio & podcast interviews

- Winifred Curran. "Can Cities Go Green Without Driving Gentrification?" 1A Program, Nation Public Radio (NPR). February 27, 2023. https://www.npr.org/2022/12/22/1145090530/can-cities-go-green-without-driving-gentrification
- Karen Bakker. "Deep Listening to Sounds of Nature." The Scholar's Circle Radio. November 22, 2022. https://scholarscircle.org/scholars-circle-deep-listening-to-sounds-of-nature-actual-history-in-american-thanksgiving-story-november-27-2022/
- Andrew Curley. "Understanding Water and Energy in the Navajo Nation." Resources Radio. November 8, 2022. https://www.resources.org/resources-radio/understanding-water-and-energy-in-the-navajo-nation-with-andrew-curley/
- Rashad Shabazz. "Spatializing Blackness." 1919Radio Black Geographies Podcast Series. June 4, 2021. https://1919mag.com/1919-radio/2021/6/2/spatializing-blackness-with-dr-rashad-shabazz

Television & documentary appearances

- Kelsey Ellis. "Tornadoes Becoming Stronger, More Common in East Tennessee." WBIR Channel 10 News. June 13, 2023. https://www.wbir.com/video/

news/local/tornadoes-becoming-stronger-more-common-in-east-tennessee/51-0609aa95-82af-4740-acba-dbbae9934d88

- Reece Jones. "Is It Time to Increase Accountability at the U.S. Border Patrol?" The Steam Program, *Al Jazeera English*. November 3, 2022. https://www.aljazeera.com/program/the-stream/2022/11/3/is-it-time-to-increase-accountability-at-the-u-s-border-patrol
- Derek Alderman. "Avenues of Dreams: Reclaiming MLK Boulevards." Xfinity Black Experience Channel. January 2022. https://play.xumo.com/free-movies/avenues-of-dreams/XM0WC5L4A3UJUY
- Joshua Inwood. "White Supremacist Threat: New Zealand's Global Call to Counter Ideology of Hate." France 24. March 20, 2019. https://www.france24.com/en/debate/20190320-white-supremacist-threat-new-zealands-global-call-counter-ideology-hate

Film and video production

- Robert Lemon is director and producer of *Transfusión* (2014), a film chronicling the struggles of an undocumented woman running a taco truck in Columbus, Ohio, to highlight relationship between food, identity, and community. https://indie-outlook.com/2014/04/15/robert-d-lemon-on-transfusion/
- Jospeh Kerski, creator of *Our Earth* YouTube Channel, has self-produced thousands of videos since 2008 that explore the importance of geography, STEM, and geotechnologies in education. https://www.youtube.com/@Our__Earth/videos
- Qiusheng Wu, creator of *Open Geospatial Solutions* YouTube Channel, has self-produced hundreds of video tutorials since 2019 to encourage use of open-source software packages for advanced geospatial analysis. https://www.youtube.com/@giswqs
- FilmGeographies, which Jessica Jacobs and Joseph Palis started as a specialty group with the American Association of Geographers, hosts an online platform for filmmaking workshops and peer reviewing and publishing of films by scholars. https://www.filmgeographies.com/

No matter how many or which strategies we choose in promoting and communicating our work publicly, one thing is certain – we will need a good "pitch." A pitch is a brief and compelling statement that persuades a wider audience that one's published research is significant and worth learning about. The pitch is closely related to another term familiar to many readers – an elevator speech. Professionals are increasingly encouraged to recognize the value of developing easily understandable summaries of their work that can be delivered in less than 30 seconds, the time of a brief elevator ride, and which can grab the attention of potential employers, student recruits, or wider civic groups. Developing a pitch or elevator speech sounds like something easily accomplished, but it is actually challenging. I learned about the art of pitching from Lola Alapo, a public relations and science communication specialist at the University of Tennessee. According to Alapo, pitching research requires summarizing its major purpose and significance in clear and enticing ways. The pitch cannot capture all that the publication is about; rather, it gets your foot in the door for longer and deeper conversations with journalists, policy agencies, and community groups. She encourages scholars to translate their latest research in three to four sentences and in terms accessible and appropriate to the target audience. In pitching publications for possible media coverage, for example, assume you are communicating with an eighth grader, extrapolate applications of your research, and speak directly to how the work advances the way we think about a pressing issue or debate.

The growing turn to public engagement across the entire academy is not just for the sake of self-promotion but out of a growing recognition of the legitimacy of alternative mediums for knowledge construction as well as dissatisfaction with more traditional academic communication channels – especially in terms of the limited visibility and comprehensibility of our writing and speaking to civic groups and decision-makers. A growing number of scientists from a wide array of fields, including several from geography, are taking advantage of a key resource in public-facing writing – *The Conversation*. *The Conversation* is a nonprofit, independent news organization that uses teams of journalists and editors to work with university scholars to publish brief but informative articles. National and international news outlets are then allowed to "pick up" or republish these articles for free under a Creative Commons license. Founded in Australia in 2011 and launched in the US soon after, the *Conversation* now publishes a number of international editions across multiple world regions.

Contributing to the *Conversation* involves making a pitch through its online portal, writing a 600–800 word essay, engaging in several rounds of writing and revision with editors, and tracking the impact of one's published item in terms of readership, comments, and social media using the *Conversation* dashboard. Unlike writing a traditional op-ed piece for a newspaper, a *Conversation* piece is more like a guest expert analysis on an issue, and authors do not have to search for media outlets willing to publish their ideas. *The Conversation* works to promote and publish one's piece through daily email newsletters and its website and network of media partners. Geographers who have authored *Conversation* pieces note how the process can clarify and sharpen their points and even provide new angles for future scholarship. Published essays can also lead to other avenues inside and outside of the media for promoting work and facilitating collaborations with nonprofits, community organizations, planning agencies, and educators.

Public speaking: creating a dialogue, not a monologue

In addition to public-facing writing, scholars are increasingly speaking outside the traditional confines of classrooms, university colloquial, and conference sessions to make public presentations at libraries, museums, teach-ins, rallies, policy institutes, city council meetings, and private industry forums. Speaking publicly about one's work is not merely a transmission of information but also creation of a place of learning that can either open up or close off certain lines of understanding. As Anja Kanngieser (2012) reminds us, voices have social and political consequences in how they affect our capacities to speak and listen to one another.

Online resources exist (e.g., Shaw 2013) that provide prescriptive advice to assist academic speakers when they go public, but one presentation style does not fit all speakers. One should develop a speaking style that fits one's strengths and temperament. Being an effective speaker is not about being a perfect speaker. One's speaking voice does not have to sound the same as one's writing voice. Ultimately, what is important is finding a presentation style that allows one to make a basic, human connection with the audience. Making eye contact with the audience is highly important. Effective speaking is not about overwhelming the audience with our intellect, but it is also not about lowering the rigor of our ideas; audiences yearn for understandable communication that challenges them to think. Many presentation coaches suggest that speakers avoid all jargon, but the use of specialist language is appropriate in some situations. Always define unfamiliar terms for audiences.

Memorable presentations evoke strong mental pictures, curiosity, and feelings from audiences. Effective speakers contextualize, and sometimes even personalize, an issue and then apply distilled ideas, research findings, and experiences to interpret that issue. This interpretation should resonate with meaning to the audience, even if audience members hold views different from the speaker. Think about how you as a speaker can tell the story of your work to demonstrate why it matters, for whom it matters, and how it will continue to matter in future discussions and debates. Do not forget to tell *your* story, how you came to do the work, why the field and topic is your passion, and what your identity is relative to the issue at hand. Your story is about bringing a human element to the conversation rather than representing the discovery process as disembodied. A detached reporting of facts, figures, and methods can be uninspiring, even if it has the feel of being scientifically objective, and it denies the socially constructed nature of all knowledge.

It is important that speakers make sure that the voices and experiences of traditionally ignored social actors and groups have a prominent place in their public remarks. I especially admire my colleagues who are able to navigate the difficult terrain of *speaking in solidarity* with these marginalized communities versus *speaking for them*. Public speaking is also an opportunity for not just vocalizing our ideas, but also creating opportunities to engage in a dialogue rather than simply a monologue with their audiences. Speakers should design presentations to leave sufficient time for public feedback and questions and take an active hand in finding and hearing from a wide range of voices in attendance. Creating this dialogue is not just about being polite; rather, it is critical for creating shared understanding between the scholar and the public.

Listening is not a passive process but actively involved in the construction of meaning. Being a good speaker is also about being a good listener. When a speaker is an active listener, she, he, or they also shares the power and authority to speak with others. Publicly engaged science is more than a one-way relationship of experts simply handing down authorized insight and research findings. Rather, by creating spaces of dialogue that blur the expert–lay divide, speakers can co-produce knowledge with public groups. At the very least, effective speaking and listening can be moments where the professional learns how people in the world outside of universities, research labs, and workplaces view and interpret science as well as how they create, use, and justify their own knowledge and expertise.

Concluding remarks

Early in my career, one of my most respected mentors advised me: "Let them hear about you, not from you." While my mentor was rightly pointing to how self-advocacy can be open to narcissistic abuse, actively promoting one's work is not inherently negative, and waiting around for other colleagues to suddenly discover our scholarship can be personally frustrating and academically and socially unproductive. Developing and carrying out thoughtful and sometimes time-consuming strategies for promoting our work is, as I have argued here, a professional responsibility. It is a responsibility we have in fulfilling our career goals and protecting our livelihoods. It is our responsibility to advocate for our disciplines and assert why our field matters in today's world and the neoliberal university. Finally, there is responsibility in disseminating and translating our scholarship widely so as to engage and offer aid to public groups, contribute to larger scientific and civic conversations, and initiate valuable dialogue with non-academic groups.

While promoting one's scholarship and ideas is a professional responsibility, it should not be treated as a one-size-fits-all approach. Rather, a scholar should discover how and where best to share their work based on their individual voice, talents, conscience, stage of career, and their institution's demands. Sweeping calls for public engagement must be balanced with an acknowledgment of the elevated vulnerability and burden that promotional work places on early career professionals and scholars from historically excluded groups (Worth 2020). While a recent survey suggests that most scientists see outreach and interacting with the media as beneficial, the risks and fears of those interactions weigh much more heavily on the minds of those who identify as women researchers given the gendered (and racialized) ways that scholarly authority is valued (SciOPS 2020). Thus, promoting one's work is about more than just learning specific techniques for sharing research findings; rather, it is also about creating a more robust culture for supporting, mentoring, and defending public intellectual work in the academy.

Note

1 This chapter is a revised and expanded republishing of three of my AAG Presidential columns: The Serious Business of Public Communication (August 2017); Moving at Twitter Speed (September 2017); On Finding One's Voice (and the Voices of Others) (October 2017); and Broadening and Caring for the Footprint of Published Scholarship (December 2017).

References

Alderman, D. H., and Inwood, J. F. 2019. The Need for Public Intellectuals in the Trump Era and Beyond: Strategies for Communication, Engagement, and Advocacy. *The Professional Geographer 71* (1): 145–151.

Brunn, S. D., and O'Lear, S. R. 1999. Research and Communication in the "Invisible College" of the Human Dimensions of Global Change. *Global Environmental Change 9* (4): 285–301.

Domosh, M. 2015. The Costs (and Benefits?) of Constant Counting. AAG Presidential Columns. 1 April, Available online at: https://www.aag.org/the-costs-and-benefits-of-constant-counting/

FLOCK, F. G. C. 2021. Making a Zine, Building a Feminist Collective: Ruptures I, Student Visionaries, and Racial Justice at the University of North Carolina at Chapel Hill. *ACME: An International Journal for Critical Geographies 20* (5): 531–561.

Jensenius, F. R., Htun, M., Samuels, D. J., Singer, D. A., Lawrence, A., and Chwe, M. 2018. The Benefits and Pitfalls of Google Scholar. *PS: Political Science & Politics 51* (4): 820–824.

Kanngieser, A. 2012. A Sonic Geography of Voice: Towards an Affective Politics. *Progress in Human Geography 36* (3): 336–353.

Kurtz, H. E. 2019. Public Intellectualism as Assemblage. *The Professional Geographer 71* (1): 179–183.

Mahtani, M., and Shafer, D. 2018. Dialogue, Discourse, Disjunctures: Building Critically Affirmative Politics in Radio. *Dialogues in Human Geography 8* (2): 156–159.

McKittrick, K. 2020. *Dear Science and Other Stories*. Durham: Duke University Press.

Moseley, W. G. 2010. Engaging the Public imagination: Geographers in the Op–Ed Pages. *Geographical Review 100* (1): 109–121.

Mott, C., and Cockayne, D. 2017. Citation Matters: Mobilizing the Politics of Citation toward a Practice of 'Conscientious Engagement'. *Gender, Place & Culture 24* (7): 954–973.

Rovira, C., Codina, L., and Lopezosa, C. 2021. Language Bias in the Google Scholar Ranking Algorithm. *Future Internet 13* (2): 31. https://doi.org/10.3390/fi13020031

Salhorta, P. 2022. A Growing Number of Americans are Questioning the Value of Going to College. *National Public Radio*. 26 July, Available online at: https://www.npr.org/2022/07/26/1113816868/more-americans-are-questioning-whether-college-has-a-positive-impact

SciOPS. 2020. "Survey Results: Supporting Scientists' Engagement with the Media." Available online at https://www.sci-ops.org/surveys/scientists-engagement-with-the-media

Shaw, C. 2013. Public speaking for academics – 10 tips. *The Guardian*. 10 May, Available at: https://www.theguardian.com/higher-education-network/2013/may/10/public-speaking-academics-10-tips

Shepherd, M. 2016. 9 Tips for Communicating Science to People Who Are Not Scientists. Forbes. 22 November, Available online at: https://www.forbes.com/sites/marshallshepherd/2016/11/22/9-tips-for-communicating-science-to-people-who-are-not-scientists/

Wethington, E., and Dunifon, R. 2012. *Research for the Public Good: Applying the Methods of Translational Research to Improve Human Health and Well Being*. Washington, DC: American Psychological Association.

Worth, N., 2020. Public Geographies and the Gendered Experience of Saying "Yes" to the Media. *The Professional Geographer 72* (4): 547–555.

SECTION IV
Building toward long-term success

Shannon O'Lear

Building toward long-term success

Graduate school has its own forms of stressors. A new faculty position is a different mode of work, with different expectations and new challenges. Learning a new position is understandably challenging, and it is quite possible that you are the only one or one of a few new members in a department or program where most of the people are already familiar with the job and settled. You are probably no longer surrounded by peers at your same stage. Where do you find support as you navigate your new surroundings in this early phase of an academic career? The chapters in this section aim at providing initial guidance to support building your long-term success.

This section addresses realities of an academic career that may seem peripheral to the research and even teaching endeavors entailed in graduate training. Yet the topics addressed in this section can help to amplify one's strengths and skills as much as they might sink one's efforts to stay on track as a successful academic. Our intention, of course, is to guide toward the former while recognizing the realities – and manageability – of pitfalls.

Overview of the chapters

In "Cultivating a Collegial Workplace," Shannon O'Lear, Kenneth Foote, and Dagmar Budikova consider the importance of intentional community-building, particularly in a workplace where students, faculty, and staff could benefit from a sense of cooperation and support. This chapter offers a variety of insights and suggested practices to cultivate an inviting workplace that values knowledge co-production and that fosters a shared sense of identity.

DOI: 10.4324/9781003342953-24

They draw on research on the development of intellectual community and the formation of scholars. They point to the importance of cultivating a shared sense of purpose among a diverse group where learning and scholarship are valued. They recognize how collegiality can be fostered immediately and developed over time through different kinds of roles. For instance, there is value in bringing students into the life of the department, not just as people enrolled in classes, but as fellow learners who bring valuable insights to your course or your program. Engaging students could be as simple as showing up early to class to be available for conversations, or it could look like involving students on departmental committees or planning professional development or mentoring events. Another approach to fostering collegiality is to generate collaborative projects such as updating the curriculum or hosting research talks that include all faculty and people in nearby programs. This chapter also discusses opening classroom doors and creating physical spaces for intellectual sharing and conversation. Collegiality can look like organizing and participating in social events that are welcoming as much as it can look like taking time to reflect, to check in on peers, and to listen. Obstacles to collegiality may come from outside the department such as how and what decisions are made by university administrators or in the form of budget cuts or financial constraints. Even in the face of challenges to collegiality, transparent decision-making that values all of the impacted individuals can help to support and build collegial places where we gather to work.

In "Valuing Mentoring, or How to Think Usefully About Mentoring," Hilda E. Kurtz makes the case that mentoring takes many forms and is usefully approached as formal and informal connections with different people through the course of one's career. She encourages us to think about how to mentor well, how to benefit as a mentee, and how to make these important connections visible. Drawing on a relatively recent body of scholarship on mentoring in academia, Kurtz considers how mentoring is both critical for student success and context-dependent. Different aspects of academic life call for different skill sets from mentors. For this reason, it is useful for emerging scholars to think about areas in which they desire mentorship and to seek mentors suited to those areas and needs. Cultures of mentorship vary across disciplines as styles of mentorship vary across individuals. Mentoring, either for career development or for psychosocial development, can happen one-on-one or in group settings, depending on the objective and people involved. In addition to utilizing any formal mentoring programs offered institutionally, it is also useful to cultivate mentors independently to support your well-being and professional success. This chapter considers stages of mentoring, the value of crafting a mentoring compact, and benefits of peer mentoring. Overall, effective mentoring is often intentional and is an investment of time that, if done with thought and reflection, can benefit everyone involved by creating supportive, professional networks.

In his chapter, "Serving With a Purpose: Thinking Strategically About Service," Michael E. Meadows discusses how service – to our department, college, university, discipline, and other professional groups or organizations – is often an expected component of an academic position. Service can take many forms, ranging from committee work focused on predetermined objectives to taking initiatives toward new, outward-facing pursuits. Meadows describes types of service that an emerging academic might face, and he discusses how some forms of service might be particularly rewarding and why it is helpful to think of service as part of being an academic citizen. Meadows considers how service roles including responsibilities within one's department and other forms of outreach or engagement can support and shape career development. The level of service and time commitment will likely increase through one's academic career. Satisfaction with service work will depend on one's personality and inclinations/values, and may also be enhanced if one's efforts and accomplishments in service areas are recognized by peers and department leadership. Service in one's department or at one's institution can involve leadership and interaction with faculty in other programs or units and can develop one's understanding of how the institution functions, which can be illuminating. Service to a community beyond one's own institution might entail engagement with professional organizations at the national or international level. It could also involve more local community-focused collaboration or service-learning. Other forms of service include peer review of journal manuscripts, book proposals, and research grant proposals. Early in one's career, invitations to review the work of others as part of an overarching, academic peer review process might be infrequent, but they become regular requests as one's research productivity increases. These opportunities to provide feedback can also be instructive as a source of insight and perspective that might not be a part of one's own research or teaching agenda. Eventually, there may be opportunities to serve on the editorial board of a journal or other publication project in one's area as well as serving as an editor for a journal or book project. Participating in professional meetings and developing one's professional network can be an excellent way to make connections within an organization and with people at other institutions, and these connections can lead to invitations to join committees, review manuscripts and proposals, and get involved in other forms of service collaborations. Often, professional connections are made not through direct planning but by making an effort to interact, participate, and even volunteer one's service at professional meetings and other gatherings. Research and teaching tend to be prioritized in academic positions, but service can be a means to learn more about both research and teaching through exposure to new ideas, the need to develop or advance particular skill sets, and the enrichment of collegiality in one's career.

In "Moving toward promotion and beyond: Understanding and connecting institutional needs and personal aspirations," Deborah G. Martin describes the intricate balance required for a successful tenure case. Navigating tenure expectations entails active involvement and teaching, research, and service duties. She emphasizes that the balance of teaching, research, and service is contingent and requires institutional context, which she goes on to detail throughout the chapter. She asserts that understanding institutional contexts is truly vital to strategizing one's progress toward tenure and promotion. This understanding is fundamental given the variable workload expectations across different types of institutions. As she details in the chapter, in the United States, higher education institutions range from community colleges to liberal arts/baccalaureate colleges to master's and doctoral degree-granting institutions. She underscores the heavier emphasis on teaching and service at community colleges and small/private liberal arts and baccalaureate colleges and, therefore, stresses the need for faculty at these institutions to focus on teaching quality and innovation. Faculty expectations at these institution types contrast with faculty appointments at universities that grant master's or both master's and doctoral degrees. These institutions are considered research-intensive and, as such, include a reduced teaching load, compared to teaching-intensive institutions, and an increased emphasis on research productivity. She discusses the pivotal role of actively engaging in departmental conversations, building a relationship with the department head, and seeking mentors to decipher tenure criteria and cultivate a community of support. She encourages seeking professional development and networking opportunities tailored for pre-tenure faculty, which can offer varying degrees of support. She discusses that every faculty member is expected to engage in some level of service, which contributes to the health and function of the department and the institution more broadly. She outlines tips for understanding the nature of a faculty member's service obligations and the possibilities for establishing camaraderie among department colleagues through service commitments. Many of Martin's strategies detailed in the chapter are underscored by her personal experiences, which she weaves into the narrative of the text. Ultimately, regardless of institution type, Martin highlights the significance of aligning individual goals with faculty expectations within an institutional context, along with developing strategies for a balanced approach toward tenure expectations in order to solidify a successful case for tenure and promotion.

In "It's OK Not Being OK: Coping with Adversity," Jennifer Ho brings to the forefront the reality that, beyond the challenges of an academic career, life can throw unforeseen and seemingly insurmountable obstacles in our path. She recognizes that sometimes, despite the pressure of academia to persevere and accomplish, we may, in fact, struggle. Ho draws from her personal circumstances to talk about what happens when life throws significant challenges, setbacks, or hardship in our way. She acknowledges that academic training tends to value independence and self-sufficiency, and that admitting to any weakness is generally discouraged even in graduate programs. Ho details events that happened in her life that upended her sense of safety and normalcy, and she describes how living in precarity felt at odds with her life as an ambitious and successful academic. She uses these examples from her life to describe how she felt and how she responded. She dares to admit that in these times, she was not OK (and revisiting these times, she may still not be entirely OK) – and that is OK. In sharing parts of her vulnerability with her students and colleagues, she discovered that they appreciated her humanness and, in turn, were able to respond more fully. A mutual recognition of the very real challenges of mental health issues turns out to have nurtured a community of care that helped her (and quite likely others) to cope with adversity. Her chapter is a powerful reminder that academic endeavors are still human endeavors, and it is OK to be fully human and therefore sometimes not to be OK.

Together, these chapters recognize aspects of an academic career that usually fall outside of formal, graduate training but commonly feature in academic positions. By discussing how these topics of collegiality, mentoring, service, and unanticipated hardship can be skillfully considered in one's own academic efforts, our intention is to ensure that your time and energy investments contribute to a workplace and career that are supportive and rewarding.

21

CULTIVATING A COLLEGIAL WORKPLACE

Shannon O'Lear, Kenneth Foote and Dagmar Budikova

Introduction

What does it feel like when you go to work in your department? Do you rush to your office to settle in and tackle your to-do list? If you see your colleagues in the hallway, do you stop to chat and share ideas? If your department organizes an event – either work-related or social – do you look forward to interacting with your colleagues, or do you dread going? Do you feel that your contributions to your department are appreciated and valued by your colleagues, or does it seem your work is overlooked?

These are questions of collegiality and the quality of interpersonal relationships in our department or university community. At its best, collegiality is expressed in a spirit of cooperation, shared decision-making, mutual respect, knowledge co-production, and support among colleagues, all of which helps us achieve our goals. Collegiality is a career-long investment in relationships with our colleagues, staff, and students that doesn't happen by accident, or overnight. But, though building such relationships is a long-term process, such efforts do offer immediate dividends in terms of open communication, constructive feedback, and timely advice.

Notably, the words *collegiality* and *college* stem from the same Latin root, *collegium*, a word whose meaning in English is close to *association* or *organization*, a body of individuals working toward common goals, while sharing decision-making power and authority. The etymology of *collegiality* is important because its root suggests three interrelated qualities: interpersonal relationships, shared governance, and a sense of identity based on shared goals.

Helping to build these qualities in your department may seem like a daunting, even over-the-top, suggestion for you at the beginning of your career.

DOI: 10.4324/9781003342953-25

But, as we suggest in this chapter, cultivating collegiality involves everyone in our community, and not only the most senior members and administrators. Fortunately, many of the qualities that will help you build collegial relationships are also covered in other chapters in this book such as those on networking, caring for ourselves, and building career momentum. This chapter suggests a range of strategies that can help you develop strong and productive collegial relationships that will help to advance your career as well as the mission of your department. Recognizing the practices that can help to cultivate collegiality is helpful at any stage of one's career.

Insights from research and theory

Collegiality is not just a feel-good topic. Collegiality is actually an important topic in organization theory that has been studied by researchers in business management, economics, business and organizational psychology, as well as in educational administration (Ashkanasy, Wilderom, and Peterson 2011). This research indicates that collegiality is associated with a wide range of positive outcomes in the workplace, including better performance of individuals and the organization as a whole, greater employee satisfaction and commitment to the organization, greater support for leadership and new initiatives within the organization, and higher levels of ethical behavior (Denison and Mishra 1995, Kotter and Heskett 1992, Schein and Schein 2017) (Figure 21.1).

FIGURE 21.1 Collegiality includes many, overlapping elements.

To understand why collegiality is so important, it is useful to see it as playing a mediating role between what organization theorists term *culture* and *climate* (Schneider, Ehrhart, and Macey 2013, Zohar and Hofmann 2012). Culture includes the shared values that are passed from generation to generation, including norms and traditions that continue in a department even when faculty retire, and new faculty and staff are hired. These may be values relating, for instance, to research and teaching specializations for which a program has developed a strong reputation; the emphasis faculty place on teaching, student engagement and fostering learning, fieldwork, or community outreach; a department's success in supporting diversity and inclusion initiatives; distinctive ceremonies and celebrations; or unique strategies for recruiting, advising, and mentoring students. These cultural values and traditions tend to change slowly.

Climate, in contrast, refers to the current mood or social character of a department. Climate encompasses attitudes, perceptions, and overall morale, and it can change from semester to semester and year to year. Climate is shaped, in part, by culture, but also responds to immediate interpersonal relationships and institutional factors which may have temporary, rather than lasting, effects. This means that changes in climate, whether positive or negative, can affect a department's culture, albeit typically more slowly, over time (Peterson and Spencer 1990).

Collegiality is a particularly important part of this dynamic because it serves as a mediator between climate and culture. It encompasses the day-to-day and week-to-week actions and activities that can transform both the cultural values of a program and the perceptions and attitudes that shape everyday interactions. However, drawing direct causative connections between culture, climate, and collegiality can be very difficult because all three factors are shaped by many personal and interpersonal factors as well as interactions. Additionally, research indicates that the cultures, climates, and standards of collegiality vary tremendously by institution type, for example between educational, business, governmental organizations, as well as by nation and national culture.

These variations also show up in higher education. One of the earliest and most influential studies in this area was Becher's *Academic Tribes and Territories,* first published in 1989, but it has since been revised (Becher and Trowler 2001). Though Becher focused his original research on universities in Britain and North America, he found tremendous differences in the culture of disciplines, even within individual universities. Other studies have addressed climate, culture, and collegiality in other ways, but all seem to underscore the complex interactions among culture, climate, and collegiality, particularly in respect to doctoral education (Gardner and Mendoza 2010, Golde 2005, Golde, Walker, and Associates 2006, Shattock 2010, Viđak et al. 2021).

The research most relevant to this chapter stems from a project begun in the 1990s by the Carnegie Foundation for the Advancement of Teaching.

The most important of the project's publications is *The Formation of Scholars: Rethinking Doctoral Education in the Twenty-First Century* (Walker et al. 2008). The sixth chapter, "Creating and Sustaining Intellectual Community" (Walker et al. 2008, 120–140), makes a compelling case that collegiality, climate, and culture are simply facets of strong, supportive academic communities. It points out that these healthy communities take many forms and are shaped by discipline, institution, student interests, and faculty expertise. What they have in common is a sense of shared purpose, diverse and multigenerational membership, a flexible and forgiving approach to learning and scholarship, and a respectful and generous attitude toward fostering student learning (Walker et al. 2008, 124–127).

Activities and practices that foster intellectual communities

Walker and his colleagues go further in their book and suggest activities and practices that help build academic communities. Here, we note some of their suggestions. We also list ideas generated during webinars and workshops we have organized and led within the American Association of Geographers for early-career faculty as well as department leaders.

We would stress from the start that community building doesn't happen by accident and involves everyone in the community, and that really means everyone, including staff, non-tenure track, undergraduate and graduate students, and visitors, a point we sometimes forget. Equally important is recognizing that community building occurs at a range of levels which may intersect and amplify each other. These include the following:

Individual roles: Collegiality at the individual scale involves civil interpersonal interactions. This looks like conducting ourselves in a civil manner, showing respect for diverse opinions, and ensuring that our interactions are inclusive and welcoming. These types of interactions help to build welcoming working and learning communities.

Team or collective roles: These interactions build on interpersonal demeanor in ways that foster collective progress and positive outcomes in mission-critical activities related to teaching, research, and service, at the individual and group levels.

Institutional roles: This level of interaction recognizes the broader environment in which a department or program operates and the importance of engaging proactively (not only reactively) on behalf of departmental interests. This kind of interaction looks like active and productive participation in shared governance such as service to the department, college, university, and professional organizations.

With these points in mind, the following sections highlight a number of activities and practices that could be integrated into the life of many departments or programs.

Engaging students fully in the life of the department

Students in our departments are not mere receptacles for knowledge; they are fellow learners who share an interest in our disciplinary view of the world. One way is by building a department culture that fosters supportive and engaged student communities. Where appropriate, seek ways to engage students in your research program. Invite them to department colloquia and student luncheons with speakers, and organize award ceremonies, club and field excursions, and other gatherings for students that help them to get to know each other as well as faculty of the program. Find ways to involve students in shared governance such as on departmental committees, in conversations about curricular pathways, recruitment of new students, and any plans to upgrade or adjust program offerings and requirements. Students can offer unique and diverse perspectives on their experiences within a program or department. Ask students what aspects about your course or program are working well and how they might be improved through either expansion in some areas or by focusing and deepening concentration in others. Check in with students occasionally through group emails or listserves. Provide useful mentorship and sound advice on course selection and professional development. This can include offering information on how to drop or add classes after a particular deadline or what courses they ought to take given their professional aspirations. Ask what needs students have. They may have seen interesting practices or activities elsewhere and might offer suggestions for helpful forms of interaction. Model for students what respectful, intellectual curiosity looks like. Be available for conversations. This might involve showing up a little early for class to engage in informal conversations or staying a bit later after class. Slow down and be available.

Collaborative work on curriculum

In addition to seeking student input on academic aspects of your program, open the conversation to include all faculty regardless of their rank or stature. How visible is a pathway through the curriculum, or how visible are program options to students? How does the curriculum play to current faculty strengths? What ideas do people have to update courses or the curriculum to bring novel and useful information, skill sets, and experiences to students? Are there courses that could be co-taught by faculty to integrate different perspectives or to model interdisciplinary inquiry? If there is a center or office on your campus that supports teaching development, what resources are available to support a curriculum update or makeover? How might your program or department work with that office or other offices on campus to collect data about popular courses, student interest, or bottlenecks in enrollment or moving through the degree requirements? What experiences do people – students, staff, or faculty – have from other programs and institutions that could bring useful perspectives to curricular or programmatic questions and challenges?

Sharing research across boundaries

Formal and informal venues to share ongoing research projects or ideas and opportunities for new projects can invigorate faculty research and inspire new collaborations involving faculty, students, and colleagues in other programs. Departmental colloquia or informal "brown bag" lunch conversations can be productive ways to foster conversations about ongoing research projects or ideas for new projects. The growing availability of webinars on a variety of topics also offers potential for group watch parties followed by discussion. To ensure that these events are productive for everyone involved, it helps for the moderator to set a tone that welcomes participation and inquiry. A thoughtful moderator will manage the floor (or microphone) to ensure different opinions and perspectives are welcomed. Encouraging a variety of people to share comments includes limiting the amount of time that any single person has to ask questions or offer comments (we have all seen faculty members take the opportunity to turn a "question" about a presentation into a mini presentation of their own work or opinions!).

There are other ways to foster conversations about research such as organizing student research teams and projects or inviting guests to classes who can speak about their research accomplishments, challenges, and insights. Research is often presented as if it is a linear process, so it can be particularly helpful for students to see that, in reality, research may zigzag, veer off course, benefit from serendipity, or hit a dead end. Informal conversations can also lead to interactions that might not be possible in the context of one-way presentations or lectures. It can be helpful to think about the next steps when a conversation about research gets interesting: Who else could be part of this conversation? How might this conversation connect to other projects or conversations in the program, on the campus, or in the broader community? What kinds of funding opportunities might support the further exploration of ideas?

Opening classroom doors

Even those of us with many years of classroom teaching experience can learn how to do it better or how to do it differently. At this time of online and hybrid instruction, there is plenty of information about best practices based on the format of a class. Regardless of whether we are teaching in person or in a hybrid format that includes online components, it can be useful to invite feedback and input to our teaching. We can benefit by literally opening the doors of our classrooms to peers and mentors who can help us understand how we are effective or less impactful in the classroom. We can also invite feedback on our syllabi, course structure, course and assignment design, and overall organization of a class through sharing our syllabi, assignments, exams, readings, and so forth. It can also be transformative to utilize a wealth of resources available on some campuses such as teaching centers, writing centers, librarians, service-learning opportunities, museum curators and exhibits, or other campus and community resources to expand how we approach subjects and assignments. Consider organizing informal opportunities in the department that allow faculty to gather and discuss ongoing challenges and future opportunities around teaching, student mentorship, and curricular matters.

Creating physical spaces for intellectual community

Departmental colloquia, informal brown bag sessions, and open classrooms are all examples of physical spaces for building a supportive intellectual community. Other ways of fostering information sharing and celebrating accomplishments and milestones in research and teaching include starting faculty meetings with a short research highlight from a faculty member, posting hard copies of recent publications or tables of contents on a physical bulletin board in a high-traffic area or break room, and sharing faculty and student research highlights online, through social media, or via hard-copy newsletters with university or college media outreach offices. Similarly, student work can be shared in a student colloquium setting, by displaying student posters or other projects in the departmental or program building, and through newsletters, online platforms, social media, and listservs. Faculty could also share their own research activity updates in any of these platforms or venues. With continued engagement and promotional efforts such as these, students and faculty will learn where they can look for up-to-date information about the kinds of work being done in the program.

Allowing risk and failure

Here is a brief anecdote about risk and failure from a seasoned university professor. Remember the summer of 2020 when many of us were facing a fall semester of online teaching due to Covid? I (O'Lear) dutifully enrolled in an online boot camp on how to teach online, offered by my university's teaching support center. Honestly, I was not looking forward to a week of online learning about online teaching. The Friday before the bootcamp was scheduled, my colleagues and I received a homework assignment that we were supposed to do over the weekend to prepare for the week of instruction. Dread and resistance were my initial responses. I finally motivated myself to watch the introductory video, and I was treated to a humorous and lighthearted welcome to the boot camp course. The homework was an enjoyable task rather than an onerous one. I got to see fun contributions from other people in the group. By the time Monday and the first day of online boot camp rolled around, I was looking forward to hearing what the organizers had to offer and to interacting more with this group. I had an A-ha moment: I could share this same kind of welcoming and exploratory tone with my own class! I, too, could invite my students to the topic of the class in a lighthearted, low-stakes way that encouraged them to be curious rather than focused on onerous or complex tasks.

This approach of inviting inquiry is open to risk and failure. It encourages people to be engaged in creative ways that may or may not work initially, and it also provides tools to assess what worked, what didn't work, and what is most important to include or achieve. Learning from failure can be effective, and we can encourage risk-taking by offering low-stakes, exploratory tasks. We can be genuinely curious about failure and utilize it creatively. Failure can be productive and exciting, especially when there are reasonable guardrails to prevent a total loss of time and energy. The value of failure – which may rely on taking a risk – is captured in this quote by Vilfredo Pareto:

> Give me the fruitful error any time, full of seeds, bursting with its own corrections. You can keep your sterile truth for yourself.

Setting aside time for reflection

"Productive" is a familiar word to many of us in academia and is a label often synonymous with "good." A productive week is a good week. A productive sabbatical is even better. Indeed, we are rewarded for being productive and for putting into the world high-quality products such as peer-reviewed publications, favorably reviewed grant proposals, and positive student evaluations of our courses. With so much emphasis on productivity, we might tend to forget the quiet power of reflection. There is value in slowing down, decolonizing the notion of time, and taking time to reflect, observe, and see connections that we might otherwise miss (Shahjahan 2015).

Prioritizing time for reflection is valuable at any stage of an academic career, and we can also model this practice for our peers and students. Valuing time for reflection may not appear to be immediately contributive to our to-do list, but it is this kind of unstructured and unguided time which allows space for new ideas to emerge and for connections across projects, between research and teaching, and from our colleagues to become apparent. We do not know what kinds of insight can be gained through reflection, and that is perhaps the best reason to spend part of each week or each day in reflection.

Social events

Departmental picnics at the start or end of the semester, graduation events, and open houses are excellent ways to bring people in a program together and to celebrate accomplishments. Building a strong and cohesive social atmosphere in a department, however, starts much smaller with civil conduct. Make every effort to treat colleagues with respect, and show your desire to listen to differing points of view. Foster healthy interpersonal relationships. As much as it is possible and appropriate, take interest in your colleagues as human beings just as much as professionals. Learn to understand their values that shape their life and professional choices and decisions. Create opportunities to socially interact and attend department-organized events on or off campus. Make yourself available and accessible for conversation and problem-solving. Taking advantage of these opportunities will make it easier to come together in a professional setting, at a faculty meeting perhaps, when critical and often challenging discussions and resolutions need to be made on behalf of our programs and departments. Entering these conversations and getting to know our colleagues professionally and personally will allow us to make more informed and mutually beneficial decisions on the collective level.

It is important to keep diversity in mind when planning social events within a department. Inviting everyone to go out for beer after colloquium may be well-intentioned, but keep in mind that not everyone drinks. Ensure that there are non-alcoholic options and that the focus of the gathering is the people, not the drinking (or eating). Similarly, not everyone drinks coffee, so an informal "coffee hour" might be more inviting to all if it is a "social hour."

The AAG sponsored a webinar series during the Covid pandemic, and one of the topics was "Healing in Leadership." Participants, who were mostly in some role of departmental leadership, shared their suggestions for best practices for cultivating a sense of community:

- Check in with people – faculty, staff, students – to see how they are doing.
- Listen carefully.
- Make sure to say hello to others and see how they are doing before you dive in right away to ask what you want.
- Ask for input during meetings and listen to people's concerns.
- Put a limit on how many people may speak so that others can talk as well.
- Disrupt unhelpful practices and policies.
- Be overtly empathetic.
- Be open to different perspectives and situations.
- Schedule wellness checks.
- Keep your office door open and be available.
- Set an office space tone with lighthearted office door tags.
- Organize social hours with students and colleagues.
- Make time to connect with each other.
- Find fun and safe activities (outside) to help people connect.
- Go for a walk.

At the departmental level, there are also multiple ways to foster collegiality:

- Focus on building inclusive and welcoming work environments.
- Encourage and reward active engagement in departmental service work so that everyone does a fair share of significant assignments appropriate to their rank.

- Understand and communicate clearly about how service is evaluated in annual reviews.
- Advocate for fair and equitable evaluation of service activities commensurate with rank.
- Monitor departmental culture through an annual survey. Identify a few predominant concerns or issues that could be meaningfully addressed.
- Welcome new colleagues (faculty, graduate students, etc.). This is especially important when colleagues arrive from outside of the country, from other institutions, or from different disciplinary departments.
- Offer (or seek) mentoring opportunities for junior faculty to help them get to know the institutional and departmental culture and available resources and procedures.
- Volunteer regularly for service opportunities at the college and university levels, as appropriate to your rank. Choose those opportunities that best align with your other interests (e.g., research proposal committee, curriculum committee). Do external service at the institution as much for your personal growth as for the department. Departments that participate in shared governance are viewed by the administration more favorably in general. Active involvement on behalf of your department may be helpful when garnering resources for your department at a later date. Those resources may impact you directly.
- Offer your expertise and institutional knowledge generously with others, now and in the future.

Challenges to collegiality

Obstacles do arise in promoting collegiality, but these needn't undermine all our efforts. The imposition of non-collaborative, top-down management systems has traditionally been the greatest challenge to collegiality (Tapper 2018, Tapper and Palfreyman 2010). Indeed, calls to reorganize universities as businesses are not uncommon. Yet, there is always a need to push back against efforts that might undercut the collaborative decision that is so important to sustaining collegiality.

Business models are often proposed to address a second challenge to collegiality: money. Currently, declining funding for higher education in some states and nations is increasing pressure to do more with less, particularly with fewer full-time staff (Hardy 1996). This can result in the hiring of part-time staff for both teaching and research: adjunct instructors, research assistants, and post-docs rather than tenure-track faculty. These temporary staff members are not always included in decision-making, and, even if they are, they may have neither the time nor inclination to participate. Again, these sorts of funding and staffing pressures should never preclude efforts to include all staff in decision-making.

These differences in participation by full and part-time staff, as well as differences related to seniority and rank, can also be a threat to collegial relationships. There are times when deference needs to be given to individuals owing to their expertise, seniority, or experience, but there is rarely a need to shape these into a permanent decision-making hierarchy that favors one group of faculty over others.

Issues of management, money, and decision-making may be perennial challenges to collegiality, but none are insurmountable. Neither are the unexpected setbacks or crises we face periodically – personal emergencies and losses, extreme weather and natural disasters, or humanitarian crises locally, nationally, or globally. These can't be predicted, but a strong sense of collegiality can help us respond quickly and effectively. Perhaps this showed up best during the Covid-19 pandemic. Working together, faculty, staff, administration, and students made a rapid transition to online education in 2020, an effort that succeeded remarkably well and, coincidentally, which has helped many of us to develop new abilities to build online communities that help support our departments' work in supportive, collegial ways.

Sometimes, even our best efforts don't build the sense of collegiality and community we most want. But this isn't a reason to give up because, as we noted earlier, collegiality happens at different levels of our personal and professional lives. We can build collegiality in areas that are open to our control in ways that increase enjoyment and productivity, relieve stress, and build a sense of shared purpose and accomplishment. Department climate and culture can change, and promoting collegiality at any level can help to move these efforts forward.

References

Ashkanasy, N., C. Wilderom, and M. Peterson. 2011. *The handbook of organizational culture and climate*, 2nd ed. Thousand Oaks, CA: Sage.

Becher, T. and P. Trowler. 2001. *Academic tribes and territories*, 2nd ed. Buckingham: Society for Research into High Education and Open University Press.

Denison, D., and A. Mishra. 1995. Toward a theory of organizational culture and effectiveness. *Organization Science* 6 (2): 204–223.

Gardner, S. and P. Mendoza. 2010. *On becoming a scholar: Socialization and development in doctoral education*. Sterling, VA: Stylus Pub.

Golde, C. 2005. The role of the department and discipline in doctoral student attrition: Lessons from four departments. *The Journal of Higher Education* 76 (6): 669–700.

Golde, C., G. Walker, and Associates. 2006. *Envisioning the future of doctoral education: Preparing stewards of the discipline*. San Francisco, CA: Jossey-Bass.

Hardy, C. 1996. *The politics of collegiality: Retrenchment strategies in Canadian universities*. Montreal: McGill-Queen's University Press.

Kotter, J., and J. Heskett. 1992. *Corporate culture and performance*. New York: Free Press.

Peterson, M., and M. Spencer. 1990. Understanding academic culture and climate. *New Directions for Institutional Research* 17 (4): 3–18.
Schein, E., and P. Schein. 2017. *Organizational culture and leadership,* 5th ed. Hoboken, NJ: Wiley.
Schneider, B., M. Ehrhart, and W. Macey. 2013. Organizational climate and culture. *Annual Review of Psychology* 64 (1): 361–388. https://doi.org/10.1146/annurev-psych-113011-143809.
Shahjahan, R. 2015. Being "lazy" and slowing down: Toward decolonizing time, our body, and pedagogy. *Educational Philosophy and Theory* 47(5): 488–501.
Shattock, M. 2010. *Managing successful universities,* 2nd ed. Maidenhead: Open University Press and McGraw-Hill Education.
Tapper, T. 2018. Collegiality in higher education. In *Encyclopedia of international higher education systems and institutions*, eds. P. Teixeira and J-Ch. Shin. Dordrecht: Springer. https://doi.org/10.1007/978-94-017-9553-1_526-1.
Tapper, T., and D. Palfreyman. 2010. *The collegial tradition in the age of mass higher education.* Dordrecht: Springer.
Viđak, M., L. Barać, R. Tokalić, I. Buljan, and A. Marušić. 2021. Interventions for organizational climate and culture in academia: A scoping review. *Science and Engineering Ethics* 27 (24). Online publication. https://doi.org/10.1007/s11948-021-00298-6.
Walker, G.E., C.M. Golde, L. Jones, A.C. Bueschel, P. Hutchings. 2008. *The formation of scholars: Rethinking doctoral education for the twenty-first century.* San Francisco, CA: Jossey-Bass.
Zohar, D., and D. Hofmann. 2012. Organizational culture and climate. In *The Oxford handbook of organizational psychology*, Vol. 1, ed. S. Kozlowski, 643–666. Oxford: Oxford University Press.

22

VALUING MENTORING, OR HOW TO THINK USEFULLY ABOUT MENTORING

Hilda E. Kurtz

Introduction

To value mentoring in academia, as elsewhere, is to recognize potential mentors in different parts of your professional world, and to recognize that no one person can or should serve as the be-all and end-all mentor. While the style, duration, and scope of mentoring relationships vary widely by context, the common thread among them is a focus on the development and success of the mentee. Mentoring in academic contexts can be formal or informal, broadly or narrowly focused, and last for varying amounts of time. Academic mentoring is understood here to refer to interpersonal relationships which provide psychosocial, role model, academic, and career-related support for students and faculty in institutions of higher education (Lunsford et al. 2017). While mentoring "serves as an essential route of transmission for the values, ethical principles, and cultural more of our various professions" (Johnson 2016, 1), it is practiced in deeply uneven ways across the academy. Mentors and mentees alike often struggle to understand how to be a good mentor or mentee (Hill et al. 2022). A 2014 Gallup survey of 30,000 adults found that no more than 22% of them had a mentor in college, an unsurprising result when reflecting on how little training would-be mentors have traditionally had, and how invisible academic mentoring tends to be in faculty evaluation mechanisms (Lambert 2018).

Systematic attention to the elements of successful mentoring is relatively recent in the sciences and social sciences. Recent years have brought more attention to graduate student and early career faculty mentoring within professional societies, such as the American Association of Geographers, American Geophysical Union, American Meteorological Society, and American

DOI: 10.4324/9781003342953-26

Psychological Association (APA 2012). As a result, a wide range of models, resources, and reviews of mentoring are available to graduate students and early career faculty, which were not offered a generation ago. This chapter will highlight key themes from the literature on academic mentoring, offer suggestions for reflexive mentoring practice, and outline the phases of student mentoring with attention to goals and resources for each phase.

Key themes from the mentoring literature

Mentoring is widely understood to be an important factor in student success. Anyone who has spent time in academic settings can appreciate the value of mentoring for student success, whether through experiencing it, being a thirdhand observer, or feeing its absence. The Council on Graduate Students (2008) lists effective mentoring as one of six key factors in graduate student success (Johnson 2016). In the *Sage Handbook of Mentoring*, Lunsford et al. (2017, 318) note that mentoring positively influences a range of student outcomes, including career-related capacities such as deep and strategic learning approaches (Chester et al., 2013), self-confidence in professional skills and abilities (Thiry et al., 2011), and psychosocial elements such as a sense of belonging in the institution and discipline and capacity for socially responsible leadership (Campbell et al., 2012).

Not all academic mentors or mentoring programs focus equally on each of these outcomes. As an early career mentor and mentee, it is worth taking some time to reflect on which possible outcomes of successful mentoring are most important to you, and how you might then prioritize them in your mentoring relationships. You may wish to continue and even deepen practice of some parts of your experience while replacing other elements with practices that better fit your temperament, working style, and career intentions.

Our needs for mentoring change over the course of an academic career. The *Sage Handbook of Mentoring* (Lunsford 2017) notes a continuum in which undergraduate student mentoring tends to be formal and programmatic, graduate student mentoring less so, and the formal and programmatic elements of mentoring diminish over academic and other career trajectories. Early career geographers and academics in related social and environmental sciences will find varying institutional opportunities and expectations regarding formal mentoring programs. That is not to say that mid-career faculty don't benefit from informal and mentoring, only that formal mentoring programs are considerably scarcer for mid-career faculty members across institutions of higher education (Baker et al., 2016; Law et al. 2014). Law et al.'s (2014) review found that early career faculty members benefit from mentors who work within their own institution, while mid-career faculty members are likely to benefit from working with mentors positioned professionally outside of their institution.

Context matters. The significance, substance, and style of mentoring varies across different professional contexts, as well as among different disciplines and fields in academia. Lunsford (2012) notes the importance of disciplinary contexts for normative models of mentoring. Citing early work on academic mentoring (Biglan 1973), Lunsford (2012) notes that

> Disciplines may be categorized along three dimensions: pure versus applied; life versus non-life systems; and hard, characterized by having a shared paradigm, versus soft, lacks a shared paradigm. Collaboration, arguably a mentoring activity, characterizes the research efforts in the hard disciplines, like chemistry, whereas individual work best describes the research enterprise in the soft disciplines, like humanities. Thus, mentor-like behaviors may be 'built-in' the research enterprise in some disciplines more so than in others.

Lunsford (2012) cites Zhao et al. (2005) who found that psychosocial support was more prevalent among mentors in the social sciences and humanities, and more strongly emphasized than career support by those mentors. The same study indicates that mentoring in the physical and biological sciences is characterized by stronger career support and weaker psychosocial support.

Geography, as a stand-alone discipline in some institutional settings, and part of an amalgamated department in others, is arguably found at an inflection point between physical sciences, social sciences, humanities, and accompanying disciplinary conventions for academic mentoring. Geography's relative location invites a wider array of models into disciplinary and departmental practice, as well as cross-fertilization between models to develop and adapt new models for graduate mentoring. As an intellectually dynamic discipline, geography is well-positioned to be more attentive to psychosocial development and evolving norms in the discipline and across the social sciences.

> The advising I received as a graduate student was conducted exclusively through one-on-one meetings, which changed in frequency according to my needs for developing and refining my research proposals and products. This model is typical of humanities and qualitative social sciences models. After following this model as an assistant professor for a number of years, I learned through casual conversation about the richness of discussions among graduate students who met in a group with their advisors. Some faculty in my broad-spectrum geography department brought team and lab advising/mentoring models with them and offered a wider array of models or modalities to draw upon for the benefit of graduate students. I adopted the group meeting model used by colleagues who were trained in that model in context of physical science lab groups, and I have found it enriching on many levels.

Mentoring is experienced differently by different students. Add to that the recognition that no one mentor or approach to mentoring fits all people, all circumstances, and all phases of career and conditions of work–life balance. Studies find that gender, age, and culture inflect the experience of mentoring. Rose (2005), for example, found that US-based female doctoral students rated a mentor's integrity higher in their definition of an ideal mentor than did male students; Ramirez (2012) discusses the specific developmental needs of undergraduate student mentees. Because one size does not fit all in the mentoring space, it is widely noted in the mentoring literature that students and early career academics often seek out informal mentors whose "knowledge, experience and/or communication abilities [resonate] with their needs" (Trube and Vanderveer 2015, 323). Informal mentors can complement or stand in for formal mentoring relationships.

Mentors with shared experiences or backgrounds can be especially important for students and faculty who are members of groups historically minoritized in the academy (Byars-Winston et al. 2015; Malone and Barabino 2009; Zambrana et al. 2015). In a study of 58 underrepresented minority (URM) faculty at 22 research-extensive institutions, Zambrana et al. (2015) found limited value in formal mentoring programs that did not include understanding (through lived experience or otherwise) the challenges faced by URM faculty. They identify paternalism and assimilationist pressures as two of many such challenges. In a more intensive study, Malone and Barabino (2009) identify challenges of invisibility, marginalization, exclusion, and undervaluation/recognition for minoritized graduate students in STEM settings. They note that when students (or early career faculty) are the "only one," myriad interactions with peers and colleagues in a community of practice can foster alienation. Malone and Barabino (2009, 506) conclude that

> such exclusions ask us to deeply reflect on how the everyday practices of laboratories [and by extension other academic settings] encourage or discourage diversity 'on the ground' and the degree to which we can remix the cognitive and social within the research process itself.

Mentoring comprises a key domain through which communities of research and teaching practice are established. Mentoring that combines the cognitive and social in research is, as Malone and Barabino (2009) describe, just one small part of the "relational approach to dismantling... multiple and intersectional barriers to inclusion and equity" avowed in the American Association of Geographers' 3-Year Justice, Equity, Diversity & Inclusion Strategic Plan (AAG 2021).

The value of multiple mentors is increasingly recognized. Hill et al. (2022, 558) note that "[a]n effective mentor is most commonly described as available, approachable, experienced, supportive, trustworthy, enthusiastic, encouraging, and an active listener." Given the uneven attention to reflexive mentoring practice across higher education, there are not highly effective mentors to meet the needs of all graduate students and early career faculty members. In addition, the variation in what different mentors can offer – due to being positioned differently in social and professional settings, and with different approaches to mentoring itself – highlights that no one person can be expected to serve a mentee equally well across all dimensions of career and psychosocial development. Thus, the value of having multiple mentors comes into clearer focus. Multiple mentors can help graduate students and early career faculty navigate a wide array of challenges, as discussed further below.

How to think usefully about mentoring

Build a mentoring constellation or network

In his foundational book, *On Being a Mentor: A Guide for Higher Education Faculty*, W. Brad Johnson observes that "it is wise to remember that most of your mentees are likely to benefit from a rich network of developmental helping relationships" (Johnson 2016, 33). Johnson dispenses with the mentor as guru model (cf. Rockquemore 2013) to applaud and encourage *mentoring constellations*, defined as "a set of relationships an individual has with people who take an active interest in and action to advance the individual's careers by assisting with his or her personal and professional development" (see also Hetty van Emmerick 2004). Working with multiple mentors can be a more or less intense proposition. The value of multiple mentors for graduate students and early career faculty navigating the various challenges of early career academia is increasingly recognized (de Janasz and Sullivan 2004; Johnson 2016; Lunsford et al. 2017; Montgomery 2017).

Team mentoring (Johnson 2016) typically occurs when a faculty member has several graduate students. Team mentoring allows for lateral learning and support among mentees, while creating conditions in which to develop professional and cultural competencies.

Peer mentoring networks are commonly organic or informal in nature (Johnson 2016), but can also be implemented formally within various academic settings (Montgomery 2017; Kroll 2016). Johnson (2016, 38) notes that "peers have an opportunity to express personal and professional dilemmas, vulnerabilities, and anxieties in ways they may be reluctant to do with their primary faculty mentors." With that in mind, consider inviting friends and colleagues to build a peer mentoring network together. This could be a loosely structured gathering, whether in person or online, a Facebook group or other digital conclave. It could be a more formalized peer mentoring group in which members hold each other accountable to progress on professional goals. It could be an even denser network such as the "critical community" described by Montgomery et al. (2014, 5), "a collective of individuals similarly dedicated to providing honest structured guidance and supporting opportunities to enhance their individual cultural capital to sustain and expand all members for their future careers."

Think reflexively about what you bring to a mentoring relationship, as mentor and mentee

Successful mentoring is a deeply rewarding aspect of professional academic relationships. As Johnson (2016, 1) notes at the very outset of *On Being a Mentor*,

> for most of us laboring behind the ivied gates of academe, the career contributions that hold greatest meaning – those that sustain us long after our careers end – are relational...I propose that more of us will measure our career impact in terms of mentees – including those who have long since become colleagues – than merely in terms of publications, grants, or courses delivered.

Early career faculty members are navigating both a new independent phase in relation to one or more graduate mentors and stepping into the role of formal mentor for students. This is an important phase in which to reflect on your emergent mentoring style, what worked and didn't work for you as a mentee, what kind of a mentor you want to be, and what kinds of mentoring practices are likely to be sustainable and work well for you in the context of work–life balance.

Seek further insight and resources to make the most of academic mentoring relationships

The mentoring literature highlights what many people working in academic settings already know, whether intuitively or experientially. Mentoring is a rich and varied practice that is meaningful to professional success. Graduate students and early career faculty should feel encouraged and emboldened to shape their own mentoring experiences, as both mentors and mentees. As an early career faculty member, your access to a formal mentoring program distinct from the mentoring you received as a doctoral student may be limited. If such opportunities or programs exist at your new institution, be sure to explore them. As an alternative or in addition to formal mentoring programs for early career faculty, take the reins and cultivate mentors on your own. Think about what you need and want as an early career faculty member. In what areas would mentoring make a difference to your well-being and professional success?

The following sections offer an overview of resources and practices that can be used to shape productive and collegial graduate mentoring relationships, in context of professional development understood more broadly.

Phases of academic mentoring with concrete strategies

Mentoring relationships unfold in phases. The stages of mentorship in a graduate program offer a useful template for understanding these phases for other mentor–mentee relationships, such as those between late and early career faculty. The phases are simplified here into early, middle and late (but see Hill et al. (2022) for a more granular differentiation of mentorship phases).

Early phase

In the early phase, mentor and mentee are getting to know one another, developing a shared understanding, and setting some norms and expectations for their professional relationship. The early phase of a mentoring relationship calls for the "self-conscious practice of articulating and developing the skills and experiences that are necessary and valuable in a profession and in other aspects of life" on the part of both mentor and mentee (Writing a Mentoring Compact, n/d). Being self-aware about strengths, weaknesses, preferences, norms, and so forth is essential for the ongoing work of aligning expectations. It is important to recognize that in a power-laden relationship such as that between a graduate mentor and mentee, the faculty member should take the lead in initiating conversation about these and other factors. I have found it important to understand how a student works. Thus, early meetings include conversation about how the student thinks and what spurs their curiosity (empirically, conceptually); how the student makes decisions (eagerly, reluctantly, carefully, intuitively), and how the student writes.

In the last several years, many universities, graduate schools, and individual departments and programs have developed tools for building healthy, transparent, productive mentor–mentee relationships. Such tools encourage self-reflection and mindfulness in developing professional mentoring relationships and more broadly support the inclusive multicultural competence to adapt and update mentoring practices. Central among these are mentoring compacts. A typical mentoring compact sets out a list of issues for mentor and mentee to discuss and agree upon, such as what is the mentee's career goals, mutual expectations for the mentorship, support needed, ground rules for meetings, notes about confidentiality, and problem resolution. Each of the elements in a mentoring compact should serve as a point of discussion before putting anything down on paper. Michigan State University Graduate School offers a more granular template to guide reflection and discussion about the nature of the mentorship, called the Student-Advisor Expectation Scale.

Middle phase

The middle phase of an academic mentoring relationship should be preparing the student or other mentee, academically and emotionally, for the rigors of academia or whatever career ambitions lie ahead for them. It typically involves close coordination between mentor and mentee through the development of the research proposal, preparing for qualifying exams, or for the early career faculty, mapping out the tenure dossier. An artifact-oriented approach that attends to building the mentee's CV through a record of accomplishment and successes offers effective scaffolding, including grantsmanship, publications, paper, poster, and presentation competitions, and small projects. Civic engagement in the department, graduate school organization, or discipline is an important element of a CV, painting a picture of a well-rounded, engaged, and productive early career academic. Both mentor and mentee should be on the lookout for such opportunities to develop the mentee's skill set and academic record.

Peer mentoring networks can become increasingly important in the middle phase, serving as vital sources of emotional support through the diverse challenges graduate students face, as well as being a source of information, advice, and accountability. Peer networks among students can deepen into collegial friendships and lead to publications, some solo- and others joint-authored (see Kohl and McCutcheon 2015; Steacy et al. 2016).

Late phase

Mentoring relationships eventually move into a late phase, where the mentee, having gained experience, confidence, and standing, evolves often into a colleague. For the early career faculty mentee, moving into this phase may have been mapped out earlier on, or it may simply evolve. The late phase of the graduate mentoring relationship is shaped more formally by the demands of fieldwork or other data collection/acquisition, writing the dissertation, and writing some manuscripts for publication. There can be a long time between the candidacy benchmarks and the start of dissertation writing, and it is not uncommon for a graduate student's confidence in their abilities to have waned somewhat. The mentor and student may need to navigate this period differently. While the student has passed important benchmarks and tests, and thus earned some credibility as a researcher, the research process itself can be emotionally taxing. Thus, some mentors may find that they are offering more emotional support in this phase than in earlier phases, as graduate students face the slog of writing their dissertations. It can be hard to predict how your mentee will respond to the particular pressures of this stage. It can be appropriate for those in mentoring relationship to revisit the mentoring compact at this stage and use it as a means to evaluate what kinds of feedback and support are appropriate to the late stage of the mentorship.

It is important and challenging to get an intermediate publication in the pipeline or other rewarding project underway while staying focused on a dissertation. It creates a sense of authorial/academic identity and keeps a project moving while the students time is freed up to focus on thesis or dissertation writing. Perhaps most importantly, it engages the student in the fundamental practices of the profession – preparing a manuscript for peer review and responding to the reviews that come back in.

In addition, as Hill et al. (2022) and others note, the late stage in which the graduate student is slogging away on the thesis or dissertation is also a time in which to envision the end of this particular phase, and what may lie beyond graduation from the academic program. As a result of hard work, dedication, and a commitment to communication on the part of both mentor and mentee, the mentee has matured into a proto-colleague and developed the confidence, knowledge base, and skill set to continue on into a relevant career, whether an academic one or other.

Conclusion

Effective mentoring is an investment of time and attention that enriches professional networks through supporting the development and success of diverse students and early career academics. As more people invest in reflexive mentoring practice in the academy, the practice is becoming better understood and better tailored to suit the needs of mentees working in different contexts. Rather than concentrate all mentoring into one dyadic relationship, students and early career academics would do well to think in terms of mentoring constellations of people who can support one another through the myriad challenges of professional advancement and personal satisfaction in the academy.

References

American Association of Geographers. 2021. American Association of Geographers 3-Year Justice, Equity, Diversity & Inclusion Strategic Plan. Retrieved from https://www.aag.org/wp-content/uploads/2022/02/AAG-3-Year-JEDI-Strategic-Plan.pdf.

American Psychological Association [APA] Presidential Task Force. 2012. Introduction to mentoring A guide for mentors and mentees. Retrieved from https://www.apa.org/education-career/grad/mentoring.

Baker, V., Pifer, M. and L. Lunsford. 2016. Faculty challenges across rank in liberal arts colleges: A human resources perspective. *The Journal of Faculty Development* 30 (1): 23–30.

Biglan, A. 1973. Relationships between subject matter characteristics and the structure and output of university departments. *Journal of Applied Psychology* 57 (3): 204–213.

Byars-Winston, A., Branchaw, J., Pfund, C., Leverett, P., and J. Newton. 2015. Culturally diverse undergraduate researchers' academic outcomes and perceptions of their research mentoring relationships. *International Journal of Science Education* 37: 2533–2554. doi:10.1080/09500693.2015.1085133.

Campbell, C., Smith, M., Dugan, J., and Komives, S. 2012. Mentors and college student leadership outcomes: The importance of position and process. *The Review of Higher Education* 35 (4): 595–625.

Chester, A., Burton, L., Xenos, S., and K. Elgar. 2013. Peer mentoring: Supporting successful transition for first year undergraduate psychology students. *Australian Journal of Psychology* (65): 30–37. doi:10.1111/ajpy.12006.

Council of Graduate Schools. 2008. PhD Completion Project. Retrieved from https://cgsnet.org/data-insights/diversity-equity-inclusiveness/degree-completion/ph-d-completion-project.

de Janasz S., and Sullivan, S. 2004. Multiple mentoring in academe: Developing the professorial network. *Journal of Vocational Behavior* 64: 263–283. doi:10.1016/j.jvb.2002.07.001.

Hetty van Emmerik, I. J. 2004. The more you can get the better: Mentoring constellations and intrinsic career success. *Career Development International* 9 (6): 578–594.

Hill, S., Ward, W., Seay, A., and Buzenski, J. 2022. The nature and evolution of the mentoring relationship in academic health centers. *Journal of Clinical Psychology in Medical Settings* 29: 557–569.

Johnson, W. B. 2016. *On Being a Mentor: A Guide for Higher Education Faculty*. London: Routledge.

Kohl, Ellen, and Priscilla McCutcheon. 2015. Kitchen table reflexivity: Negotiating positionality through everyday talk. *Gender, Place and Culture* 22 (6): 747–763. doi:10.1080/0966369X.2014.958063.

Kroll, J. 2016. What is meant by the term *group* mentoring? *Mentoring & Tutoring: Partnership in Learning* 24 (1): 44–58. doi:10.1080/13611267.2016.1165488.

Lambert, L. 2018. The Importance of Helping Students Find Mentors in College. Internet: https://news.gallup.com/opinion/gallup/245048/importance-helping-students-find-mentors-college.aspx.

Law, A. V., Bottenberg, M. M., Brozick, A. H., Currie, J. D., DiVall, M. V., Haines, S. T., and Yablonski, E. 2014. A checklist for the development of faculty mentorship programs. *American Journal of Pharmaceutical Education* 78 (5): 98.

Lunsford, L. 2012. Doctoral advising or mentoring? Effects on student outcomes. *Mentoring and Tutoring: Partnership in Learning* 20 (2): 251–270. doi:10.1080/13611267.2012.678974.

Lunsford, L., Crisp, G., Dolan, E., and Wuetherick, B. 2017. Mentoring in higher education. Ch. 20 In *Sage Handbook of Mentoring*. Thousand Oaks, CA: Sage Publishing, pp. 316–334.

Malone, K. R., and Barabino, G. 2009. Narrations of race in STEM research settings: Identity formation and its discontents. *Science Education* 93: 485–510.

Montgomery, B., Dodson, J. and S. Johnson. 2014. Guiding the way: Mentoring graduate students and junior faculty for sustainable academic careers. *Sage Open* (October–December): 1–11.

Montgomery, B. 2017. Mapping a mentoring road map and developing a supportive network for strategic career advancement. *Sage Open* (April–June): 1–13.

Ramirez, J. 2012. The intentional mentor: Effective mentorship of undergraduate science students. *Journal of Undergraduate Neuroscience Education* 11: A55–A63.

Rockquemore, K. 2013, July 22. A new model of mentoring. *Inside Higher Ed*. Retrieved from https://www.insidehighered.com/advice/2013/07/22/essay-calling-senior-faculty-embrace new-style-mentoring.

Rose, G. 2005. Group differences in graduate students' concepts of the ideal mentor. *Research in Higher Education* 46 (1): 53–80.

Steacy, C., Williams, B., Pettersen, C. and H. Kurtz. 2016. Placing the "analyst" in discourse analysis: Iteration, emergence and dialogicality as situated process. *The Professional Geographer* 68 (1): 166–173.

Thiry, H., Laursen, S. L., and A. Hunter. 2011. Student-advisor interactions in apprenticing undergraduate researchers into a scientific community of practice. *Journal of Science Education and Technology* 20: 771–784.

Trube, M. B., and B. VanDerveer. 2015. Support for engaged scholars: The role of mentoring networks with diverse faculty. *Mentoring & Tutoring: Partnership in Learning* 23: 311–327. doi:10.1080/13611267.2015.1099869.

Zambrana, R. E., Ray, R., Espino, M. M., Castro, C., Douthirt Cohen, B., and J. Eliason. 2015. "Don't leave us behind": The importance of mentoring for under-represented minority faculty. *American Educational Research Journal* 52: 40–72. doi:10.3102/0002831214563063.

Zhao, C., Golde, C. M., and McCormick, A. C. 2005, April. *More than a signature: How advisor choice and advisor behavior affect doctoral student satisfaction.* Paper presented at the American Educational Research Association, Montreal, Canada.

23

SERVING WITH A PURPOSE

Thinking strategically about service

Michael E. Meadows

Introduction: a personal journey

In considering how to approach this topic I initially hesitated to make any mention of my personal experiences. However, given that my own academic career has been illuminated by such deep engagement with 'service' at all levels – departmental, university, national and international – I decided to reflect on this personal involvement at the risk of seeming self-indulgent, and in the hope that it may better expose the benefits (and costs!) of serving both the academic institution and the discipline in general. Ultimately, an academic career is shaped by one's personality and how it interfaces with the personalities of others – whether they be students, peers, or senior line managers. Indeed, the personal perspective is fundamentally the only genuine one.

While many, if not all, academic tertiary institutions globally state 'service', or 'outreach' or 'engagement' as criteria to support promotion, I have the strong impression that research and teaching are by far the most significant requirements for career development. Such a situation may act as a disincentive to early career academics to involve themselves in activities that may be seen as peripheral, and that may consume time and energy that would otherwise be spent on producing research outputs or developing their teaching portfolio. Performance review, tenure and promotion processes are seen to place greater value on research in particular, at the expense of teaching and service (Schimanski & Alperin 2018). Service is integral to what has been termed 'academic citizenship' (Macfarlane 2007), although there have been worrying signs for a while that it may be increasingly underappreciated and undervalued (Macfarlane 2005; Beatson et al. 2021) and younger academics may therefore be drawn away from engaging in what may be considered as

DOI: 10.4324/9781003342953-27

more collegial activities. However, in this chapter, I aim to demonstrate that a deep engagement with service in the academic context can be a richly rewarding experience and, indeed, can also provide excellent support for career development. Such service, which can mean anything from simply contributing to departmental administration to playing a lead role in national or international academic bodies, offers opportunities to obtain a broader perspective on the academic enterprise and learn from more experienced colleagues.

I cannot recall at what point in my career I really started to commit to service, but I think that I fell in love with the discipline at secondary school and perhaps that underlies the desire to 'give back' in the form of service. The trajectory of my career testifies to the fact that I have found myself consistently in positions of service, quite a number of them in a leadership capacity. I do not recall immersing myself in this very early on in my first academic appointment, which was as lecturer at what was then called the City of Liverpool College of Higher Education but was eventually incorporated into what is now Liverpool John Moores University. There is so much to do in these early days – new lectures to write, practical and fieldwork exercises to design, research papers based on the recently completed PhD to complete. It is a hectic time, indeed, and consolidating the two key components, research and teaching, must be the priorities. Eventually, however, there is more confidence in one's own ability, fewer new lectures to prepare from scratch and a little more time to start thinking about contributing more broadly to the academic project.

To be honest, I am not sure how 'strategic' my own contributions have been, since they often came about through random encounters, or because no one else was available to fill the role in question – but I will endeavour at least to demonstrate that service can certainly support, rather than constrain, an academic career. I think it is useful to consider service on a continuum – from that which relates to the department, faculty and university, to external service for the community, and for the wider discipline through involvement with national and international organizations.

Service to the departmental, faculty and university

My first substantial service experience came during the early days of my appointment to the Department of Environmental & Geographical Science at the University of Cape Town (UCT). The department was relatively new, in the sense that although geography as a subject had been taught at UCT since 1936 on the appointment of Professor Bill Talbot (see Meadows 2018), the School of Environmental Studies had recently been incorporated. Such mergers can be stressful, of course, but in this case the underlying administrative and governance issues had largely been resolved before my arrival, and so the situation was really more of an opportunity to shape an exciting new curriculum.

For most academics, their first service assignment is most likely to be at the departmental level. The nature of such duties will vary, of course, depending on the governance structure in the unit concerned, but it is highly likely that, at some stage, whoever is head of the department (HOD) will approach a younger staff member to take on one or more portfolios. These too will differ from school to school, but they are likely to relate to the administrative functions such as course convening, curriculum design and re-design, participating in the examinations process and student advising. Initially, engagement may take the form simply of participation, which may be somewhat passive or more active as required. The duty may involve only attendance at regular or occasional meetings but is most likely to require a deeper commitment, and, ultimately, the individual may end up as chair of the relevant committee or subcommittee. In my case, I started with convening a semester course, a reasonably straightforward expectation. I also recall working with a small team on a major curriculum planning and review exercise. As challenging and time-consuming as this was, it was an excellent opportunity to learn about how my own courses articulated with others, and it was rewarding to help develop a vision for teaching and learning in the department that had a lasting impact. Ultimately, of course, the level of engagement is likely to increase, either because the HOD requests it or because the individual academic sees a particular opportunity and volunteers. Strategically, it makes sense to offer to work on issues that one has a greater interest in, and so discussion with line management indicating a willingness to participate in a particular initiative is likely to bear fruit. In my experience, managers rarely turn down offers of assistance! Either way, the level of responsibility will doubtless increase with time, and a successful academic will find themselves acting as chair of one or other of the various departmental committees, or perhaps in charge of undergraduate or postgraduate matters. The degree of satisfaction one gains from this is to some extent personality-dependent but made much easier if there is acknowledgement and support from peers and in particular from the HOD. Sadly, although the latter definitely cannot always be taken for granted, contributing positively to the departmental culture (see elsewhere in this volume) through taking on one's fair share of administrative responsibility is definitely worthwhile.

The ultimate level of departmental service would be as HOD. A worst-case scenario is that this turns out to be a 'poison chalice'. Here, 'service' merges into 'leadership' and will demand a great deal more in terms of time, energy and expertise. A short digression seems warranted here for a few reflections on leadership because, for one reason or another, I spent a substantial part of my career as HOD, some 17 years in all. The department at UCT had always been characterized by stability and continuity in terms of headship (see Meadows & Fuggle 2016) and, I think, is somewhat unusual in that regard. Indeed, when I stepped down from the headship at the end of 2018, the department had been led by only four professors since its foundation in 1936 (just four HODs in 82 years!). Seventeen years is a long time (although it went by quickly enough), and I am not entirely sure why it so transpired. In the past, at most universities I know, there would have been only one professor in a department who was *de facto* appointed for life, and the chair would typically become available only upon the incumbent's retirement or death. But by the time I joined UCT, there was a five-year rotational arrangement for the headship and an *ad hominem* promotion system in place. Accordingly, the headship issue was resolved through a semi-democratic process conducted at least every five years, whereby the Dean consulted with members of the department and then made a recommendation to Council for appointment. That I was elected and then reelected three times thereafter probably had more to do with the fact that nobody else wanted to take the task on! Leadership stability is obviously valuable, although it is separated only by a fine line from stagnation, so the trick is to keep moving, inventing and reinventing. This only pays off if it is achieved through good communication and teamwork. My own style of leadership has never been of the autocratic variety, although I know of many departments where that has worked perfectly well. I took solace and advice by heeding the wisdom of my peers and then, perhaps, acting as a kind of 'shepherd' to see that the good ideas were developed and bore fruit. I was lucky enough to be able to place a great deal of trust in my colleagues over the years, and the spirit of mutual support was very much part of our departmental culture. It may be over-romantic, but I really felt that we were part of a large family (it is the largest department of its type in South Africa – and probably Africa as a whole for that matter). Anecdotally, I am pleased to report that this atmosphere still prevails there, despite many recent challenges due to political protests, the pandemic, academic staff retirements and so forth.

Depending on the governance structure of the institution in question, the next administrative level up from the department is usually the faculty (sometimes school – the semantics here are highly variable), where there are several departments representing different, sometimes diverse, disciplines. In the case of UCT, the department falls under the Faculty of Science, along with others from the natural, earth and physical sciences. Here, higher level decisions are made regarding, *inter alia*, the approval of departmental course curricula, award of undergraduate and postgraduate degrees and appointment of new staff. Engagement here brings with it opportunities to interact with colleagues from other academic fields and learn more about how decisions are made that keep the academic project as a whole on track. For one reason or another, more especially when I was HOD, I found myself sitting on a good number of faculty committees and chairing several of them. However, there are rewards here too that, I believe, far outweigh the costs in terms of time and effort. A more detailed understanding of how different disciplines interface with each other, and how each fit within the overall curriculum structure, adds considerable value and provides a useful conceptual context for one's own work. There is also a great deal to be learnt from those in other departments, and the casual interactions that take place before or after the inevitable meetings are especially worthwhile.

At the university level, responsibilities become more diverse. Again, these duties offer a wider perspective on the higher-level operation of an institution, bring even more connections and contacts, this time with more senior colleagues, and are all the more rewarding – and quite possibly strategically valuable – as a result. The highest decision-making bodies at universities are usually the preserve of the professoriate, and are unlikely to be accessible to early-career academics, although at UCT there are several members of the senate who are elected from among the lower academic ranks. Nevertheless, other university committees may be available, and membership may well turn out to be a bonus in terms of a future promotion application. At various stages before I was HOD, I served on university committees dealing with issues such as environmental management, academic conditions of employment, extension service (i.e., outreach to the community, specifically the disadvantaged and marginalized) and several non-statutory advisory boards of research institutes across the campus. Universities typically operate through countless committees of this type and, while nobody really likes to sit in meetings, being an active participant in decisions at this level can be both illuminating and empowering.

Service at the local, national and international levels

Service to the wider community at the local or national level has become very important, particularly as universities and funding bodies strive to overcome their 'ivory tower' reputation (Pfeifer 2016) and ensure that society more readily appreciates the value of higher education. There is ever more pressure to demonstrate the relevance of research to real-world issues (Ozias & Pasque 2019), and community engagement is often now a requirement in many research funding calls. Indeed, such work has become imperative to secure government funding for higher education in general. As is the case at UCT, more and more institutions of higher education are incorporating such activities in their criteria for appointments and promotion. Undoubtedly there are significant benefits to incorporating 'community geography' into curricula (Fischer et al. 2021), and so-called service learning is a potentially valuable tool in promoting social transformation (Liu et al. 2020). Accordingly, introducing a service-learning element to the curriculum offers some particular advantages, especially given that universities are increasingly called upon to justify the relevance of their teaching to societal needs (Cantor & Englot, 2014). For example, in the case of geography, local environmental organizations, government or otherwise, may prove to be enthusiastic partners if the teaching activity can be designed in such a way as to support their work.

Whatever the nature, applied or otherwise, of scientific research, the major unit of currency for academic appointments and promotion continues to be the peer-reviewed publication. Within the social sciences and humanities, the single-authored monograph has been in decline (Shaw et al. 2022), and the rise of bibliometrics has shifted the balance in favour of journal articles. The entire community of academic research is underpinned by the system of peer review, a process that relies on the independent assessment of submitted articles by experts in the field and is a key element of academic citizenship (Wei 2016). The reviews are solicited by journal editors and are intended as a means of quality assurance that is undertaken without remuneration or other incentive in order to minimize bias. An early-career researcher's experience of this process begins with submission of a manuscript, probably based on the work completed for their master's or PhD research, and receiving reviewer responses from the journal editor with several possible outcomes that range from 'reject' (hopefully not!) through to 'major' or 'minor' revision and resubmission or (very, very rarely) 'accept'. Since the reviewers are themselves research peers, it is logical that we are all, sooner or later, approached by a journal editor, frequently these days by automated means, to review a submission. This can be time-consuming, but it is a collaborative process involving constructive dialogue with peers in the field and may well, in giving early insights into new research in the area of interest, help to improve one's own work. Notwithstanding the undoubted hegemony of academic publishing that is today dominated by a small number of major publishing houses (see Meadows et al. 2016), serving as a reviewer or, perhaps eventually on the editorial board, even as editor, of a scholarly journal is extremely valuable in yielding a more detailed understanding of the academic publication ecosystem. Moreover, such duties can pave the way to fresh collaborative partnerships and open up future job opportunities (Pfeifer 2016).

For most early-career academics, the first opportunity to network with fellow researchers is at the national, or perhaps regional, disciplinary conference. These meetings are intended to facilitate engagement with, in real-time, current research best practice, and they offer a chance for aspiring academics to promote their work and receive (hopefully) constructive criticism. At the national scale, such conferences and congresses are typically organized and hosted annually or biennially in different localities around the country, for example, the annual meetings of the American Association of Geographers (AAG https://www.aag.org/) or Royal Geographical Society (RGS https://www.rgs.org/). In most countries, the national academic body (sometimes more than one) has elected representatives that form a committee or council that aims to foster interest in the discipline locally. These are honorary (i.e. unpaid) positions, usually comprising more experienced academics, but many have an 'early-career' portfolio that would allow a younger person to gain valuable insights into the operations of such bodies. Again, the knowledge gained, and connections made, may well be highly advantageous in the long run. For larger organizations, such as the AAG and RGS, there may be a full-time secretariat to take on the substantial administrative burden, in particular that associated with holding such meetings. In many countries, however, the task falls to the committee members of the relevant national or regional geographical association to organize the meetings. Having been involved in hosting more than one of these meetings, I can attest to the fact that this involves months, if not years, of preparation, and sleepless nights during the period of the meeting itself! However, there are considerable trade-off benefits that again include an improved network of research contacts, administrative experience and the warm appreciation of colleagues. Strategically, involvement in the national geographical association is definitely worth considering. Given that these are usually elected positions, however, probably the best way to get involved is simply by volunteering to help out through contacting the conference chairperson or secretary. Offers of such assistance are unlikely to be flatly refused!

Internationally, a number of organizations exist at the continental scale, for example, the European Association of Geographers, Eurogeo (https://www.eurogeography.eu/) and the Association of Asian Geographers, AGA (www.asiageo.net). However, it is the International Geographical Union, IGU (https://igu-online.org), that officially represents the discipline of geography globally. The IGU is an international, non-governmental, professional organization, the purposes of which are primarily to promote geography through initiating and coordinating geographical research and teaching in all countries of the world. Its work is carried out through its National Committees, Commissions and Task Forces. The IGU hosts the International Geographical Congress every four years, and Superfluous - and clumsy celebrated its centennial 2022 in Paris, having been founded in Brussels in 1922 (see Kolosov et al. 2022). The IGU also organizes regional conferences and other meetings that further its objectives (i.e. promoting high-quality research and teaching in the discipline) and facilitates the participation of geographers in the global community of scientists through its formal affiliation as a Category I Member Union within the International Science Council, ISC (https://council.science/) and the International Council for Philosophy and Human Sciences, CIPSH (https://cipsh.one).

Allow me again to digress to reflect on my own association with the IGU, because I think it is illustrative of how, sometimes in life, 'things just happen'. My personal involvement with the organization is really a lesson in serendipity – being in the right place at the right time. My first IGU Congress was in 1992 in Washington DC, where I presented a paper. However, it was four years later in The Hague, Netherlands, that my life took a different turn. While there, I was approached by the chair of what was then known as the Mediterranean Erosion and Desertification Study Group to join its steering committee – largely on the basis of the fact that I was based in South Africa and there was nobody at all representing Africa on the committee. The chair was also keen to visit Cape Town and asked me to host a field-based meeting there, which I subsequently did. From then on, there was a succession of connected – but in a sense quite accidental – encounters and connections that eventually led to my being nominated by the South African National Committee for the IGU to the position of IGU Vice-President; I was then duly elected at the Tunis Congress in 2008. Once again, I believe that the African affiliation was a factor – I would doubtless not have been elected if I had been based in the country of my birth. Then, just two years later, the position of Secretary-General and Treasurer unexpectedly became vacant. Professor Ron Abler, then IGU President, suggested I should make myself available, and I was in due course elected unopposed. Eight years in that role followed, punctuated by a re-election (again, I suspect, because nobody else really wanted the position!), and then a nomination and election to the presidency in 2020 for a four-year term. I have to be honest – there was never really a strategic 'plan' involved in any of this, and I have never thought of myself as ambitious in regard to this kind of service. But I do have a vocation of commitment to the discipline that has given me so much joy, and I feel a strong sense of loyalty to the people who drive it. Indeed, the entire IGU experience has been extraordinarily rich and rewarding and has introduced me to people and places I could only have dreamt of visiting as a child (albeit rather heavy on the carbon footprint!).

While the IGU is a global organization, its reputation differs from place to place. Arguably the two major Anglophone conferences for geographers are those hosted annually by the AAG and RGS, and the IGU probably has a relatively modest profile in both countries. In India and China, which between them posssiby account for almost half of the world's academic geographers, the situation is different, and the IGU is held in high esteem. Accordingly, IGU has global reach that the national bodies are unable to claim and, in hosting its Congresses, Regional and Thematic Conferences in different countries around the world, IGU is well-situated to nurture the discipline in places that perhaps the others cannot.

To return more directly to the issue of service, it is worth noting the opportunities the IGU offers in this regard. Although formal membership of the IGU is by country, individual affiliation is possible via one of its many commissions. Joining them is actually very easy; it is totally free and it is without any obligation. The IGU now has more than 40 'commissions' – these are groups of geographers interested in particular aspects of the discipline. IGU Commissions, supplemented by specialized task forces that are formed from time to time, span the whole range of topics and themes that constitute the discipline. Each Commission has a steering committee, led by a chair, that oversees the organization of conferences and workshops (often associated with major IGU events). Each Commission also has a website that distributes newsletters, including information on any upcoming meetings or other activities, and publishes books and special issues of journals related to the theme. Becoming an affiliated member of any one (or more) IGU Commission is as easy as writing an email to the chairperson indicating willingness to be included on the relevant email contact list. A full list of all IGU Commissions, with details of contact persons and website addresses, is available at: https://igu-online.org/organization/commissions/#. One of the IGU commissions is aimed primarily at attracting young and early-career geographers. This is a really easy way to make a connection with the global network of geographers, and IGU commission events, including conferences and workshops, sometimes held online or in hybrid format, are available to reinforce the contacts that may further deeper engagement and provide opportunities to serve the community more directly. The IGU also co-hosts the annual *GeoNight* (https://www.geonight.net/) in which, each year on a date in early April, events are organized in many countries around the world to showcase geography and the work of geographers. These events give the greater public a chance to familiarize themselves with geographical concepts and studies and make geographical research more accessible. Assisting with the organization of one of these *GeoNight* activities is a really exciting way of meeting and working with likeminded geographers.

In conclusion: the costs and benefits of service

Given that research and, to a lesser extent, teaching are frequently regarded as the most important elements of an academic position, it is hardly surprising that newly appointed lecturers may well regard the prospects of involvement in service with some trepidation. Certainly, there are perceived 'costs'. No doubt such service can be time-consuming and, particularly if the newly appointed are still engaged in completing their doctoral dissertation, contribute to the considerable pressures of academic life and may be seen as compromising that effort. There is a sense too that, in a university environment increasingly characterized by managerialism (see van Houtem & van Uden 2022, for a critique of this trend), and where research outputs are frequently evaluated principally through bibliometrics, time spent on such activities may easily be seen as time conflicting with the research endeavour. There is also a perception, rightly or wrongly, that the appointment and promotion process may undervalue service. Moreover, let's face it, if service involves – as it frequently does – attending lengthy committee meetings, this is hardly a motivation!

However, there are also many benefits that far outweigh these costs. Service is a critical component of academic citizenship and offers opportunities to improve an emerging academic's knowledge and skills, as well as to expand their network of connections. Engagement at any of the aforementioned levels is obviously beneficial to building the *curriculum vitae*, yields opportunities to learn from other more experienced academics (or one's peers for that matter), fosters collegiality and creates a positive impression as a 'team player'. In short, service should not really be seen as a supplement to the workload, but one that is essential to the development of an academic citizen, and which may well benefit both teaching and research.

I can certainly attest to the fact that service adds value to an academic career. Pfeifer (2016) offers some sound words of advice for developing what she calls the service agenda and highlights two key factors to take account of before accepting or seeking a service role, namely, relevance to the research focus and the amount of time required by the activity. While particular departmental tasks may not be geared necessarily to satisfying the first of these factors, joining the national geographical society or association and attending its conferences certainly will. Membership and active participation place the emerging academic in direct contact with the relevant network of scholars in the field, and this may, in turn, lead to a deeper form of involvement, such as contributing to conference organization, or even election to the committee or council of the association. Managing one's time for the competing roles of an academic life is dealt with elsewhere in this volume, suffice to say here that it is probably a good idea to consult with senior colleagues as to what a particular service role may entail and carefully weigh up the expectations with other commitments.

Given my own rather serendipitous, if not exactly random, journey through service and leadership, I am quite possibly the last person to be offering guidance as to how to strategize in this regard! What I can say definitively is that service at every level in which I have been involved, from the department through to the coalface of a major international academic organization, has given me enormous satisfaction, and that the people I have met along the route have substantially enriched my life in ways that go far beyond the mere sense of fulfilling a duty. I also have no doubt that the diverse array of engagements that I have enjoyed has materially greased the wheels of progress through the academic ranks. Accordingly, I can emphatically recommend that any aspiring academic should seek out and grasp any appropriate service opportunities as may come their way, thereby reaping rewards that are substantial compared to the risk.

References

Beatson, Nicola J., Tharapos, M., O'Connell, Brendan T., De Lange, Paul, Carr, Sarah, and Copeland, Scott. 2021. The gradual retreat from academic citizenship. *Higher Education Quarterly* 76 (4): 715–725. https://doi.org/10.1111/hequ.12341

Cantor, N. and Englot, P. 2014. Civic renewal of higher education through renewed commitment to the public good. In: J. N. Reich (Ed.) *Civic Engagement, Civic Development, and Higher Education*. Washington, DC: Bringing Theory to Practice, pp. 1–11.

Fischer, H., Block, D., Bosse, A., Hawthorne, T. L., Jung, J. K., Pearsall, H., Rees, A. and Shannon, K. 2021. Doing community geography. *GeoJournal* 87: 293–306. https://doi.org/10.1007/s10708-021-10457-8

Kolosov, V., Garcia-Alvarez, J., Heffernan, M., Schelhaas, B, Eds. 2022. *A Geographical Century: Essays for the Centenary of the International Geographical Union*. Cham, Switzerland: Springer Nature.

Liu, R., Fisch, S., Hong, A. and Melville, K. 2020. Spatializing community-based learning: how a critical geography framework can foster understandings of structural inequality and egalitarian relationships. *Journal of Community Engagement and Higher Education* 12 (1): 16–28.

Macfarlane, B. 2005. The disengaged academic: The retreat from citizenship. *Higher Education Quarterly* 59 (4): 296–312. https://doi.org/10.1111/j.1468-2273.2005.00299.x

Macfarlane, B. 2007b. Defining and rewarding academic citizenship: The implications for university promotions policy. *Journal of Higher Education Policy and Management* 29 (3): 261–273. https://doi.org/10.1080/13600800701457863

Meadows, M. E. 2018. William John Talbot. *Geographers' Biobibliographical Studies* 32: 124–136.

Meadows, M., Dietz, T. and Vandermotten, C. 2016. A perspective on problems and prospects for academic publishing in Geography. *Geo: Geography and Environment* 3 (1): e00015. https://doi.org/10.1002/geo2.16

Meadows, M. E. and Fuggle, R. F. 2016. Geography and environmental science on the slopes of Devil's Peak. In G. Visser, R. Donaldson and C. Seethal (Eds.) *The Origin and Growth of Geography as a Discipline at South African Universities*. Stellenbosch: Sun Press, pp. 117–142.

Ozias, M. and Pasque, P. 2019. Critical geography as theory and praxis: the community-university imperative for social change. *The Journal of Higher Education* 20 (1): 85–110. https://doi.org/10.1080/00221546.2018.1449082

Pfeifer, H. 2016. How to be a good academic citizen: the role and importance of service in academia. *Journal of Criminal Justice Education* 27 (2): 238–154. https://doi.org/10.1080/10511253.2015.1128706

Schimanski, L. A. and Alperin, J. P. 2018. The evaluation of scholarship in academic promotion and tenure processes: past, present, and future. *F1000Research* 2018 (7): 1605. https://doi.org/10.12688/f1000research.16493.1

Shaw, P., Phillips, A. and Gutiérrez, M. B. 2022. The death of the monograph? *Publishing Research Quarterly* 38: 382–395. https://doi.org/10.1007/s12109-022-09885-2

Van Houtem, H. and van Uden, A. 2022. The autoimmunity of the modern university: How its managerialism is self-harming what it claims to protect. *Organization* 29 (1): 197–208. https://doi.org/10.1177/1350508420975347

Wei, P. 2016. Journal work as community service for the common good and academic citizenship. *Mass Communication and Society* 19 (1): 1–6. https://doi.org/10.1080/15205436.2016.1114824

24

MOVING TOWARD PROMOTION AND BEYOND

Understanding and connecting institutional needs and personal aspirations

Deborah G. Martin

Introduction

Achieving a tenure-track academic position might feel like winning a lottery. For people whose career goals are oriented to the practices of teaching, research, and publication, as well as serving one's discipline and institution, the tenure-track is a kind of gift. Yet, it arrives with a very clear timeline and an often-fuzzy set of expectations and metrics for demonstrating satisfactory job performance for tenure. This chapter aims to demystify the essential components of a successful trajectory toward tenure and promotion. I suggest guidelines for clarifying expectations for tenure and for achieving progress in research, teaching, and service in order to reach the tenure stage successfully. The chapter points to the importance of institutional context as well as personal strategizing and ways to balance personal goals with job expectations.

In most higher education settings, getting to tenure involves successful progress in three general areas of the professoriate: research, teaching, and service. However, the workload and type expectations in each of these three areas are far from universal. Indeed, the most important aspect of meeting tenure expectations is understanding the landscape of higher education for the institution that will be granting tenure. Institutions of higher education vary tremendously in their character and primary missions, regardless of country. Each country has a different system of institutions. This chapter focuses initially on the United States to illustrate institutional context, but the advice herein can and should be adapted to different national settings. Fundamentally, success in tenure hinges on knowing your context and adapting your own behavior and work to fit the essential elements of your institution's mission while keeping an eye on what helps you to be happy and successful in your work.

DOI: 10.4324/9781003342953-28

Institution types

In the United States, the landscape for higher education is somewhat complicated, as the Federal Department of Education relies upon non-governmental agencies to verify that colleges and universities meet certain standards of quality (American Council on Education (ACE) 2023). These standards are developed and recognized by various accrediting organizations or state educational accrediting agencies, both of which are recognized by the federal government. While specific to one country, this framework has analogues in other places. In all higher educational contexts, there are a number of different types of educational institutions. In the US, these types range from technical and community colleges, four-year public (state-funded) and private (usually non-profit) institutions offering bachelor's degrees, and universities offering bachelor's and graduate degrees. Universities can be further divided into those offering primarily master's degrees in addition to bachelor's degrees, or those offering a full range of degrees that includes doctorates. To make sense of the wide variety of higher education institutions in the United States, the non-profit American Council on Education manages a classification system that was originally developed in 1973 to group similar types of institutions together (ACE 2023).

It is not necessary to understand all the permutations of institution type in order to figure out how tenure expectations vary across institutions and to use this knowledge to apply to one's own career. The next section clarifies some of the key elements in different types of institutions in the United States, and, in doing so, it points to the distinct types of work, and the varying balance expected among them, that are important for progressing toward tenure.

Community colleges and liberal arts/baccalaureate colleges: mostly (but not entirely) teaching-oriented

Community colleges and colleges (the latter are four-year bachelor's degree institutions in the US system) differ significantly from one another but share the commonality that teaching is the signature activity for faculty and the work that must be satisfactorily performed for a successful tenure case. Community colleges often have geography departments or social science departments that hire geographers. The workload in these institutions is heavily focused on teaching and service, with far less orientation to research. Faculty may have teaching loads of up to four classes per semester (eight per year) and commensurately low research expectations. Faculty in community colleges need to focus on teaching quality and accessibility to students. Their research expectations might focus on or include conference participation, publications in technical manuals, or community-oriented publications such as policy reports.

The other institution type that leans significantly toward teaching is a college primarily or exclusively offering bachelor's degrees, designed as a four-year degree for full-time students. Students in liberal arts/baccalaureate institutions might take anywhere from three and a half to six years to earn a degree, depending on their ability to continuously maintain a full-time schedule of classes over four years. Although these "liberal arts" colleges, like community colleges, prioritize teaching students, they often also promote "research-active" faculty to those students and expect faculty to develop research programs that can engage undergraduate students as lab or other kinds of research assistants. Four-year institutions vary significantly, from state-funded settings where teaching loads might be six courses per year (three per semester), to private, more exclusive institutions where professors teach two classes per semester and are expected to maintain research activity on a par with, or at least aspiring to, institutions with PhD-degree programs.

Faculty in private liberal arts colleges, in particular, should be expected to find ways to nurture a research agenda that involves undergraduate students in data collection and sometimes analysis. Doing so can involve developing regularly offered classes that help students to learn skills that will support data collection or analysis related to the research program. In a setting where all or a large portion of the students are undergraduates, it can be helpful to cultivate a research agenda that is friendly to new researchers adding to a database over time. In this type of environment, advising loads may be heavy, including for student-driven research such as honors theses. The burden of advising is lessened if student projects are oriented to your research agenda, so having a project that appeals to students at your university can enhance the synergies of teaching, advising, and research.

If you are on the tenure-track in a community college or baccalaureate college, it's crucial to understand that your effectiveness in teaching and your accessibility to a wide range of undergraduate degree-seeking students, will be the primary lens through which your tenure case is judged. Two- and four-year institutions in the "college" category in the United States – with counterparts in other countries including technical colleges and undergraduate-focused universities – are so varied that it's important to pay attention to specific messages about tenure expectations and values in each institution.

Universities, including master's colleges and doctoral universities: high research expectations and wide-ranging student goals

Universities in the United States grant some level of graduate degrees, whether primarily master's degrees, or master's and doctoral (PhD) degrees. In such institutions, it's important to understand the role of your department. Even in departments with large numbers of PhD degree-seeking students, faculty may still teach a significant number of undergraduate students within a teaching load of three or four courses per year. Indeed, one of the workload challenges of faculty in large doctoral institutions is balancing large courses that involve management of teaching assistants (TAs) with smaller PhD seminars, while maintaining high research productivity. By contrast, some faculty in master's-oriented universities might be expected to teach professional oriented master's degree-seeking students who value professors who have links to potential employers, making research in applied or policy-related themes particularly valuable. Publication expectations in graduate degree-granting institutions is likely to range from 1–2 articles per year in master's-oriented institutions, to 2–4 articles in traditional academic journals in doctoral institutions, with an emphasis also on grant-getting. The specific expectations of a given department in a given university depend significantly on the institutional mission and values; these are important to learn and integrate into your mindset and work plan in the years leading up to the promotion and tenure review.

Discerning institutional values, finding personal support

Knowing institutional type shapes the academic job and tenure expectations in a number of ways. It helps you better understand the range of students you will encounter, and to contextualize department expectations about the types and sizes of courses you will be asked to teach, the range of research you should aim to produce, and even the types of service you can best provide to contribute to your department. Knowing the type of institution where you are seeking tenure is only a starting point for situating your job tasks and expectations, however. Far more important for understanding the journey to tenure will be the formal and informal conversations that you have with colleagues in and outside of your department.

Department chair

Conversations with your department head or chair (that title depends in part on institutional type and leadership structure) will be a primary source of information about tenure expectations. Department chairs will likely have a number of conversations with you, some of which might feel quite informal, about what courses you should teach and how you are settling in. It is a good idea to ask for a formal meeting each term with your department chair, with the primary agenda item as your goals for the term. These can be short meetings, but, in them, it's important to listen to how your chair frames the teaching load and tasks that semester and to ask specifically, "What should I be doing this term to work towards tenure?" Take notes, and be sure that teaching, research, and service are part of the conversation.

Other mentors and professional development

While the department chair is a formal conduit of information about tenure expectations, that person shouldn't be the only source of knowledge about the process. Casual conversations with colleagues about workload, work–life balance, and also specifically about tenure expectations will be invaluable for better understanding the department and its culture. Additionally, be sure to reach beyond the department to develop an institutional network for yourself. While departments are the front line of tenure decisions, they are not the end of the line, and having allies and knowledge sources from outside of the departmental context provides a fuller basis for understanding the tenure process, as well as the parameters of the job overall, both before and after promotion.

Most institutions now offer, and sometimes require, programming for pre-tenure faculty, including providing individual mentors to junior faculty within one to three years of hiring and cohort-based programming on topics relating to teaching, publication, and grant-getting. Universities sometimes offer programs on managing a social media presence, using mentors effectively, and developing collaborative research. These meetings may feel like one more drag on your calendar, taking you away from precious research and writing time. But they can pay huge dividends and should be a priority on your schedule, at least for the first year.

Official, institution-led programs for new faculty provide an opportunity to develop a supportive group of colleagues who are at the same general stage – most (but not all) new hires invited to this programming will be pre-tenure and on the tenure-track. The people you meet in these programs, even if they don't become close friends – and some likely will! – are a source of support and a wealth of knowledge about what they are hearing in their own departments about the process to getting tenure. It's always informative to compare stories and work to reconcile differing messages about tenure expectations. These conversations can provide more understanding about the institution as a whole and can put departmental messages in a different light. That said, it is the department that votes first on a tenure case, so departmental priorities have to be, to some degree, also your priorities, or at least your framework for building a work plan for getting to and through tenure.

Formal, college/university-led programs also expose you to decision-makers and leaders of different kinds in your institution. The people who create the programs and invite speakers are those who have been recognized for their skills and expertise by the institution; otherwise, they would not be trusted to help acculturate new faculty. Regularly attending the meetings that these folks plan as part of a faculty orientation and development series will provide you with an opportunity to regularly interact with recognized campus leaders. Their programs will give you explicit and implicit signals about campus priorities; their knowledge will help you understand and synthesize a wide variety of messages that you are hearing about your job responsibilities and the path to tenure. These leaders will also serve as informal mentors and may introduce you to other campus leaders and more senior faculty whom you may not otherwise meet. These connections can serve you well as you move toward tenure and beyond, and they may be instrumental in helping you to navigate especially the teaching and service responsibilities of your role.

Spending some of your first year and beyond learning about the institution and connecting with colleagues within and outside of your department will provide important context and support as you navigate the day-to-day work in academia. Fundamentally, however, the day-to-day core of the job revolves around three distinct arenas of work, each of which forms an explicit component of the tenure evaluation: research, teaching, and service.

Strategizing the work: teaching, research, and service

The doctoral degree signifies expertise in doing research, but the explicit, scheduled, and most consistent component of the tenure-track position is teaching. While many faculty new to the tenure-track will have one or sometimes more teaching releases in the first or second year, teaching will place the most regular demands on the junior faculty's calendar. Balancing research and service usually means figuring out how to set goals and parameters on the teaching role, and then connecting ways to produce research and contribute service to institutional expectations, as well as one's own personal goals.

Teaching

Every institution of higher education has students, and teaching those students is an imperative for all faculty, regardless of institution type. One of the most important early conversations with one's department chair, usually even before officially starting a tenure-track job, is about the teaching schedule and courses in the first year. It's important to try to balance the course needs of the department (often for large and/or popular introductory classes that serve as prerequisites to other courses) with the needs of the new faculty member. Teaching smaller, more specialized classes for advanced undergraduates or graduate students is usually more oriented to the faculty member's expertise, and sometimes involves far less time to design and execute than larger introductory courses. At the same time, having a reliable, popular introductory course that serves the department also serves the needs of the individual faculty member; the class can be repeated, every year or even every term, and the flow of satisfied students through multiple sections/offerings of the course makes the faculty member a highly valued member of the department.

Important jargon in the teaching arena is the "prep," or preparatory, class. A new prep signifies a class that someone has never before taught and, therefore, must completely plan from scratch. Sometimes, a new prep is truly a course from scratch; other times, the faculty member might have a previous syllabus from someone else and sometimes even course materials from which to build their own class. Paying attention to preps is particularly important in high-teaching-load institutions; a faculty member teaching 6–8 courses per year might have fewer preps in that same year if they teach the same material to different groups of students across two or even three terms.

Most new faculty, particularly those in a faculty appointment at a baccalaureate college or university, get some sort of teaching release in the first year. A teaching release is a reduction in the standard number of courses the faculty member is expected to teach per term/year. Typically, the release is granted in the first teaching term. Sometimes, however, the job contract allows a faculty member to choose when to take one or two course releases within a given time period, usually the in first one to three years of the appointment. It's important to consider how individual teaching prep and department needs over the first years may be brought into alignment, and sometimes taking a release after a semester or two of teaching can make that time more meaningful for pursuing research goals.

Writing teaching syllabi and turning those outlines into 10-, 12-, or 15-week instruction is a craft that takes time to develop: plan to take more time than you might want to devote to the task in order to feel confident and well-prepared going into your classes every week. When I started teaching, I spent one weekend day each week planning my course materials for that upcoming week, which meant a good five to eight hours every Sunday. But after that first year, I never spent that kind of time on a single course again: I got better at knowing my own material, and what worked well (and didn't) for my lectures. Individual life circumstances will shape the time you have available for teaching development; in my case, I didn't yet have children, and I could carve out half my weekend for course preparation. For those who have household/family/community obligations that don't allow such concentrated time, they may need to spend one to two hours four nights (or mornings, or whatever) per week on course prep in the first year. But it won't be necessary to spend such time every year on course development.

It's absolutely true that courses eventually become stale and need wholesale revising. But, extra time spent in the first term or year of teaching a given course pays off the next time that same course is taught. Minor tweaks can be made each term to fix parts of a course that didn't work the first time, and it won't need extensive development again for several years. While you might want to take a larger chunk of time to revise a course after teaching it a few years, it's best to save that task for the first sabbatical after getting tenure. Sometimes, faculty can use course development opportunities from their institution, gaining a release from teaching (a "course release") in return for revising a course syllabus and learning strategies. In the meantime, take the time in the summer before you start teaching, or the December-January break in between terms, to carefully select texts and supplemental materials so that the bulk of a class can remain the same for a few years, at least past reappointment if not through gaining tenure.

When I first started teaching, a senior colleague advised me that my courses did not need to be perfect – only "good enough." In the moment, I was appalled, because I thought of myself as someone who wanted to do the very best for my students. Over time, however, I have come to see the wisdom of this advice – no course is perfect, and a junior faculty member trying to make their class the best it possibly can will always need to devote more time and will never be completely satisfied. Seek to do well by your students, and respond when the term doesn't seem to be going well. But also, don't expect to be perfect or the most popular teacher your first time with a class. Know that you can do better, and you will.

Time spent on improving and developing teaching is not always a zero-sum game against other areas of responsibility. Some faculty are able to strategically use teaching in ways that support a publication agenda, or they find that time devoted to pedagogical training can have unexpected research dividends. For example, my first year teaching an upper-division undergraduate course in my subfield allowed me to create a course assignment that required extensive fieldwork on the part of students. Developing the assignment involved extra reading about the city where I assigned the fieldwork and driving a significant portion of the area in order to create a self-guided field project. But I was able to then write about the development of that course guide and publish it in a journal about pedagogy in my field (Martin 2003). Thus, I was able to use the extensive time that I put into developing course material toward a research productivity outcome.

In my early years as an assistant professor at my first institution, I applied to a junior faculty teaching-mentoring opportunity. At the time, I was in a large, state-funded doctoral university which had a highly regarded and well-resourced teaching development and mentoring "Teaching Fellows" program. A bit counterintuitively, this program – putatively focusing on teaching – offered a course release in exchange for committing a year to regular meetings and workshops, and two out-of-town retreats. It was a significant time commitment but involved both one-on-one mentoring and the opportunity to develop skills and share a cohort experience with other junior faculty. It might seem that such a program would lead me away from focusing on my research. It was a risk in that regard, but I also knew that the program was highly valued in my department. I viewed the course release as a reasonable trade-off; given I would be teaching less if I was in the program, it would probably not take too much of my time.

What I did not expect was that participation in the program vastly enhanced my professional networks within the university. Furthermore, I enjoyed the time in the workshops immensely, as well as benefited from all that I learned. The time spent with nine other junior faculty from across the university enabled me to learn about different departments and the many strategies that my peers pursued to balance their teaching, research, and service. Finally, I never anticipated that I would connect with one colleague who, while on the surface from a totally different research context, actually shared substantive intellectual interests with me. After creating a teaching assignment through which I brought my expertise to his classroom, we ended up developing a research agenda together, which ultimately culminated in a National Science Foundation-funded collaborative research project and a number of publications (Martin and Scherr 2005; Martin, Scherr, and City 2010; Pierce et al. 2012; Martin 2013; Martin and Pierce 2013).

Research

Research productivity may be a consistent part of tenure expectations, but the type of research and what counts as productive vary considerably by institutional type. While it's crucial, as noted above, to understand your institutional context, it's also crucial to understand that you have to find the time to devote to research, so that you can meet the expectations of your department and institution.

Coming into a tenure-track job, you should have a planned strategy for the first few years of publication. For most people, this involves publishing the results of the dissertation research or recent postdoctoral work. In the case of a dissertation, ideally you have already thought about three distinct arguments – three articles, really – that you can package into articles from the dissertation. At the same time, it's entirely possible that some of the dissertation articles were published before you got the job. So it is particularly helpful to be able to conceptualize products from your dissertation beyond the initial "three articles"; either you focus on getting three dissertation articles published and/or you focus on other arguments, analyses, or insights from your dissertation beyond the three you have published. Sometimes, for example, you can develop a methodological article from dissertation work.

It is especially efficient and effective for progress to tenure to be able to publish from the dissertation – even if you already have some articles published from it – because it represents data that you have in hand. You do not need to spend time on data collection, and, in many cases, much of your analysis will have already been completed. So, two of the significant time investments in research are already done. Regardless of institutional type, disseminating as much as possible from a dissertation is a particularly effective way to demonstrate productivity. Even if you are in the most teaching-intensive type of institution, finding outlets for already completed research is an optimal way to demonstrate research productivity. Rather than a research article, maybe you present various aspects of your work at conferences over the first few years. Maybe you develop an insightful policy paper that you publish through a think tank, or present to an industry or policy group in your community. Regardless of the particular outlet, think about ways to use your dissertation for research productivity beyond the defense of the PhD degree, and try to align this research dissemination with the values and priorities of your institution.

You might be focused on developing your dissertation into a book as the main part of your tenure case. If you are in a doctoral institution in a social or physical sciences department, it's very important to have conversations with your department chair, and others in your department, about how books are regarded in a tenure case. You might seek advice on how to develop some articles to accompany a book-based tenure case. Try to balance your personal research dissemination goals with the goals of your institution, and consult with mentors to find ways to balance these priorities if they seem different.

Getting to tenure from the research perspective also involves demonstrating that you can develop a new research program beyond the dissertation. But you do not have to start from scratch. It is completely legitimate to build off your dissertation work and data collection. In my case, I developed a small, local research project that built off my dissertation work, but in a new context. I used similar methods, and applied for internal, institutional funds to support the new work. I also took advantage of opportunities to collaborate with colleagues, adding my insights to their data or collecting new data with them. Each of these modest, small projects helped me to develop insights that fostered the grant application that I wrote with my friend and colleague in the teaching fellows program I was doing at my institution. Synergies abounded, even ones I hadn't anticipated and certainly could not have planned.

One of my new faculty peers when I started my first job was a more senior person in a different discipline who had already worked for a few years as an assistant professor at a different institution, and was coming into the new job with a shorter tenure clock. He already had an established publication record, whereas I was just beginning; I had a few publications on my CV when I started, but just one co-authored peer-reviewed piece in the pipeline. In a casual conversation about writing, he told me he woke up early every day and spent one to two hours writing; that way no matter what happened during the day, he got some writing done. His goal in those one to two hours was relatively modest: 300–700 words. I remember being awed and shocked at the discipline and knowing I probably would never replicate that type of schedule. But, I did take some lessons that have served me well: (1) just one to a few hours a day can produce a good amount of writing over time; (2) when you stop writing, always know where you're going next (leave a note in the text for yourself); and (3) even a few hundred words in one sitting will pay off over time (even if some of those words get edited the next time you sit down to write).

One strategy I did pursue was to always be actively writing one article each term. (I could not sustain focus on more than one thing at a time until well into my tenure-track; one article per term, two terms per year, plus a summer, enabled me to slowly develop a cycle of writing, revising, new research, writing, etc., that turned into a publication track record over time.) I rarely wrote every day. But, each month I had a part of an article I was trying to write, and each small vacation – such as a four-day weekend – had one or two half-days cleared to work on some part of a paper. Longer breaks had multiple sequential days blocked off, with deadlines set for article submissions. I planned conference papers around future publications, writing out my conference papers in long form. I didn't do that for conference presentation purposes, but in order to have a rough draft of a paper which I would finish as soon as the term ended after that conference had occurred. Through these strategies, I slowly developed a pipeline of papers submitted, then returned – usually with eviscerating comments, but most often with the opportunity to resubmit. Then I created a new set of deadlines and worked on revisions until I could resubmit, then turned to the next partly-written conference paper or very rough article idea based on data from my dissertation that I had collected but never used (Martin 2004).

It might seem as though I was always writing, but I wasn't. I always gave myself some days off, especially on breaks. And, I always maintained my personal exercise schedule. It helped me stay grounded, and gave me time to reflect and think. One of the most significant aspects of my research schedule in the days of being a junior faculty was experimentation. I tried a number of different schedules until I found the right rhythm and timing in my writing. I never could write a little bit every day like my friend did, but I found ways to take chunks of two to three days every week to write. And as life and its demands changed, I had to change my expectations and self-organization.

While both teaching and research represent aspects of being an academic that allows for creativity and experimentation, one aspect of the tenure track that tends to be relatively inflexible is the service expectation and burden.

Service

Every faculty member is expected to help manage the business of the department and, sometimes, to help manage the work of the institution as a whole. Departments usually organize their work into committees, addressing everything from diversity and inclusion in the work of the department, to curriculum design, personnel reviews, graduate admissions (particularly for PhD programs), and colloquium events. The specific committees and their scope of work can actually be another valuable source of knowledge about the institution, department, and their values. But they are also sites for additional workload expectations.

It's generally the chair's job to assign a junior faculty member a role that allows them to contribute without the service assignment taking too much time. It's very legitimate, even if intimidating, to ask questions about how much time a suggested committee assignment should take. Additionally, a junior faculty member should not typically be assigned a leadership role on a committee; service work in the early tenure years is as much about observation and learning as it is about substantive contribution. But, it is important nonetheless to contribute, offer opinions, make suggestions, and be prepared to put in some extra time to bring a suggestion to fruition. Service assignments are opportunities to shape the business of the department; colleagues generally hope that junior faculty will be engaged in the work while respecting existing systems. Sometimes, these systems need to be changed, but it is prudent to learn about them before suggesting that they don't work.

Service to a department and institution provides an opportunity to understand oneself in relation to administrative work; is it interesting, mind-numbing, or fun? These reactions can signal dimensions of one's workplace aptitudes and skills that we don't often get to explore in a PhD program. (Some PhD programs do offer their students substantive governance experience, and people who have had such experiences will generally be able to engage in service in ways that can be especially rewarding, and not overly time burdensome.)

Committee work is also a way to learn about colleagues, including what they do, and how it might connect to your own work or teaching. Academic life can feel isolating at times; take advantage of the small group setting of many department committees to ask questions and learn how colleagues have approached teaching or writing challenges. These moments of camaraderie can be nourishing.

Conclusion: self-care and time

Embarking on a tenure-track position is exciting, intimidating, and sometimes overwhelming. There is a reason the tenure clock lasts about six years before the tenure review. The work takes time. Some semesters will feel as if you've done nothing but teaching. But, that teaching might have been really important – certainly to the students, and often to one's development as an instructor (even a bad semester can be a significant learning experience!).

It's important to find ways to remember what brought you into the field in the first place, and to notice what brings you small (or big!) joys in your day. We all have personal pleasures that have nothing to do with work; don't forget to practice those rewarding activities with people who support and sustain you. Get into the habit of reminding yourself of your current goals for the term, and to recalibrate when it seems your work–life and teaching–research–service balances seem out of whack. Ask for advice and help when you need it, and look for insights occasionally from unexpected quarters, especially from your junior faculty peers and your growing networks in the department, and the larger institution.

References

American Council on Education (ACE). 2023. *Carnegie Classifications of Institutions of Higher Education*. Accessed 22 November 2023. https://carnegieclassifications.acenet.edu/.

Martin, D. G. 2003. Observing Atlanta, Georgia: Using an urban field study to enhance student experiences and instructor knowledge in urban geography. *Journal of Geography* 102(1): 35–41.

Martin, D. G. 2004. Non-profit foundations and grassroots organizing: reshaping urban governance. *Professional Geographer* 56(3): 394–405.

Martin, D. G., and A. Scherr. 2005. Landscapes: Lawyers as constituents of landscape. *Landscape Research* 30(3): 379–394. [Reprinted in K. Olwig and D. Mitchell, eds. *Justice, Power, and the Political Landscape*. Routledge, 2008.]

Martin, D. G., A. Scherr, and C. City. 2010. Making law, making place: Lawyers and the production of space. *Progress in Human Geography* 34(2): 175–192.

Martin, D. G. 2013. Up against the law: legal structuring of political opportunities in neighborhood opposition to group home siting in Massachusetts. *Urban Geography* 34(4): 523–540.

Martin, D. G. and J. Pierce. 2013. Reconceptualizing Resistance: residuals of the state and democratic radical pluralism. *Antipode* 45(1): 67–79.

Pierce, J., D. G. Martin, A. Greiner, and A. Scherr. 2012. The spatial justice of mental health: How local development politics shape public health outcomes. *Annals of the Association of American Geographers* 102(5): 1084–1092.

25

IT'S OK NOT BEING OK

Coping with adversity

Jennifer Ho

2010. This was supposed to be my year – a year where I would be on an upward trajectory of success, both personal and (hopefully) professional. I was going to turn 40, which I thought was great (I'm in the camp of age is a number, though being on the other side of 50, my knees would disagree with that statement). I was trying to get pregnant for the first time with my fiancé, who will appear later in the essay as my husband, a second marriage for me. I was going up for tenure and promotion to Associate Professor, and all my mentors at UNC Chapel Hill assured me that I had nothing to worry about. I had just been asked to co-chair the annual meeting of the Association for Asian American Studies (AAAS), a testament to my growing national reputation in Asian American studies.

Then I got a phone call on April 15 from a radiologist at the Lineberger Cancer Center that my biopsy had come back positive for breast cancer. In the space of that ten-minute phone call, I felt a flutter of sheer and utter panic start to engulf me – which I envisioned as the shutters of a window opening up to a dark unending landscape. Because I knew I wasn't ready to face the reality of my mortality, I mentally closed the shutters of the window, walked down a set of stairs, and then closed a door behind me. Thirty minutes later, I actually walked into my graduate seminar on cultural studies, acting as if everything was fine. It was only later that night that I started to think through the things I'd need to do: the phone calls I'd have to make, who to contact in my department about a potential leave of absence, how to talk with my fiancé about this diagnosis, and how this would impact our upcoming wedding and summer plans. When my best friend later asked why I continued to teach my grad seminar, I said that I wasn't ready to share this news with anyone, to which she pointed out that I could have simply

DOI: 10.4324/9781003342953-29

canceled class by putting a note on the door. But I wasn't able to write a note to put on the door – writing the note would have been admitting that I was not OK. And I was not ready to admit that I was not OK.

Academics are mostly over-achieving, intensely focused, and high functioning people – we approach problems as things to be solved, challenges as things to be overcome, and failures as things to learn from on a path to success. But not all problems, challenges, or failures can be solved, overcome, or lead to success. Sometimes adversity is adversity: we fail, and we are not OK, before, during, or after the adversity. I've written about the various failures I've experienced in and out of academia, in "The Accidental Academic, or How to Succeed in Academia through Failure and Doubt." Inspired by Jack Halberstam's thought-provoking *The Queer Art of Failure*, a work that destabilizes our notions of normativity through success by offering an interpretation of failure as queerly progressive and generative, I wanted to share the many moments of academic failure that I've experienced. I end "Accidental Academic" with my cancer diagnosis because I wanted to address what happens when we are not OK.

Not being OK and coping with adversity is the subject of this essay because, like Halberstam, I want to destabilize our idea of what is normative in academia and challenge us to rethink an academic culture of productivity and accomplishment. I want to create spaces where I can share the many times I have not been OK in my life because admitting I was not OK was part of how I coped with adversity. I do not intend this essay to be a narrative of triumph through hardship nor to discount the ways that sometimes you are not OK; and you ask for help, you don't get the help, and things continue to be awful, and the adversity does not end; or people try to help, but you are still not OK. There are so many things in the world, currently, that feel awful and never ending: gun violence, white supremacy, climate change, to name a few. This essay will be a continuation of my "Accidental Academic" article in sharing times when I was not OK in my personal life and how I managed to cope – and by cope I mean that I survived them but didn't necessarily get over them or even move past the hurt and pain entirely.

I'm writing this essay as a personal reflection because I believe we need to share stories about professional setbacks in order to normalize adversity and find ways to endure – and professional setbacks are often intimately tied to the personal hardships we navigate in the complexities of our all too human lives. I've done a formal presentation about one of my key experiences with professional adversity: how I nearly gave up on my second book, *Racial Ambiguity in Asian American Culture*. I submitted my book proposal and sample chapters to three university presses, received three rejections, and finally hired a book editor (again, something we rarely talk about in academia – that you can hire people who can help make your book proposal legible and your introductory chapter more discernible to readers). The fourth press, Rutgers, said yes, and the book eventually won a best book prize, which did feel like vindication for all the negative reader reports I received. This story is likely one you're familiar with: the grant application that was rejected several times, and then eventually was reworked to win a lucrative prize; striking out on the job market two years in a row until finally getting a dream job in year three; the article that couldn't find a home, but eventually was published in a journal that led to academic acclaim. This is the narrative of triumph over misfortune that circulates as a means of instilling scholarly resilience.

What we don't have, in academia, is a language for the personal – perhaps because we desire a clear demarcation of the personal and the professional.[1] Yet so many of us have a certain porousness of our work and personal lives – we develop friendships with our departmental colleagues, and some of our students, especially our graduate students, will transition to being our friends. And for some of us (for me especially), our areas of research are deeply tied to our personal identities.[2] I don't want to make light of the academic and scholarly setbacks most of us have experienced – and the difficulties of admitting those failures and being vulnerable. But even less visible and legible are the personal afflictions we often hide from our colleagues that impact our professional lives in profound and lasting ways. So, I will share a few personal adversities and the ways they impacted me professionally in the hopes that my sharing can normalize the ways it's OK to not be OK.

As I alluded to above and discussed in my "Accidental Academic" essay, my breast cancer treatment and recovery was a central trauma in my life, though oddly enough not one that derailed my professional life, largely because I didn't delay going up for tenure and promotion to Associate Professor, though the chair of my department encouraged me to do so. She wanted to make sure I felt no pressure to turn in my file in the midst of my chemotherapy treatments, but I wanted to push through because I was in the fortunate position of having my first book out and a second book underway, as well as several articles. It is also true (and no one talks about this) that chemotherapy is boring: there is a lot of empty time to fill. When I wasn't sleeping, I had a 30-minute attention span and truly could not focus my mind on anything that wasn't rote. Organizing my tenure file into six physical binders was a welcome distraction that felt satisfying – like I was accomplishing something meaningful rather than dealing with the tedium of my days between chemotherapy treatments.

Making sense of my health after four months of chemotherapy, a bilateral mastectomy, along with a daily regime of Tamoxifen that lasted for five years meant that I tried not to worry too much about recurrence. But the reality of having a cancer diagnosis means that the worry is always there.[3] Three years after my breast cancer treatment, in July 2013, there was a 36-hour period in which I thought I might have liver cancer. During a physical examination, my primary care physician felt a mass in my abdomen and sent me to the hospital for an ultrasound later that week. After the ultrasound, I received a phone call from my physician that there was a mass on my liver that they suspected might be cancer given my previous breast cancer diagnosis, so I would have to do a CT scan at the hospital the following day. My husband and I went to the hospital on Saturday morning, thinking I might have a year to live, and during dinner with friends I cracked jokes, coming up with a top ten list of things I wouldn't have to do if I had liver cancer, like take over as director of graduate studies in my department or meet my book deadline. I even named the mass "Tommy Tumor" in an effort to control the panic that was just under the surface of my dark humor – a nickname I revised when my doctor called that evening to share that the mass was a benign hemangioma – thus, Tommy became Hank Hemangioma.[4]

Though I have not had another cancer scare as acute as this 36-hour period, I have had other masses appear on my body – lumps that I can feel and that remind me of the lump in my left breast that led to my breast cancer diagnosis. Each time I find a lump, I see my primary care physician about whether it is a lipoma (a benign mass) or a potential cancer recurrence. In October 2022, I found myself in a breast imaging center surrounded by pink ribbon imagery, with a tech doing an ultrasound of the mass in my left armpit. Putting on the hospital gown and seeing the pink ribbons plunged me back in time to my cancer diagnosis and treatment. As I lay on the table waiting for the radiologist to look over my screens, I felt the familiar fear recur. I wondered how I would cope with another cancer diagnosis. I thought about having to find a new network of doctors and cancer support now that I had moved from Chapel Hill to Boulder. I dreaded having to tell people that my cancer had come back. When the tech returned, she shared that the radiologist was 99% certain that it was a benign lipoma and that I was still cancer-free as far as they could tell.

Living with a cancer diagnosis means living with uncertainty – not one that impairs my ability to live but one that I am reminded of, acutely, when I find a mass on my body or have to fill out medical forms giving my full health history. One of the things I researched before accepting a position at the University of Colorado Boulder was the proximity of a comprehensive cancer center to Boulder County. This is now part of the professional calculation I make – to know what the health care system is like if I have another cancer diagnosis. These personal factors related to my health influence my professional decision-making.

The July 2013 physical that led to the brief liver cancer scare happened because my husband and I were trying to adopt a baby, and the adoption agency required us to submit medical exams. Our adoption story began as many do: with an inability to get pregnant. We had tried to get pregnant in 2008 and again in 2009 – the first time was derailed due to a fibroid diagnosis, one that led me to lose so much blood that they pushed up the surgery by a month and had to prep me with three bags of blood before the procedure. When we resumed trying, I was diagnosed with breast cancer six months later, which then threw me into early menopause post-chemotherapy. And after my cancer treatments, I learned that the back-up plan I always assumed would happen if I wasn't able to get pregnant – adoption from China – was off the table: China does not let people who have had cancer adopt.[5]

Indeed, most countries have restrictions when it comes to a cancer diagnosis, and for a variety of reasons, my husband and I decided that domestic private open adoption was the right choice for us. We found an agency in North Carolina, created a profile, underwent health screenings, had friends write recommendation letters for us, and then waited. We had been told that it can take years to make a match, so when Cindy contacted us just under a month from when our profile went live, we were very surprised, especially because Cindy wasn't pregnant: she had a nine-month-old daughter, Chloe.[6] As much as we fell in love with Chloe, we fell in love with Cindy, a young woman who reminded me of the many undergraduate students I had taught over the years. Because we needed to do the placement before Chloe turned one, everything happened quickly once Cindy chose us. We had several visits, including two overnight stays, one where Cindy and Chloe stayed with us and another where we had Chloe alone without Cindy.

It was after this last overnight visit that the reality of what was about to happen in a month set in for Cindy – at least this is what I'm guessing because a day after our overnight stay, we received a message from the agency that Cindy needed to take a step back and wouldn't be communicating with us for a while. This was mid-September, and we had a projected date of mid-October 2013 for when Chloe would join our family. We had prepared a baby room, collecting clothes, a crib, a stroller, and I had arranged to have friends cover the rest of my classes for the semester while taking the following spring off. We had also been communicating steadily with Cindy since we first met her and Chloe – in the last few weeks it was at least a text a day, if not more. But two weeks after the agency said she needed a break, they called me when I was on campus and told me Cindy had changed her mind.

Like with my breast cancer diagnosis, I had 30 minutes to digest this news before walking into a faculty meeting in which I was doing a presentation as Director of Graduate Studies. I texted my husband, attended the meeting, then went home and wept. It felt like someone had thrown me down a set of stairs and kicked me in the teeth, so great was the shock and pain. I sent a mass email to friends and family asking them to share the news widely because I didn't have the energy to tell people that we would not be welcoming Chloe to our home, and that weekend my husband and I dismantled the baby room. I fell into a depression during which I barely got off the couch and watched reruns of innocuous sitcoms. I stopped grading assignments, and the only thing I immediately told my students was that we would be reverting back to the original syllabi I had shared at the beginning of the semester – there would be no guest teachers coming to substitute for me.

I want to be very clear here: Cindy had every right to make the decision not to move forward with an open adoption with us. I had only known Chloe for eight weeks and felt her loss acutely. I can only imagine the pain and heartache that overwhelmed Cindy when considering relinquishing Chloe to us – even in a situation where she could have contact with her, if she chose to. We were very open to Cindy having as much or as little contact as she wanted. Perhaps this was unrealistic, but because I had taken Cindy into my heart as much as Chloe, I thought we would be this messy blended family and we'd grapple with the challenges of this new family structure as they came up. While Cindy's reasons for wanting an initial adoption for Chloe are not for me to share, I know there were structural barriers that didn't immediately allow her to keep Chloe. Friends of ours, out of solidarity and love, said that Cindy was wrong and selfish to not move forward with the adoption because Chloe would have a better life with us. I'm not sure that's true. Maybe we could have provided, more easily and readily, a material existence that Cindy couldn't have, but the loss of her first mother would always be a loss for Chloe to navigate, even in an open adoption situation where she'd hopefully get to see Cindy on a regular basis.

My husband and I decided not to pursue another adoption. We pulled our application from the agency, went to therapy, and decided we would find other ways to have fulfilling relationships with young people – future nieces and nephews and the children of our friends. We are risk-averse people, and we knew we didn't have it in us to experience this type of loss again. This is the first time that I've written about our failed adoption, and I'm surprised at how hard it has been to revisit these scenes from a decade ago – that a wound I thought had completely healed and scarred over, is actually still tender to the touch. My failed fertility and failed adoption ended a vision I had of myself as a mother. Losing Chloe felt more painful than my cancer diagnosis and treatment, perhaps because it was the loss of a future I had wanted and hoped for myself – to be a mother. Reconceiving our family as just two instead of three was a mental and emotional adjustment for myself and my husband that felt more acute than navigating my breast cancer diagnosis and treatment. I'm not entirely sure why that is. Perhaps one reason is the amount of support and care I received from my cancer treatment – from medical authorities at the Lineberger Center to friends and family who provided meals and sat with me while I received chemo or who dropped off gifts and sent words of support and love. Whereas losing Chloe was a more private affair – not because we were less public but because I think it was hard for people to know how to show care for us and hard for us to know what would make us feel better, other than time. Prior to writing this essay I would have said that I don't think of Chloe or Cindy much anymore – that we've moved past this period in our lives. I think this is largely true, though I also think there's a reason I don't think of this period very often and why it has taken me a decade to write about this experience: it is still, as I've just discovered and disclosed, a tender area, reminding me that I'm not as OK about this as I thought I was.

And that's OK. It's OK for me not to be OK about this, even a decade later. Not being OK – admitting that I'm not OK – is something I'm trying to model in this essay and that I modeled for my students when it was over eight weeks before I was able to return their essays to them. In fall 2013, I was teaching two classes but only had grading responsibilities for one class since I had two teaching assistants who handled assignments in the large lecture class. Six weeks after the adoption fell through, I walked into class and told them about our failed adoption.[7] I told them I was depressed and found myself unable to do anything other than watch Netflix on my couch. I apologized for not getting their papers to them and understood that there were consequences for my negligence, such as having negative course evaluations. I also shared that I was going to therapy for my depression and hoped that if any of them were also feeling depressed that they consider getting help – that asking for help, especially mental health services, should be destigmatized, which is one reason I was telling them about my own struggles.

I was nervous sharing such intimate details with my students, but I felt I owed them an honest explanation. What I didn't anticipate was the overwhelmingly positive reaction of my students. Some immediately stayed after class to express their condolences for what my husband and I were going through. Others came to see me in office hours to thank me for sharing that I was seeing a therapist and confided their own mental health struggles. I had been afraid that my vulnerability would either be seen as too intimate and unprofessional or would undermine my authority. My fears were wholly unrealized because I found that students appreciated my vulnerability. My vulnerability allowed students to express empathy with me and affirmed students struggling with mental health that their issues were validated. Together, we could say we were not OK and that being not OK was OK.

Very recently I had an opportunity to put this approach of personal vulnerability and openness with my students into practice again. On a Sunday morning in early October 2022, a former graduate student contacted me and shared that Allison, the last of the PhD students I was working with at UNC Chapel Hill, had killed herself.[8] I had just been texting with Allison the week before – we exchanged texts on Tuesday, and on Thursday she was gone. The Monday after I received this news I walked into the large lecture course on social justice that I was teaching and stopped half-way through my PowerPoint slides and let students know what I was going through – that I had just received this news and was not OK. I told them that I was sharing this with them because if any of them were not OK – and not OK to the point where they were thinking about harming themselves – then there were resources I had just put on our course management site. I told them I cared about them as humans first and students second – that grades in my class or any other class were not as important as their overall well-being. I said I would stay after class if any of them wanted to talk or could be reached by email or during office hours if they wanted help finding help. I told them I was sharing the news of my loss with them because I didn't know how I would be in the weeks and months that would follow. Perhaps I'd burst into tears one day in class, and I didn't want them to freak out if this happened, so I wanted them to know that I was experiencing pain and grief over Allison's death.

Like when I disclosed my depression to students in fall 2013, telling them about Allison's suicide and saying I wanted them to be OK was a moment of profound vulnerability for me, especially because I was so numb and raw. One student ran out of the room as I started to talk about Allison's suicide. Other students began to cry. I wondered if I had done the right thing, watching the students' faces crumble in front of me. But after I awkwardly ended class, about 20 students stayed to talk with me, sharing that they also had painful experiences with death and loss. Students hugged me; one dropped off a card and flowers to my office the next day; the student who ran out of class contacted me that night by email to thank me for sharing and disclosed their own mental health struggles. I heard from numerous students on email and in office hours that disclosing my pain over Allison's death was a profound moment for them, one they appreciated because it made myself human-sized and relatable to them. I don't know if I felt a noticeable shift in class after that moment, but I experienced this class as one where students were incredibly kind to me and to one another. Perhaps it was just that we were all dealing with so much loss already due to the global pandemic. Perhaps it was me admitting just how not OK I was and that it was OK if they, too, were not OK. Perhaps it was stressing that I wanted them to find help if they needed help. Because we all need help.

This is the biggest takeaway of this essay, and the one I will end on. It is OK not being OK, but it is especially OK to ask for help, both in the moment of crisis and in the days, weeks, months, and years afterward. Trauma, grief, and mourning have their own timeline, and each trauma and loss are unique: how I mourned the loss of my breasts was different from how I grieved the loss of Chloe, and both are different from the pain I feel over Allison's death. For me, writing and talking about my losses has brought solace. I shared drafts of this essay with a new friend who didn't know the details of what I had been through. After reading the piece, she immediately reached out to me, providing space for me to talk about Chloe and the loss of motherhood – which was not something I thought I needed until I wrote this piece. My friend had a similar story about fertility, cancer, and the loss of motherhood, which I hadn't known. In sharing my own struggles I created space to hold her losses and hear her stories of trauma and grief. There is no set deadline when it comes to coping with loss, and while there is a temptation to feel like something that happened years ago shouldn't still be tender to the touch, the reality of our all too human lives is that there are so many tender parts of ourselves that need tending to. Being open to receive help from others can also be a way for people to open themselves up to you, and for people, together, to find solace – true mutual aid.

Mutual aid is not engrained into the culture of academia. As academics and scholars we are trained to be independent and self-sufficient. We are living in an academic culture where we are told not to admit our vulnerabilities, either professionally or personally. Across disciplines, the message that so many of us absorb, from the time we are graduate students to our first academic appointments, is not to admit weakness. Because if we are weak, we are not cut out for the life of an academic. Like athletes, academics are given messages, overtly and covertly, that we should push through our pain. We should be stoic in the face of adversity, and we should work longer and harder to receive the rewards of academia.

However, this culture of stoicism is neither realistic nor sustainable. To endure, in academia as in our personal lives, we require aid and care, which should be mutual and communal. When I admitted weakness and vulnerability, I was generally met with kindness: from colleagues during my cancer treatment – who dropped off food, flowers, and messages of support – and from students who expressed their gratitude at my willingness to make myself human-sized. When I learned of colleagues who had also been diagnosed with cancer, who had lost loved ones, who were struggling with fertility, I reached out and shared some of my story – I made myself vulnerable to allow space for their vulnerability. And that space of communal care is capacious and expansive, allowing us to extend that same care to our students and our loved ones. Because this is all a gift: to ask for help and to receive it is as much a gift as providing help to someone. In the receiving, as in the giving, we are admitting people to our intimate spaces, which requires trust. And the trust helps solidify respect, cementing relationships of mutual aid. So, if any of you readers are not OK, please know it's OK. And if you have it in you to make yourself vulnerable and human-sized, to your colleagues, students, family, and friends, then I hope you will find what I found: that a community of care is necessary to help cope with the adversities that will inevitably find us.

If you are having thoughts of suicide, call or text 988 to reach the 988 Suicide and Crisis Lifeline or go to SpeakingOfSuicide.com/resources for a list of additional resources.

Notes

1 One excellent exception to the idea that the personal narrative is not centered in academia is the volume *Counternarratives from Women of Color Academics: Bravery, Vulnerability, and Resistance*. All of the essays in this volume are written in the first person (whether singular or plural) and express the intersection of research and the personal (or as some authors describe, autoethnography). The existence of this volume also feels like resistance to white supremacy and patriarchy in its affirmation of the personal and its assertion of the political.

2 Two edited collections, *Written/Unwritten: Diversity and the Hidden Truths of Tenure* and *Presumed Incompetent: The Intersections of Race and Class for Women in Academia*, discuss the challenges for scholars who are women and/or people of color and/or queer, many of whom do research that touches on their personal identities.

3 Tamoxifen is a medication used to treat breast cancer recurrence.
4 A hemangioma is a non-cancerous vascular mass that is congenital; it developed in the womb at the time of birth. In my case it was considered an incidental finding, meaning I had no symptoms that would have led them to look for a mass, were it not for the fact that my doctor felt something and the ultrasound showed something on my liver. My particular hemangioma is very large — about half the size of my liver.
5 China's restrictions on who can adopt also include people who have not sought mental health treatment and have a BMI index that is below a certain threshold, and they will not allow LGBTQ+ people or single people to adopt (queer people had been adopting as single parents and not disclosing their partner relationship, but China caught on).
6 Cindy and Chloe are pseudonyms, and our adoption agency mostly placed expectant mothers with adoptive parents, though 10% of their placements were with babies under one year.
7 I assumed students had guessed that the adoption had fallen through since I had told them they would be having guest instructors because I was adopting a baby.
8 Allison is a pseudonym.

Further Reading

American Association of Geographers. 2023. Mental Health in the Academy Affinity Group, https://www.aag.org/groups/mental-health-in-the-academy-affinity-group/.

Gutiérrez y Muhs, G., Y. Niemann, C. González, A. Harris, eds. 2012. *Presumed incompetent: The intersections of race and class for women in academia*. Boulder: The University Press of Colorado.

Halberstam, J. 2011. *The queer art of failure*. Durham, NC: Duke University Press.

Ho, J. 2017. The Accidental Academic, or how to succeed in academia through failure and doubt. In *Staging women's lives in academia: Gendered life stages in language and literature workplaces*, eds. M. Massé and N. Bauer-Maglin, 29–39. Albany: SUNY Press.

Ho, J. 2015. *Racial ambiguity in Asian American culture*. New Brunswick: Rutgers University Press.

Matthew, P., ed. 2016. *Written/unwritten: Diversity and the hidden truths of tenure*. Chapel Hill: UNC Press.

Peake, L., B. Mullings, K. Parizeau, G. Thornburg, J. Magee, D. Metzel, V. Wadhwa, K. England, N. Worth, A. Mountz, J. Finlay, B. Hawkins, and L. Pulsipher. 2018. Mental health and well-being in geography: Creating a healthy discipline report of the AAG Task Force on Mental Health (2015–2018), https://www.aag.org/wp-content/uploads/2021/12/AAG_Mental_Health_Task_Force_Report.pdf.

Puig de La Bellacasa, M. 2017. *Matters of care: Speculative ethics in more than human worlds*. Minneapolis: University of Minnesota Press.

Skop, E., M. Caretta , C. Faria and J. Smith. 2021. An ethos of care. *Inside Higher Education*, https://www.insidehighered.com/advice/2021/10/15/pledge-foster-more-equitable-research-collaborations-opinion.

Whitaker, M., and E. Grollman, eds. 2019. *Counternarratives from women of color academics: Bravery, vulnerability, and resistance*. New York: Routledge.

INDEX

Note: **Bold** page numbers refer to boxes and tables; *italic* page numbers refer to figures.

FSC
www.fsc.org
MIX
Papier | Fördert
gute Waldnutzung
FSC® C083411

Kolibri 360 GmbH
Ferdinand-Jühlke-Straße 7
99095 Erfurt, Deutschland
produktsicherheit@kolibri360.de